T0174495

BIRDS and FORESTRY

Birds and Forestry

by
Mark Avery
and
Roderick Leslie

Illustrated by
PHILIP SNOW

T & A D POYSER

London

First published 1990 by T & AD Poyser Ltd
Print-on-demand and digital editions published 2010 by T & AD Poyser,
an imprint of A&C Black Publishers Ltd, 36 Soho Square, London W1D 3QY

www.acblack.com

Copyright © 1990 by Mark Avery and Roderick Leslie

The right of Mark Avery and Roderick Leslie to be identified as the authors of this
work has been asserted by them in accordance with the Copyright, Designs and
Patents Act 1988.

ISBN (print) 978-1-4081-3768-0
ISBN (epub) 978-1-4081-3767-3
ISBN (e-pdf) 978-1-4081-3766-6

A CIP catalogue record for this book is available from the British Library

All rights reserved. No part of this publication may be reproduced or used in any form
or by any means – photographic, electronic or mechanical, including photocopying,
recording, taping or information storage or retrieval systems – without permission
of the publishers.

This is a print-on-demand edition produced from an original copy.

It is produced using paper that is made from wood grown in managed sustainable
forests. It is natural, renewable and recyclable. The logging and manufacturing processes
conform to the environmental regulations of the country of origin.

Typeset by Paston Press, Loddon, Norfolk

Printed and bound by CPI Group (UK) Ltd, Croydon CR0 4YY

Contents

List of Photographs

List of Figures

List of Tables

Foreword

One day, and in the not too far distant future, 20%, perhaps more, of Britain will be covered with forest. At least 14% of it will be *new*, in the sense that it will have been planted since 1919.

This is, I believe, a fact of life and one which we all should welcome.

Forests are, or at least should be, renewable resources providing vital raw materials for a number of industries and worthwhile employment in what are often remote, job poor areas. And when planted and managed with care and foresight they should provide good habitat for an important cross section of our wildlife.

This book is a reasoned argument to that end. An end, or rather a new green beginning which could see much of our urban sprawl contained and interleaved with coppice woodland, and farmers sharing at least their grade 3 land with woodlots managed for wildlife, fieldsports and home grown fenceposts. Also our uplands could become part covered (those parts which have been so misused in the past that they are now biological deserts) with 'Rotational Coppice', landscaped mixtures of conifers of all ages complete with streamside corridors and broadleaf buffer zones. A new vibrant, living Britain held together as nature intended by a diversity of working woodland.

Although written with an undenied bias on the side of forestry, this book in no way shirks the spectre of sterile black forests striding across our hillsides, nor the unforgivable destruction of key parts of the Flow Country, Europe's last great untouched wilderness. Indeed it poses a number of key questions which cry out to be answered. Why are there no National Parks in Scotland? Why did the Nature Conservancy not buy the wet heart of the Flow Country for the same ridiculously low price paid by Fountain Forestry? Why, in a country abundant with universities and field naturalists, haven't the surveys been done, the maps drawn and the information regarding our most important wildlife areas been published? Why do the foresters regard the conservationists as enemies and vice versa?

Well, now this excellent book has been published the answers can be found and there is no turning back. The directives for cooperation are there. No more threats to SSSIs, no more money wasted in the hot-headed acrimony of public enquiries. A rational landuse policy based on scientific fact and economic planning which has no need to hide behind subsidies and half truths. You cannot hide a forest, but hiding within a well managed forest there can be a wealth of wildlife and a wealth of opportunity for all our futures.

David Bellamy
The Conservation Foundation

Preface

This book is a processed tree, a tree which was once part of a habitat, and which has now provided paper for about 30 books of this size. Did these pages once form part of a British conifer plantation or part of a Scandinavian or American forest? Perhaps they were once branches supporting a Goshawk nest. Did a Three-toed Woodpecker pass through them or was there once a male Spruce Grouse displaying below them?

These flights of fancy signal our intention to discuss timber production both as an industry and as an influence on the environment. In Britain, forestry has become a major environmental issue due to the apparently unrelenting afforestation of open hills. Here we look critically at the impacts of British forestry on nature conservation. We have had to take a very narrow view of what constitutes nature conservation and have only considered birds. This is simply because most of the public arguments and published information have dealt with birds, and because our own interests are centred on birds.

This book has three main themes. The first is to describe the British forestry industry and how it arrived at its present state. We consider this to be a fascinating story in its own right, but also a prerequisite for informed comment on the relationship between forestry and nature conservation. The second theme is to describe the effects of forestry on birds. This involves not only describing the birds which live in commercial forests but also the birds of treeless areas which are affected by afforestation. These data are at the heart of any assessment of the net effect of forestry on birds. The third theme of the book is that of measuring conservation value. This theme grew during the writing of the book and plays an important part in it. Even if we knew, perfectly, what changes occurred in Britain due to forestry how would we decide whether or not they represented a net conservation gain or a net loss? We do not claim to know all the answers.

To anyone outside the acrimonious debate, or slanging match, between foresters and conservationists, this book may seem to cover fairly tame ground. However, we know that we risk stirring up a hornets' nest. We expect to be criticized from both sides of the current divide. So, for the record, we regard this book as being both pro-forestry and pro-conservation and it is a sad comment on the current situation that we know many conservationists and many foresters who regard such a position as impossible. We regard further afforestation in Britain as both desirable and inevitable, but what is also desirable, and on past performance far from inevitable, is that forestry should take full account of conservation value in both the planning stage and as part of routine forest management.

This book is, therefore, a personal account of our own views. It does not necessarily coincide with the views or policies of either the Royal Society for the Protection of Birds or the Forestry Commission. We would like to thank our colleagues in both organizations for providing the stimulating jumble of different views that we have tried to assess here.

Mark Avery
Roderick Leslie

Acknowledgements

The authors would like to thank the following for their help and advice: Chris Bowden, Charles Critchley, Ron Hoblyn, John Mather, Bob McIntosh, Barbara and Rick Mearns, Steve Petty, Bernard Pleasance, Dr Phil Ratcliffe and Dick Roxburgh.

The photographs on pages 3, 19, 20, 25, 27, 109, 119, 133, 137, 142, 163, 171 and 225 were kindly supplied by David Whitaker, and the photograph on page 106 was supplied by Edmund Fellowes.

CHAPTER 1

Introduction

The conflict between forestry and nature conservation became one of the major environmental issues in Britain during the 1980s. At a time when worldwide concern focuses on the destruction of forests, this dispute is paradoxically over the planting of new forests on open hills which were deforested centuries ago. Although afforestation is a way that Britain can replenish natural timber resources depleted by the longest continuous industrial history in the world, conservationists are concerned at the scale of what is the country's largest deliberate land use change of the century. In contrast to agriculture and urban or industrial development, which alter in response to changing economic conditions and government policy, national policies for forestry have remained unusually consistent since the creation of a national forest service (the Forestry Commission) in 1919. Government funding, initially to the forest service and latterly to private landowners and forestry companies as well, has supported stated targets for areas of afforestation. Today the target for afforestation of open land, 'new planting', is 33 000 ha (1 ha = 2.47 acres) per annum, although less than 25 000 ha per annum have actually been planted over the last few years.

The new forests created by this programme now cover over 1.5 million ha and are characterized by their poor soils and largely non-native coniferous trees. This book is about these new forests, the birds that live in them and the birds of the open uplands that the forests replaced. Bird populations have been at the centre of concern over the environmental impact of afforestation of open land. There are good reasons for this. The effects of change on birds are obvious because birds are highly visible, and bird conservation organizations

1

also receive more support than all the other voluntary nature conservation organizations in Britain put together. The land that is being afforested does not generally have as rich and varied a flora or entomology as it has bird fauna, which though at low density is of high diversity and interest. Because man destroyed most of the natural forests, Britain has an unusually large area of open ground at lower latitude and elevation compared with countries where natural forests still exist.

The issues raised by afforestation are complicated compared to others in conservation. Human activities often result in an indisputable loss of wildlife, and the only question is whether the benefit to people offsets the environmental loss. A petrochemical works replacing bird-rich estuarine mudflats, or oiled seabirds, are fairly clear-cut losses to wildlife. This is not the case with afforestation; the bird assemblages of open ground are lost but are replaced by a forest bird fauna that includes a number of rare and interesting species. Because of this the argument over afforestation is about relative values, in contrast to the absolute values attached to many conservation issues. This factor is apparent in the public debate; people opposed to afforestation try to minimize the value of forest birds, describing them as 'common garden birds', which, of course, many are. Foresters present the forest species as a more than adequate substitute for what was there before, stressing the higher density and species diversity of forest bird assemblages, but ignoring the fact that this does not amount to an absolute conservation value. This question of relative values, far too little considered in assessing the implications for bird conservation of extensive rural land use change in Britain, will be a persistent theme of this book.

The main benefits of forestry to people, in contrast to developments like agriculture, housing or industry that may affect birds, are greatly delayed. It is a very long-term process. Only two things are certain; that people will still be using a lot of wood at the end of the 50 years it takes to grow a tree, and that the ways in which they are using it will be significantly different from today. The reasons for which forests are planted have changed over the years, and the outside observer might be forgiven for thinking that the justifications for forestry put forward by the forest industry and its supporters tend to follow rather than precede the decision to plant trees. Wood products are a massive import, but Britain could never grow enough trees to be self-supporting (Forestry Commission, 1977a), which means that arguments over self-sufficiency must also be of degree rather than absolute.

Similarly, whilst forestry certainly does create jobs, it is not so different from other rural land uses for rural employment to be an absolute justification for afforestation. Economic return on investment is inevitably low because of the long-term nature of the business, but to place too much emphasis on this effectively endorses a short-term view of world resources at odds with present-day conservation thinking. On balance there is sufficient uncertainty attached to forestry to invalidate any claim for a dominant priority in land use, and the case for forestry, like the relative values of forest and moorland birds, must be one of degree rather than absolute.

Although on the surface public debate remains determinedly logical and factually based, there are strong emotional undertones on both sides of this argument. The Forestry Commission (FC) was almost stillborn, coming close to extinction within a decade of its founding, due to urgent government economies. Subsequently, forestry has faced many attacks, but has always managed to attract more support than the logic of opposing arguments might suggest would be likely. This is because underlying all the rational arguments is a basic feeling that planting trees is a good thing. Equally irrational and at least as powerful is the opposition of many people, including bird conservationists, to large-scale change in the uplands. There is a deep sense of loss when familiar landscapes and the peculiar wildness of the unconfined birds of the open moors are irreversibly transformed. The obvious artificiality and insensitivity of much afforestation has deepened this conviction. Also contributing to the heat of the argument is that, in contrast to most conservation issues, where at least one side is arguing largely from short-term financial self-interest, both foresters and conservationists have an almost religious conviction in their cause, believing their actions to be justified not only for themselves, but also through their legacy to succeeding generations.

The central controversy is the afforestation of open land which started in 1919 and continues today. These new forests are very different from the native British forests wiped out by man centuries before.

A bird of open land, Hen Harriers were persecuted to extinction on the British mainland. Finding safe nesting sites in young plantations they re-established themselves as far south as Wales. In England and Wales Hen Harriers are now very rare again as breeding birds, perhaps partly due to shortage of young forestry.

THE FORESTS

The new forests are only the latest stage in man's impact on Britain's vegetation stretching back well over 3000 years, and the almost total clearance of the native forest was the most important effect. Woodland is the climax vegetation of most of mainland Britain, broadleaved deciduous species in England, Wales and south and central Scotland merging into evergreen Scots Pine forest in the Highlands of Scotland. The story of the ancestral 'wildwood' is of continuous decline, paralleling the development of a heavily populated and, more recently, industrialized country. Woodland clearance began on a small scale about 2500 BC. By the Middle Ages the most densely settled and productive land of East Anglia was almost totally deforested, and remnant woodland was intensively managed by a variety of silvicultural systems supplying local needs (Rackham, 1976, 1980; Peterken, 1981). Woodland area in England at least remained remarkably constant from the Middle Ages until the 20th century. In complete contrast to traditional thinking, we now also know that industry, especially iron smelting and leather tanning, were powerful forces for the conservation, not the destruction, of woodland.

It is surprising how well Britain was able to meet its timber needs from woodlands covering only a tiny proportion of its land surface. Wood was a scarce resource in much of England by the end of the 19th century but concern over shortage of timber for ship building is not quite what it might seem. Rackham (1986) points out that whilst the navy complained bitterly about the

The beauty of the Scottish Highlands near Loch Lomond. The man-made origins of the uplands are often overlooked but these hillsides have been actively deforested over a period of thousands of years.

difficulty of obtaining timber during the Napoleonic wars, there were no equivalent problems recorded from the private dockyards building a far larger quantity of merchant shipping. It was not shortage of timber the navy suffered from, but shortage of money; they expected to buy special shapes and large sizes for the going rate for ordinary oak. The Royal Forests were used as a cheap source of timber, and the 'Napoleonic Oak' of the Forest of Dean, the planting of which began in 1808 in response to a 1792 government committee report recommendation that 70 000 acres of new woodland be planted, is a monument to this. Some timber was imported for ship-building in 1809, but most of the 19th-century construction, which accounts for over half the wooden shipping ever built in Britain, was from home-produced timber (Rackham, 1986). Nothing like as much timber was ever used for shipping as for bulk uses like domestic and industrial firewood or tanning. Although coal replaced wood for fuel, and metals for fencing, traditional woodland management continued well into the 20th century. Some new woodland was also created, for landscaping large estates and as commercial plantings by a few Scottish estates.

Although the remaining woodland was in a better state than generally appreciated, there was by the start of the 20th century a growing feeling that something needed to be done about Britain's forest cover. It took the 1914–1918 World War to create the urgency that gave life to the idea of reforestation. Although the war did not permanently destroy many woods, it did demonstrate conclusively that Britain could no longer supply all the timber needs for a vastly increased level of economic activity for even a short period. In contrast to previous false starts, this attempt has persisted through many ups and downs to the present day.

The First World War both ate up the timber resources of the tiny remaining forest area and brought home the dangers of dependence on imports. Timber was of great strategic importance for its use as structural props in the coal mines that provided the main energy source for the war effort, and the high volume that had to be imported tied up a disproportionate amount of shipping. During the German submarine campaign of 1917, the European allies suffered one of the worst crises of the war, and timber imports played a major part.

A turning point for British woodland came in 1919; tree cover had reached its all-time low, at less than 5% of the country, and, on the recommendation of a committee led by Sir Richard Acland, a new government body, the Forestry Commission, was set up under the 1919 Forestry Act with a remit of reforestation. In constitution it was effectively the national forest service that Britain had never previously had. Although totally funded by government, the FC was not a conventional department, in that it reported to ministers through a board of commissioners rather than directly. It also had the dual role of growing trees as a business in its own right and administering and promoting forestry carried out by private landowners. This latter function has only been widely appreciated in the last five years, as a higher proportion of new planting has been done by the private sector.

The objective which the government set the FC was to build up a resource of industrial timber for use in the mines as quickly as possible. The requirement to do this as cheaply as possible was soon added. The forests that resulted reflected this objective and were inevitably very different from the multi-use native broadleaved woodlands handed down from the Middle Ages. Relatively small size, straight, round timber was required in quantity. In Britain conifers are generally faster growing, straighter and more tolerant of poor soils than broadleaved trees. Large blocks created economies of scale for an industrial-style tree-growing operation, whilst intensive management of plantations through close spacing and systematic thinning was adopted to produce the highest yield of straight trees of a consistent size. From the beginning, non-native conifers dominated planting; Britain has only one major commercial native conifer, Scots Pine, which is well suited to dryer ground but slow growing. European species such as European Larch and Norway Spruce have increasingly given way over the years to Sitka Spruce, Douglas Fir and Lodgepole Pine, species from the temperate American Pacific coast better suited to Britain's oceanic moist, warm climate than those from the hotter, drier summers and harsher winters of continental Europe.

Twenty years later, by 1939, the FC had planted 149 000 hectares, 75% of the 1919 target. The immediacy of the depression years and the receding memory of the 1917 emergency resulted in reduced money for the FC from 1932 to 1936. Private planting barely kept pace with felling. The Battle of the Atlantic recreated the 1917 crisis, and refuelled enthusiasm for afforestation. Immediate post-war forest policy was developed by a 1943 wartime reconstruction committee (Forestry Commission, 1943). The strategic imperative was the basis of the 1951 Forestry Act, which also introduced the 'Dedication Scheme' that succeeded in motivating more private owners to manage their woods. A target of five million acres of 'effective' forest was set, to be reached by planting new ground and rehabilitating woods devastated by wartime felling.

Up to this point, although sometimes diverted from its main aim (for example, to provide work for the unemployed), the underlying reason for afforestation had been constant. The mid-1950s, and the rather belated realization that a nuclear war was likely to be so short that planning for a long siege was pointless, saw the first major change in thinking. The 1957 Zuckerman report, *Forestry, Agriculture and Marginal Land* (Natural Resources Committee, 1957), preceded a change in emphasis that by 1963 meant that the concept of establishing a strategic timber reserve had been almost totally replaced by economic and social priorities, and timber production for indigenous industry remains the prime objective of British forestry today. From 1972, when its accounts were reorganized, the FC was charged with making a 3% return on investment in its own plantations. This change in objective, combined with technological progress, has had a profound effect on the way forests are established and managed. By far the greater part of the new forests have been established in the post-strategic era; total conifer planting, most of which was new planting, from 1931 to 1940 was 87 581 ha, leaping to 310 271 ha in the decade 1951–1960 and to 365 806 ha between 1961 and 1970 (Locke, 1987).

However, the balance of objectives continued to change; during the 1950s and 1960s the cost-effective production of wood became dominant over all other considerations, but by the 1970s the significance of environmental objectives was increasing rapidly. FC forests were opened to the public in the 1960s, access having previously been restricted from fear of fire, and recreational use grew rapidly with the affluence and increased car ownership of the period. During the 1970s a more concentrated awareness of the environment developed, and forestry has come under increasing pressure from both landscape and nature conservation, leading to rapid change during the 1980s. At the same time, both the nature of the forests and the work of foresters have changed radically as more and more forests reach maturity. The 1980s were the first decade since 1919 when the job of the forester has been dominated by harvesting rather than new planting of open ground.

The FC has, since 1947, carried out censuses of all woodland in Britain (Forestry Commission, 1952; Locke, 1970), and the most recent (Locke, 1987) sampled individual hedgerow trees as well as covering woodlands. The census provides a fascinating picture of Britain's changing forests. Both the location of new afforestation and the organizations carrying it out have changed over the years. From 1920 to 1940, the areas of conifers planted in England and Scotland were broadly similar, but from 1960 onwards, planting in Scotland has always been more than double that in England. Between 1950 and 1970, the greatest concentration of conifer planting proportionate to country area was in Wales. By 1980, upland afforestation in England and Wales was of minimal importance compared to Scotland; in financial year 1985–1986 new conifer planting was 654 ha in England, 749 ha in Wales and 21 044 ha in Scotland.

By 1980 the woodland area of Britain was 2 108 397 ha (5 207 740 acres), 947 688 ha, 7.3% of the country's area in England, 240 784 ha (11.6%) in Wales and 919 925 ha (12.6%) in Scotland. Of this, 1 316 808, or 62% was coniferous high forest, most of the balance being lowland broadleaved woodland. FC woodland amounted to 873 435 ha, 41.42% of total woodland. Although precise definition is difficult, roughly 1.5 million ha (70%) of the total woodland area is 'new' upland forest (Balfour and Steele, 1980). To put this in perspective, the uplands as a whole cover about 6.6 million ha, 29% of Britain's land surface, so that about 23% of the uplands are now forest. No region or even local authority district in the uplands has over 50% forest cover: Dumfries and Galloway Region is highest at 21%, its Stewartry District topping 30%. West Glamorgan has 20% forest cover, but lowland, predominantly broadleaved; Surrey (19%) and West Sussex (17%) have higher woodland cover than the Scottish Borders (14%), or Northumberland (15%). This is, in part at least, because nearly a third of the uplands are physically unsuitable for forestry, because of exposure, or too great altitude for economic tree growth, or lack of soil.

Although double the area of woodland in 1919, Britain's present-day forest cover, at 10%, is still amongst the lowest in Europe; the EEC as a whole has 25% forest, and only Ireland (6%) has significantly less than Britain, whilst Germany has 30% cover and France 28%. Scandinavian countries are very heavily forested, with 76% forest in Finland and 68% in Sweden.

Fifty-nine percent of England's woodland is broadleaved, but the conifers of the new upland forests predominate in Wales (70% conifer) and Scotland (83% conifer). These are young forests; in 1980, 77% of conifers were under 30 years old and only 3% originated before 1910. More than half the new forests are still over a decade away from completing their first rotation.

The proportion of different conifer species planted has changed greatly over the years: averaged from 1919 to 1980, the area of Sitka Spruce planted has almost doubled every decade. However, it was not until 1971–1980 that it first made up more than half the conifer planting, and a higher proportion (though not area) than Scots Pine had done in the FC's first decade. In fact, pines made up over half the planting of that first decade, to be overtaken by spruce in 1931–1940, when Norway Spruce was only just behind Sitka. The actual area of Norway Spruce planted in each decade increased up to 1960, but proportionate to Sitka Spruce, it fell from 1940 onwards.

Lodgepole Pine planting rose dramatically from only 346 ha in 1921–1930 to 50 240 in 1961–1970 (out of a total conifer planting of 365 469 ha, and a Sitka Spruce area of 153 888 ha). Its use is an index of the declining quality of land available to forestry from 1950 onwards. However, planting of Lodgepole Pine as a pure crop dropped as quickly as it had risen, partly because of its low yield but mainly because methods of growing Sitka Spruce on heather ground were developed.

Larches and Douglas Fir are now a far smaller proportion of the total planting than in the early days, but actual areas planted (including restocking) are consistently higher post- than pre-1945, except for European Larch, which has been largely replaced by Japanese Larch and Hybrid Larch. The only major commercial conifer to have declined in actual area planted is Scots Pine, the mainstay of the initial heathland plantings; in 1971–1980 15 378 ha were planted as against 27 605 ha from 1921 to 1930.

What does this tell us about the composition of the present-day forests? As well as the highest proportion of broadleaves, the greater range of conifers in England, including the greatest area of Douglas Fir, reflect better soils. There is also more pine than spruce in England because of the heathland forests of the east coast. The forests of Scotland and Wales are today dominated by spruce, over half the conifer area in both countries. In Wales, Douglas Fir and larch are as important as pines, but in Scotland, Scots Pine and Lodgepole Pine make up the bulk of the non-spruce forest. The younger a conifer stand is, the more likely it is to be Sitka Spruce.

The planters of the new forests have changed, too. Although it was always envisaged that the FC and private landowners or companies would share the job of forest planting, most large-scale planting up to 1939 was by the FC. After 1945, a number of companies planting forests as an investment for private clients were established, and by the mid-1970s were planting about 25% of the national programme. Major private upland planting took advantage of the now much publicized tax arrangements which were originally introduced in 1916: it was not until the 1950s that the potential benefits of changing ownership and tax schedules were appreciated, and they formed a key element in the growth of the management companies, including the

Economic Forestry Group, Tilhill and Fountain Forestry (see Chapter 8). From 1979 onwards, government policy favoured private rather than state investment, so that in 1982 private planting exceeded state planting for the first time since 1919, and now accounts for 75% of new planting. Significant areas of woodland, with a majority in the lowlands, are still managed by traditional private estates.

Different owners have different management objectives, both financial and environmental, which have profound effects on the value of forests for wildlife. The requirement to make a return on initial investment limits the length of FC rotations, whilst the rate at which land is planted in forests managed for private clients has often been determined largely by the individual's taxable income for that year. Inheritance tax relief on timber means that felling often follows a traditional estate changing hands, and if an owner is long-lived, rotation length may be extended far beyond the average for FC forests. The needs of Pheasant shooting have a profound effect on the management of lowland estate woods, but in the uplands, design for deer stalking and fishing are recent inclusions in both private and state forests. However, private owners do have greater freedom to diversify their activities than the FC, which acquires land on behalf of the nation on the strict understanding that it is to be used primarily for forestry.

The FC remains centre stage in the forestry debate because it administers government grant aid to private sector planting. It is not compulsory to claim a grant, but there has been an unwritten convention that application for a grant will be made for large forestry schemes. This brings them within the FC's statutory consultation process, which involves allowing comment from a range of other organizations, including local authorities and the agriculture departments. Environmental bodies are generally consulted within designated areas only, but local authorities may subconsult other organizations as they wish. This consultation process acts at present as the main planning medium for the location of afforestation and has been increasingly criticized as concern over the environmental impact of afforestation has grown in recent years.

CONSERVATION

Environmental concern dates back almost to the inception of the new forests. The earliest real environmental restraint came in the Lake District, where opposition to planting in some of England's most famous beauty spots was orchestrated by the Council for the Protection of Rural England, eventually leading to a voluntary agreement with the FC to limit further planting in what is now the core of the National Park. Up to 1939, interest in the biological effects of afforestation was limited, perhaps not surprisingly as there was little awareness of habitat loss as a conservation problem at the time. The restoration of tree cover was generally seen as a good thing, and the few biologists seriously involved in afforestation, such as Lack, tended to comment on the changes they observed without critically assessing its overall impact on bird populations.

Landscape continued for many years to be the main source of complaint against forestry; based not just on a natural resistance to change in familiar, though in reality often fairly young, landscapes, but also the nature of that change. It is interesting to speculate how people might today feel about forestry had the best of modern landscape design been applied from the start. It was not, and regardless of arguments over the relative monotony of conifers versus broadleaves versus heather, many forest edges today cut like knives straight through naturally rounded terrain, inexorably drawing the eye. Inside the forest, as the young planting grows into thicket stage, dark walls of trees crowd up to forest roads, cutting across walking routes to the open hill.

The first post-1945 opposition to forestry on nature conservation grounds seems largely to have been inspired by a gut reaction to the artificiality, the scale and the change brought about by forest planting. However, in contrast to concern over the landscape, it has not until recently been particularly intense or effective. The exact reasons for this are not clear, but a number of factors were probably important. Prior to the Wildlife and Countryside Act of 1981, there were far more obvious and arguably more serious changes continuing to threaten some of the best conservation sites in Britain, with several lowland habitat types facing virtual extinction. Conservation efforts were concentrated on protecting the very best sites, and a very high proportion of resources went on purchase of land for nature reserves. The potential impact on wildlife populations of afforestation was probably not fully appreciated, and linked to this there was probably less awareness of the scale of change forestry was causing.

Sustained and effective criticism of forestry really dates from the late 1970s, and in the 1980s it has become the centre of the continuing debate over afforestation (e.g. Andrews, 1979; Grove, 1983; Nature Conservancy Council, 1986). For the first time, the real concerns of nature conservation are being clearly stated in contrast to the often muddled arguments of the past. Attention has focused on birds and the effect on them of the total ecological change from open moor to high forest. As the trees grow, the habitat changes from open ground to something equivalent to scrub, to tall, dense trees. Wildlife species change too, the birds of open ground gradually being eliminated and replaced by species characteristic of the later successional stages of forest. Despite the fact that the vegetation of the British uplands is to a large extent an artificial climax arrested by fire and grazing, it has been in its present state for a considerable time and has built up an interesting and characteristic bird fauna. Raptors, waders and grouse are particularly important terrestrial species, and divers and wildfowl are important in aquatic habitats.

The very artificiality of the uplands gives them special value for birds; a tundra or montane-type habitat occurs at lower elevation and on a larger scale than in comparable latitudes and elevations elsewhere. The moorland bird assemblage is not in itself extraordinary on an international level, but there is a good representation of a variety of species, there are southern subspecies of the Golden Plover and Dunlin, and a British form of the Willow Grouse, all of which can be identified in the field. The British uplands have become internationally important for raptors such as the Golden Eagle and Peregrine during the 20th century because of their decline elsewhere caused by land use change, persecution and pesticides. In the British uplands these birds are relatively protected from human interference, and populations in the wilder parts of the country have now largely recovered from the decline caused by pesticides in the 1950s and 1960s.

Concern over afforestation is exacerbated by the decline of a number of the important moorland species for other reasons, which have been reasonably well documented. These birds are an indicator of the economic fragility of moorland land use which has been an important factor in the growth of forestry. Management of large areas has declined, both through the abandonment or modification of traditional management and through increased pressure on vegetation. Carefully controlled rotational burning of heather (most frequently for Red Grouse shooting) favours many important bird species, especially Golden Plovers and Merlins. In many places the decline in Red Grouse and higher cost of labour have resulted in the abandonment of burning, or of careless burning of very large areas, which does not create the desirable mosaic of small patches of heather of different ages and may result in moorland vegetation changing from heather to grass or bracken. Overgrazing, too, can lead to a change in vegetation from heather towards grass: sheep and Red Deer numbers have increased on virtually all moorland areas since 1945. Land use change may also affect aquatic habitats through changes to runoff into watercourses and water bodies, in particular acidification. As well as commoner birds like Dipper, rarer species such as divers could also be affected.

The uplands pose special problems because bird distribution is on a different scale to that of the lowlands. The sort of clearly defined boundaries that estuaries or lowland heaths now have do not exist in extensive tracts of moorland, meaning that the site-based Site of Special Scientific Interest (SSSI) system does not work as well in the uplands as in the lowlands. Golden Eagles range over vast areas, and a difference of opinion over whether a territory size of 10 000 or 20 000 acres is required represents a difference equal to several substantial farms or a large forest. To the outside observer it may often seem impossible that such vast areas can be needed for bird conservation, or for that matter forestry, but in the uplands both interests depend on scale for their viability, and the environment in which they work is vast, unconfined, and of low fertility and productivity, meaning low densities of everything whether it be sheep production, timber production or Golden Eagle production.

The splendour and majesty of the uplands and their birds, underpin strong emotional commitments which make it difficult for many to accept any change. The haunting cry of the Curlew was cited by Ian Armstrong of the Royal Society for the Protection of Birds (RSPB) as a component of natural beauty at the North Pennines Area of Outstanding Natural Beauty enquiry, and who could dispute that idea? So too is a soaring Golden Eagle, and for a slightly different group of upland devotees so is a whirring Red Grouse on 12 August, or a September Red Deer stag.

Whilst most general countryside conservation organizations, both statutory and voluntary, and many individuals or institutions have contributed to the debate, two have played the leading role in putting the case for birds. These are the Nature Conservancy Council (NCC), the government nature conservation body, and the RSPB, the largest voluntary nature conservation body in Europe, with over 500 000 members. Afforestation has served to create a common cause out of two very different approaches to nature conservation.

The NCC's conservation philosophy stresses the scientific importance of wildlife, and has in recent years concentrated on 'naturalness' as a key criterion of conservation value. In a country as comprehensively altered by man as Britain, this often looks like the pursuit of the impossible and the new forests cause a particular problem (e.g. Nature Conservancy Council, 1986). The NCC is concerned with all groups of wildlife, but the way in which the RSPB has grabbed the limelight for birds has perhaps contributed to the NCC's rather elitist public image.

In contrast, the RSPB's spectacular growth during the 1970s has been dependent on the marketing of birds and conservation as a popular activity and concern. The RSPB's approach to conservation is simpler than the NCC's: it is to protect birds (at a population rather than individual level) wherever they are threatened by whatever means available, whether by land purchase for reserves, or obtaining changes to the law or to local or national policies. The RSPB has used the media to great effect in a country with a long tradition of sympathy to, and interest in, wildlife. Its magazine *Birds* has an estimated readership of over one million people.

The two organizations have been drawn onto parallel courses because afforestation is changing both semi-natural habitats and habitats for rare and interesting birds on a large scale.

It was two reports in 1977 and 1980, *The Wood Production Outlook in Britain* (Forestry Commission, 1977a) and *Strategy for the UK Forest Industry* (Centre for Agricultural Strategy, 1980), advocating major increases in upland afforestation on the grounds of timber production for industry, which really drew conservationists' attention to the scale of land use change that forestry could cause. At the same time, the work of Rackham (1976, 1980) and Peterken (1981) caused a reappraisal of the conservation and historic importance of broadleaved woodland, large areas of which had been converted to conifer forest or farmland since 1947.

Although the attack on forestry published in *Birds* (Andrews, 1979) had many factual inaccuracies, it made a number of telling points. The titles of the editorial and article, 'Forestry: a conflict of claims' and 'Who can impress the forests?', prophesy the subsequent events of the dispute. It has been characterized by arguments over factual accuracy on both sides and growing frustration amongst conservationists at their inability to influence the progress of upland afforestation. The article put the bird conservation argument to a wide audience with considerable clarity, and our broad assessment of the value of birds of the open uplands is little different 10 years on. In contrast, Balfour and Steele (1980), speaking at the British Association in the same year, did not give a clear exposition of the conservation case for the open uplands beyond the special sites, and no real warning of the storm to follow.

At the same time the NCC were preparing a paper on afforestation, something a number of scientific staff had felt a need to do for some time. The failure by the NCC to publish its 1979 paper has been an important contributory factor to the intensity of the present dispute, because most foresters were allowed to carry on unaware of the growing opposition. If anything, the determination amongst this paper's supporters to put the conservation case was increased, resulting eventually in 1986 in the publication of *Nature Conservation and Afforestation in Britain* (Nature Conservancy Council, 1986). Though weak on basic forestry knowledge, the 1979 paper made most of the key points of the 1986 publication but without the same antiafforestation intensity. It is unfortunate and predictable that the NCC spent the intervening years building up the case against forestry, but learned little about the forests themselves. Ironically, their case would have been the stronger for it. This general course of events was predictable by 1979; it was unlikely that conservation concern would diminish, and the seven-year delay did no good for either side.

Within the FC, increasing public access in the 1960s was closely linked to growing interest in wildlife; many early forest walks were 'nature trails'. However, despite the interest of many individual managers, consideration of wildlife at a national level was slow to develop. The 1970s were the locust years of missed opportunities for conservation in the national forests: there was enthusiasm for it, and so much to be gained for forestry in the long term, but the harsh rationalism of economics that dominated foresters' thinking found no place for wildlife that did not damage the forest. Johnston *et al.* (1967) discuss only resource, not nature, conservation, and the 1978 edition of the FC's standard setting publication *Forestry Practice* (Blatchford, 1978) devotes virtually the whole of its two pages on wildlife to species that damage the forest. By 1979 the forestry profession, in general, was almost as ignorant of conservation (other than on nature reserves or SSSIs) as the conservation profession was of forest management other than coppice broadleaves. Andrews (1979) was really rather mild in describing the statement in *Wood*

Production Outlook that 'in the uplands large-scale coniferous woodlands have a quite dramatic beneficial effect on wildlife' as 'an assertion which is very far from the whole truth.' It must be stressed that many foresters were interested in wildlife, and a few very knowledgeable, and there were also some conservationists who knew a lot about forestry. They have played a crucial role in shaping the new ideas on forest management described in Chapter 4. However, even today many foresters cannot understand that the public standing of forestry, including its environmental image, is as important a commercial force as fencing against deer or the number of kilograms of phosphate applied per hectare.

By the end of the 1970s, the opportunity was lost and the forest industry was plunged into a very real crisis by economic recession which reduced timber prices and led to the complete collapse of the British market for small-sized conifer wood. Neither time nor money were available for anything but the essential business first of damage limitation and then of recovery. Initially, contracts to export pulp to Scandinavia at starvation prices were organized, followed by a modernization of harvesting methods which in a few years transformed the industry and cut costs dramatically. But the main priority was to attract new processing industry, and in doing this forestry has created one of the major industrial successes of the 1980s. The last major element in the overall plan to use home-grown timber, the Scandinavian-owned Kymene company's mill at Irvine, is the biggest single foreign investment ever in Scotland. The crash in the timber market was important for wildlife in the short term in that it diverted attention away from the growing conservation concern, but the industry's response to its problems demonstrated once again its ability to get things done at a practical level, something which hopefully may increasingly be used to the benefit of the environment.

The first half of the 1980s were dominated by two main forestry issues: the dispute over the Creag Meagaidh SSSI in 1983, and the protection of ancient woodland, which culminated in the 1985 government broadleaves policy. A number of reports critical of forestry were published, which by 1986 had become an increasingly repetitive flood.

It was not until 1985 (Royal Society for the Protection of Birds, 1985) that the 'Flow Country' of Sutherland and Caithness really became a major public issue. Although awareness that there was a problem had been growing for some time amongst conservationists, up to 1984 it had remained a fairly well kept secret from the 'culprits'. That the Flow Country issue arose when it did was not by chance: it was a disaster waiting to happen somewhere. What, as we will show in Chapter 7, was unfortunate for the birds but fortunate for the conservationists arguing the case, was that out of all the British uplands, the Flow Country is the richest place for moorland birds. Finally the conservation concern over new afforestation that hatched in the 1970s came home to roost and upland afforestation became the public issue it is today.

The attitude of conservationists towards the new forests has contributed to a serious lack of knowledge both of the birds living in them and of the effects of forestry on birds of the open uplands. Important studies have been widely spread in time, starting with Lack's work in the Breckland in the late 1920s.

Like this study, much of the most valuable information has come from work directed at questions other than the actual change brought about by forestry, Newton's work on the Sparrowhawk being a good example. It has been central to a whole range of work on birds of prey and related subjects, including that of Moss *et al.* (1979a) on forest songbirds and Petty (1979, 1983, 1987b) on birds of prey of developing upland forests.

Over the years there have been sporadic attempts to look directly at the birds of the developing forests, Williamson's work (1972, 1975) being a particularly notable example. He addresses key questions of forest management in relation to birds more effectively even than many later studies, but his work is unfortunate in two respects. Firstly, although involving planted conifers, most of his studies were undertaken in unrepresentative situations, like lowland woods or the NCC planting on the island of Rhum. Secondly, in contrast to much of his work in other fields, his own enthusiasm failed to inspire amateurs to expand on the foundation he created. Recently, research has increased considerably, but nowhere near to the extent that general levels of interest would suggest. The work of Bibby *et al.* (1985, 1989a) is particularly notable in its effective identification of questions of high practical and conceptual importance, and in the economy of the approaches developed to answer them.

However, in contrast to this, extraordinary and unjustified assumptions have not only put people off working on forests, but have also damaged from the start a number of projects. The idea that there are no birds in the new forests, or that those that are there are 'common garden species', seems to have effectively stifled interest in research, and the important contribution of amateur ornithologists in characterizing and monitoring the range of habitats is missing. O'Connor and Shrubb (1986) give a wealth of detailed information on the effects of changing agriculture on birds, not only over geographical regions but also over time. A large number of consistently worked study areas made this possible: there are none for upland forests. Throughout this book we have time and again to admit to vast gaps in knowledge of direct and crucial importance both to forest management for wildlife and the broader conservation issues. We do now have just about enough information to show that the suggestion that 'one spruce forest is much like another' is wrong, but we certainly cannot characterize these differences in the way that it is possible to characterize the birds of predominantly cereal farms from different parts of the country. Had it not been for the need to monitor the effects of insecticide spraying (Crick and Spray, 1987), we would have no systematic information whatsoever on the birds of the vast 1960s and 1970s Lodgepole Pine plantations of northern Scotland, a habitat not readily comparable with pine further south.

In this book we therefore try to present the facts as they are at present available; we identify gaps in knowledge and try to reach some conclusions on the future of the forests, the moors and the conflict between the opposing parties. In the next three chapters we look at the birds of the new forests, how the forests have been established and managed in the past, and recent developments that aim to solve some of the problems of earlier management methods, and we speculate on the future potential of British forests. Alternative options for broad silvicultural systems are considered, as well as the development of smaller-scale management techniques and the possibility of new breeding species for Britain coming into the new habitat provided by the forests.

Chapter 5 is about the birds of the moorlands. We identify which species are dependent on the true moorland habitats on which afforestation is at present concentrated and consider how our existing knowledge of the distribution of moorland birds can help in identifying the most important areas for them. The next chapter considers the contentious subject of effects of forestry beyond the forest edge, including possible effects on vegetation, predation patterns and water quality. Drawing on these general ideas, we then discuss four case studies of areas that have seen above-average afforestation since 1919 to try and determine what has really happened at a regional level.

The last part of the book of necessity strays rather further from birds. It is easy to get the feeling from recent publications (e.g. Stroud *et al.*, 1987) that forestry has quite intentionally set out to seek conflict with conservationists. Of course, this is not so and to find out why things have turned out the way they have it is essential to look at the land use and fiscal structures within which forestry and conservation operate in the British uplands. This is equally the case in considering the future, where we discuss the prospects for the forests that already exist, for new forests and for moorland birds.

Most contributions to the forestry debate claim to be objective, balanced or scientific, but unfortunately this is sometimes a transparent cover for the transmission of half-truths and personal opinions. We do not claim to be objective in the opinions we put forward in this book: anyone who has been closely involved in the affray over forestry of the last few years carries too much emotional baggage, resentment at distorted claims, and antipathy based on supposed or real slights, to claim to see a whole picture clearly. What we have tried to do is present the facts such as they are as clearly as possible so that people have the information to make up their own minds about the ideas and conclusions we put forward.

CHAPTER 2

Birds in forests

The work of foresters during the past 70 years has added huge plantations to the British landscape, often in areas where few trees grew before. This change has resulted in increases in habitat for a large number of species of woodland and forest birds. In this chapter we discuss the birds of the forests. We take an approach which some will regard as blinkered but seems eminently sensible to us; we treat commercial plantations as an interesting habitat and assess the importance of the birds which inhabit it. Here we do not consider the birds which might have been present in the area if the forest did not exist; those will be considered in Chapters 5 and 6.

We take this line because the value of British commercial forests for birds has been ignored and underestimated in some of the arguments about upland afforestation. This is probably because afforestation has often, particularly in recent years, occurred in areas which have important or relatively highly regarded upland bird communities, and conservationists have tried to play down the value and potential value of the new forests because they have been opposing their existence. Because of this there have been few rational examinations of the value of commercial forestry plantations for birds in Britain. This question has been ducked by conservationists in recent years; here we provide the beginnings of a more rational basis for the arguments about forestry.

In assessing the richness of commercial plantations for birds we are handicapped by a surprising lack of information. The BTO Common Birds Census is the only national scheme which is suitable for monitoring the numbers of common British birds. Its plots are all either farmland or woodland but very few are in commercial coniferous plantations. It is remarkable that whereas Common Birds Census plots were established partly as a means of measuring the changes caused by agriculture in the British countryside, no such system has arisen as a result of the threat to moorland birds which is posed by afforestation even though the threats posed by the forester are every bit as great as those posed by the farmer. This lack of hard data has allowed the arguments about the birds which live in forestry plantations to take place in an atmosphere of claim and counter-claim which is potentially never-ending.

Few detailed studies have looked at the birds of commercial forests, and those which have taken place have not been on anything like the wide scale which is necessary to generalize. Without a network of representative study plots it is difficult to assess the value of the bird communities in commercial plantations; indeed it is difficult to know anything about them. We have thus been forced to rely mainly upon the information from a few published studies supplemented by that published in the reports of the Rare Breeding Birds Panel. In the case of the Rare Breeding Birds Panel reports, much of the information is given confidentially to the Panel and therefore is published in a deliberately noninformative manner in order to protect the breeding localities of rare species. This limits the usefulness of the information to people such as ourselves. It is also unfortunately true that an unknown, but in some cases probably very large, proportion of breeding records are not given to the Panel at all and may never be published.

Here we examine several questions. What are the ecological factors which determine the birds to be found in plantations? Which are the common birds of commercial forests? Which rare birds have important populations in commercial forests?

Before describing which birds are found in commercial forests it is useful to discuss the general principles which lie behind bird distribution in woodland habitats. A thorough understanding of why birds occur where they do would be very useful in assessing the impact of forestry on birds. In particular, if we could fully understand the *status quo* we would be in a much stronger position to recommend how to change forestry practice in ways which would be suitable for particular bird species. Similar reviews can be found in Newton (1986), Newton and Moss (1981), Fuller (1982) and Petty and Avery (1990).

FOREST SUCCESSION

Commercial forests are always changing so it is meaningless to talk about 'typical' birds of plantations without stating the age of the trees. The birds which are present in each successive stage of a commercial coniferous plantation are different, just as they are in a commercially coppiced broadleaf stand (Fuller and Moreton, 1989). It is often overlooked in the arguments about the value of commercial plantations that the birds which inhabit the early stages of the forest cycle are every bit as much 'forest birds' as those which live in mature trees. And when the crop is harvested, the colonizers of the restock site again have to be regarded as 'forest birds'. Thus, to assess the conservation value of a commercial plantation, one needs information on exactly what the forest will hold at different ages. Unfortunately we are not in a perfect position to answer these questions. Other handicaps include the fact that only a relatively small proportion of British plantations have reached felling stage so that there is even less information on the birds of restocked sites than there is on the birds of the first rotation. However, some information does exist and has been widely used.

One of the very first studies of the birds at different stages of a woodland succession was that of the Lacks (Lack, 1933, 1939; Lack and Lack, 1951) in the Breckland pine forests of East Anglia. The unplanted sandy heaths held rather few bird species; Skylarks, Meadow Pipits, Stonechats and Whinchats. As trees grew, these open-country birds gradually fell in numbers and disappeared, and were replaced by birds whose characteristic habitat is scrub. At this stage the dominant species was the Willow Warbler. As the trees grew taller and the canopy closed, Willow Warblers were replaced by species

A male Whinchat perched on a conifer leader is a characteristic sight in young plantations. Whinchats are birds of open land or sparse scrub.

characteristic of true woodland: Chaffinches, Goldcrests and Coal Tits. The Lacks did not use any of the census methods which are standard today, so that their findings must be regarded as rather impressionistic, although the general patterns which they found have been replicated by many studies of succession in many parts of the world.

Leslie *et al.* (in preparation) studied 20 restocked sites in the Brecklands over a four-year period and found changes which mirrored to a large extent those found by the Lacks 50 years earlier when those same areas were originally planted. In the first year of their four-year study, Leslie *et al.* found that the restocks were dominated by open-country species such as Skylarks and Red-legged Partridges which gave way to scrub species such as Whitethroats and Willow Warblers with time.

Moss (Moss, 1978, 1979; Moss *et al.*, 1979a, Newton and Moss, 1981) is the most frequently quoted study of succession in upland conifer plantations. Moss studied a succession in upland Galloway. He chose several small plots and surveyed them using the BTO Common Birds Census technique. This method has several disadvantages in the upland situation. First, it is labour/time-intensive, so that only relatively small areas can be covered. Second, it is unsuitable for studying the numbers of non-passerines such as waders or raptors, which have large home ranges or territories, because these species are unlikely to occur in sufficient numbers to be analysed if only a small area is

Using the 'scrub' part of the succession, Whitethroats will nest in young Sitka Spruce at about 2 m high. Whitethroat numbers dropped dramatically in the late 1970s, probably because of drought in their African wintering grounds, but lack of studies in the new forests means that we know very little about what could be a large part of the British breeding population.

covered. Although this piece of work was a good start to quantifying the changes which occur throughout the forest succession, its scope (in terms of the small number of plots and the one geographical location) was very limited. It is certainly not a sufficiently detailed study on which to base any sort of case either for or against the conservation importance of plantations, although the lack of other data has tended to result in this happening.

Moss's findings were that as the trees grew, the bird community changed in a fairly predictable way. Open-country species such as Meadow Pipits and Skylarks persisted in the young plantations for a few years but were absent from plantations which had reached the thicket stage, where scrub-loving birds such as Wrens and Willow Warblers predominated. After canopy closure typical forest birds predominated. Coal Tits, Chaffinches, Wrens, Robins and Goldcrests were common. These findings closely replicated those of Lack's Breckland study and have since been replicated by other less detailed studies (Jessop, 1982; Sykes *et al.*, 1985).

A more recent study (Bibby *et al.*, 1985) investigated the apparent succession of birds on clearfelled sites in Wales. This is the most detailed study to date and goes a long way to confirm the findings of Moss and the Lacks that densities of birds increase with successional stage. The method used in this study was the point count; all birds encountered within a given radius of a static observer are counted during a short, fixed length of time. Although not suitable for all purposes, this method has the great advantage that many points can be visited and those points can be chosen by rigorous sampling procedures which make the interpretation of the results much easier than with the more intensive methods. An important difference in this study of restocked sites was that densities of birds were much higher than those found in studies of newly planted land. It remains to be seen whether this will emerge as a general rule for studies of restocked sites but such a difference in density could be explained in at least two ways. First, the habitat will be different in many respects on restocks than on recently planted moorland; there will be less ground disturbance, drainage may be better and the debris of the last rotation will have changed. Second, the pool of potential colonists may be very different. With present forestry techniques, large areas tend to be planted at the same time. This means that the nearest woodlands may be very distant. At the beginning of the second rotation the restock is likely to be surrounded by blocks of forest which themselves are nearing felling age and/or other blocks which themselves were restocked a few years previously. Thus there is a much bigger pool of potential colonists. Thirty-one passerine species were using Welsh restocks, of which the Willow Warbler was by far the most numerous.

A study within another type of commercially managed forest, Sweet Chestnut coppice, also adds to the knowledge of succession in British woods (Fuller and Moreton, 1987). Sweet Chestnut coppices can be thought of as being similar to very small restocks with a much shorter rotation length. In this habitat the familiar pattern emerged with some species (Tree Pipit, Yellowhammer) being characteristic of the early open stages of the succession, others gaining dominance in the scrub stages (particularly the familiar Willow Warbler) and other species (e.g. Robin, Blackbird) being characteristic of the older coppices. Species richness declined with coppice age, particularly after canopy closure, and the ratio of migrant to resident birds was also lowest in the oldest coppices. Fuller and Moreton show that species richness is maximized by opting for a short coppice cycle of about eight years, which maximizes the numbers of open-country species in the woodland mixture.

As well as these British studies considerable research has been carried out elsewhere in the world. In Europe, studies in France (Ferry and Frochot, 1970), Scandinavia (Helle, 1984; 1985a,b) and Poland (Turcek, 1957; Glowaci'nski, 1972, 1975, 1979, 1981a,b; Glowaci'nski and Jarvinen, 1975; Glowaci'nski and Weiner, 1977) have documented rises in species richness and density through the forest succession, as well as documenting the change in the characteristic species at different stages. In the USA the number of studies is even greater. Important works include Odum (1950) and Engstrom *et al.* (1984). Overall, it is fair to generalize and say that both species richness and overall bird density tend to rise through the course of woodland succession. In some cases there is evidence that the peak in one or other or both of these measures occurs in an intermediate stage (scrub) rather than in the oldest vegetation. The second generalization is that the identity of the most abundant species changes through the succession. In Britain, depending on the original nature of the unplanted ground open-country species such as Skylarks, Meadow Pipits and Yellowhammers may dominate. In the intermediate stages of scrub, it is always likely to be the Willow Warbler, which is the commonest species, but that in turn is replaced by thrushes, tits, Chaffinches, Robins and (in conifers) Coal Tits and Goldcrests.

All these studies advance our knowledge of the succession of passerine bird communities in different situations but what do they tell us of the conservation importance of British plantations? Rather little, since they concentrated on passerine birds, and were largely concerned with the commoner species, whereas the species of greatest conservation interest are often the rarer species of non-passerines. However, it is clear that almost every species of bird which might realistically have been expected to nest in a commercial plantation has at some time or other been found there. There are very few woodland or scrub birds which cannot nest in conifer forests at some stage or other in the forest succession.

FOREST AREA

Ecologists have shown great interest in recent years in the relationship between the size of an area and the wildlife which it holds. This interest was stimulated by the theory of island biogeography (MacArthur and Wilson, 1967), which provided a theoretical background to the observations that larger islands (and less isolated ones) tended to have more species of breeding landbirds than did small islands (and more isolated ones). Many authors have treated woodlands as 'habitat islands', and have shown that large woods tend to have more nesting species than small woods. Although such a result indicates that given the choice it would usually be better to protect a single large wood than a single small wood, even that is not unquestionably true. For example, a wood whose bird community consisted solely of Goshawks which preyed upon Capercaillies (an admittedly unlikely scenario) would be rated by most people as of high conservation interest, whereas a community with many more species, all of which were very common, widespread species, would almost certainly be rated more moderately. The point is that species-area relationships predict the number of species which will occur in a wood, but do not of themselves predict which species they will be, and conservation value is critically related to the identity of the species present. The precision with which even species number can be predicted is in fact sufficiently low that it would be rash to use the general relationship between area and species richness, for example, to formulate a strategy for reserve acquisition.

Despite this, the species–area relationship itself is of ecological interest. The earliest study in Britain was that of Moore and Hooper (1975), which used data from a very large sample of 433 woods. Larger woods were found to contain a larger number of species than smaller woods, and as a rough guide it was suggested that to double the number of species to be found in a wood it was necessary to increase the size of a wood 10-fold. Despite its very large sample size, this study can be criticized on numerous grounds, so that the applicability of the 10-fold rule is very doubtful. The sample of woods was a complex amalgam of large and small; coniferous, deciduous and mixed; nature reserves and unprotected sites; northern and southern sites. And the methods used to collect the data were a similar mixture of fieldwork and literature search. Given this haphazard assortment of data, it is impossible to decide whether the observed species–area relationship is a real effect of area or whether it is confounded by some other effect. For example, it is not clear whether coniferous and deciduous woods of equal sizes were selected from each part of the country. Thus if coniferous woods tended (whatever their size) to have smaller numbers of bird species than deciduous woods, then if the average size of coniferous woods in the sample was smaller than that of deciduous woods, a spurious species–area relationship would emerge which was really due to differences between tree species. The work of Woolhouse (1983) showed that biases in surveying effort can greatly affect the observed species–area relationship, and Fuller (1982) suggested that the form of the species–area relationship differs from region to region in the British Isles, so an uncritical acceptance of Moore's and Hooper's 10-fold rule would be unwise.

A finding which is related to the existence of species–area relationships is that some particular species are found in a greater proportion of large woods than of small woods. This suggests that certain species require woods of above a certain critical size before they will occupy them. If true, this could be of great conservation importance; it would suggest that steps should be taken to protect the integrity of current sites and perhaps to manage unused sites in such a way that their size increased. The concept of 'minimum area' has been an influential one in shaping conservationists' attitudes to woodland area, particularly in the USA. There, where the natural broadleaved and mixed forests were felled and burned by European settlers so that the woodland cover of many New England states is a mere remnant of that of two centuries ago, there has been considerable interest in the idea that some species need a minimum size of wood to exist. Interest has been particularly focused on a few rare species, such as the Worm-eating Warbler, which rarely occur in the smaller plots.

It takes quite a lot of ingenuity to come up with convincing ecological reasons why a small warbler needs to live in a forest which has to be several times as big as its territory. One of the more convincing suggestions is that near the edge of the forest open-country or farmland species can penetrate into the wood. If these species outcompete the forest-interior species for either nest sites or food, then this might provide a mechanism for restricting certain species to the interior of the forests. In the USA, nest parasitism by Brown-headed Cowbirds is likely to reduce the reproductive success of some species near the edge of forest plots. However, there is a much more prosaic potential explanation for the apparent restriction of some species to the larger forests, which is as follows. If a species is rare then it will not occur in many woods. If birds were distributed at random with respect to wood size, then there would be a much greater chance of a rare species occurring in a big wood than in a small wood. This would tend to increase the species number for big woods and reduce it for small woods simply by chance. It can easily be the case with a rare species that most large woods contain the species and few small woods do, even if the species shows no true preference for the larger woods. The more important comparison to make is to see whether, given the total area of woodland in woods of all sizes, any particular species tends to have a higher probability of being found in a given area of a big wood than of small woods. This requires data of high quality, since it depends on knowing the total numbers of the bird species one is considering, not just presence or absence.

There is no evidence that any British species of birds need a minimum area of woodland much greater than their territory size for their survival. After all, the main criteria for a species to exist in any area are that it can find a nest site and enough food in the area of its territory. Therefore, the species most likely to show minimum area effects will be those which have large territories and hunt only inside single woods. Most of the species which have large territories in Britain which might fall into this category are raptors, such as Sparrowhawks, which spend much of their time hunting for prey outside the single wood in which they nest. Another reason for not expecting a very convincing demonstration of species–area effects is that in many parts of Britain the distance to the nearest neighbouring wood is actually short. This will allow species to appear to exist in small woods even if this is only due to their commuting between several small woods. An example of such a species might be the Great-spotted Woodpecker which is known to commute between woods (Smith, 1987). It would be interesting to know whether any species would show a strong preference for woods of above a particular size if only isolated woods were considered. Our guess would be that there would not be many such species, since woods would have to be very far apart to be truly isolated for large species and it is unlikely that many small species will show a strong effect, because they need only small territories.

TREE SPECIES

It is quite likely that different species of tree will support different species of bird; both because they provide different nesting opportunities, and, probably much more importantly, because they contain different feeding opportunities. However, there appear to be very few species whose habitat requirements are so specific that they are met only by one tree species. Excellent evidence of this comes from the large number of common species which can use Sitka Spruce for nesting and feeding even though this is a recent addition to the potential tree species on offer to the British avifauna.

The most obvious division between different tree species is between broad-leaves and conifers. Most species will nest in both types of wood. It is clear that a number of species prefer broadleaves (Marsh Tit, Nuthatch, Lesser-spotted Woodpecker, Turtle Dove, Hawfinch, Garden Warbler, Wood Warbler, Nightingale) and a smaller group of species prefer conifers (Crossbill, Scottish Crossbill, Goldcrest, Coal Tit, Siskin, Crested Tit, Capercaillie). The best evidence which shows that particular species of British birds prefer broad-leaves to conifers comes from the work of Bibby *et al.* In Welsh conifer forests, it was shown that certain species, notably Wood Warblers, Chiffchaffs and Pied Flycatchers, nested in the conifer plantations but only in areas where there

A Scottish Crossbill nest on the lichen covered branch of an aged 'granny' Scots Pine. Native Scots Pine, but also old plantations, are vital to Britain's one endemic bird. The highly specialized mandibles do not cross until after the young have left the nest.

were at least a few broadleaved trees. Small patches of broadleaves were enough to attract these species.

Petty and Avery (1990) reviewed the evidence on the effects of different tree species and came to the conclusion that there were surprisingly few studies which had surveyed several types of tree in the same area. It is this type of study which will give us the most useful information on tree–species preference of birds. There is little point in, for example, comparing the birds of a New Forest oak wood with those of a Galloway thicket-stage Sitka Spruce plantation and deciding that oak woods are better than Sitka plantations because that will not really be what the data show. Any differences between the two studies might be due to geography, altitude or any number of other confounding variables. Petty and Avery also drew attention to the fact that the existing studies have looked at a small number of woods which probably were not typical at all. It is surprising that we really know so little about the typical bird communities of different woodland types.

The larger number of species which prefer broadleaves to conifers might suggest that upland broadleaved plantations would have richer bird populations than conifer plantations but this would be stretching the evidence too far. Geography and latitude may be more important than tree species in predicting woodland bird populations. Upland broadleaved woods do not exist on a very large scale in Britain. One of the best studied types is Birch wood, which has been surveyed by Simms (1971), Yapp (1962, 1974) and Bibby *et al.* (1989a). Natural Birch woods are interesting, because in upland Scotland Birch is the main naturally occurring broadleaved tree, and therefore is the species most likely to be included as a broadleaved element in primarily coniferous plantations. Despite the fact that natural Birch woods are beautiful places, their typical bird communities are rather unimpressive. Simms (1971) and Yapp (1962, 1974) used transect methods, and their Scottish samples each showed that the three commonest species in Birch woods were Willow Warblers, Chaffinches and Tree Pipits. Bibby *et al.* (1989a) studied Highland Birch woods using point counts and found that Chaffinches and Willow Warblers were the commonest two species. Tree Pipits were the fourth most often recorded species, but once this species' high detectability was taken into account, its densities were calculated to be lower than those of Wrens, Blue Tits, Robins and Spotted Flycatchers. Twenty-eight species of songbird were recorded in all from a total of 4630 bird records. Apart from the expected common woodland species, few more unusual species were recorded. Some Redwings were noted, but Bibby *et al.* point out that Birch woods do not seem to be a very important habitat for this species in Britain. Small numbers of Pied Flycatchers, Wood Warblers and Redstarts were recorded, and single Wryneck and Brambling were located outside the study plots. There was rather little variation in bird community between different Birch woods, but what variation did exist was related primarily to longitude, with eastern woods being better in terms of species richness and densities. The low diversity of the Birch wood bird communities, with more than half of the total bird density being contributed by just three species, resembles the rather uniform communities associated with conifer forests (Bibby *et al.*, 1989a).

GEOGRAPHY

Identical woods in different places would be likely to hold different assemblages of birds. For example, a Scots Pine plantation in northern Scotland might possibly hold Crested Tits, Capercaillies, Scottish Crossbills and Buzzards, yet it is certain that all of these species would be missing from a similar plantation in Kent. In general, in Britain more woodland species show a tendency to be restricted to the southeast of the country than to any other region. Examples of species which are absent from, or barely reach, Scotland are Marsh Tits, Willow Tits, Nightingales, Lesser-Spotted Woodpeckers and Turtle Doves. It is presumably often climatic factors which limit bird distribution, although it would be difficult to point to any particular woodland species and state with confidence that the species' range is largely determined by the British climate.

The geographical constraints which appear to exist suggest that planting broadleaves all over Scotland would do little to increase the range of Nightingales, Hawfinches, Turtle Doves, Marsh Tits etc. These species have not ever been known to have lived in northern Scotland and there is no evidence that their ranges are constrained by lack of habitat. However, such evidence

With perhaps 40% of British Crested Tits living in plantations future management could have a critical effect on this rare British bird. Older Scots Pine plantations with some dead wood are favoured, but the trend in the central Highlands is is to plant more Sitka Spruce instead of Scots Pine. This change in practice could undo any benefits forestry has created for the species in the past.

must be treated with care because the ranges of many conifer specialists (e.g. Siskin, Crossbill) have expanded considerably with the provision of new habitat (see below).

(see below)

ALTITUDE

Few British studies have looked directly at the effects of altitude on bird distribution. In general terms, a rise in altitude has similar climatic effects to a rise in latitude, so we would expect conditions to get harsher in woods at higher altitudes. Newton (1986) did look at songbird densities at different altitudes in upland conifer plantations and found that passerines were more abundant at lower altitudes and on better soils. This is just what might have been predicted; but it had not before been demonstrated in Britain.

The tetrad atlas of the birds of Devon (Sitters, 1988) analyses all breeding species by their percentage occupancy of tetrads at different altitudes. It is difficult to interpret this information easily because the distribution of available habitat is unknown for each species. However, the typical woodland species can be divided roughly into three groups. The first small group is composed of those species which appear to be showing a preference for the higher-altitude areas of Dartmoor, and include Siskins, Crossbills and Redpolls. A larger group appears to be showing a preference for low-lying woods; Lesser-spotted Woodpeckers, Great-spotted Woodpeckers, Turtle Doves, Bullfinches, Greenfinches and Blackcaps. The last group includes species whose preference seems to lie in what are, for Devon, mid-altitudes, and includes species such as Tree Pipits, Redstarts, Pied Flycatchers, Nightjars, Garden Warblers and Wood Warblers. The fact that two congeners with very similar ecologies, Blackcaps and Garden Warblers, have different altitudinal distributions, suggests that there is a real effect of altitude, since the distribution of available gross habitat must, of course, be identical for both of them.

The altitudes of individual plantations cannot be manipulated, but by changing forestry policy it would be possible to affect where new plantations would go. Information such as that in Sitters (1988) cannot be reliably used to predict what would happen if more planting were to occur in Devon unless the new plantations were to be exactly similar at all altitudes to the existing woodlands. However, these data reinforce what seems to be a defendable view; that the upland plantations which lack such species as Turtle Doves and Lesser-spotted Woodpecker are unsuitable for them, not just because the plantations consist of exotic conifer species which are closely packed together, but also because of their locations at high altitudes. If correct this will limit which species can be attracted into the new upland forests by sensitive management, particularly planting of broadleaves. The data suggest that putting nest boxes up for Redstarts and Pied Flycatchers might be worthwhile in upland plantations in Devon but that Marsh Tits are less likely occupants.

STRUCTURE

A factor which has been investigated in the search for an understanding of which birds would be expected to occur in different woods is whether the physical structure of the wood might be important. This could either be because some species of bird might be influenced by the structure itself, e.g. flycatchers cannot make flycatching sallies in impenetrable thickets, or perhaps through an effect that structure might have on the plant community of a wood, which may often be determined by the degree of penetration into the wood by sunlight.

One of the difficulties of studying this subject is in finding a way to describe the structure of a wood. One of the commonest methods that has been used is to calculate a mathematical index of Foliage Height Diversity (FHD) (Mac-Arthur and MacArthur, 1961; MacArthur *et al.*, 1962, 1966). To do this the amount of vegetation which is present in a number of defined horizontal layers is estimated and these are combined in a mathematical relationship which takes a maximum value when the vegetation is equally distributed between the different layers but is at a minimum when all of the vegetation occurs in just one layer. Two main British studies have shown that woods of high FHD tend to have a high diversity of woodland bird species (Moss 1978a; Newton and Moss, 1981; French *et al.*, 1986), as have many more in other parts of the world (Beedy, 1981; Haila *et al.*, 1980; Hino, 1985; Røv, 1975; Ulfstrand, 1975; Willson, 1974). However, the results of these studies are difficult to interpret clearly. For one thing a number of studies have failed to find any such relationship (Balda, 1975; Pearson, 1975; Willson, 1974). In some cases the relationships which were found are only true if taken across very wide habitat differences and were not detected within woodlands of different types but between woodlands and other habitats. In the study of Moss, various subdivisions of the vegetation profile were tried in calculating the relationship between FHD and bird species diversity, and the best relationship was chosen. Another problem in interpreting the results is that differences between tree species may underly the differences between FHDs. For example, in the study of Moss, 18 sites were chosen in the Spey Valley and Dumfries. These covered a very wide range of woodland types, from plantations of spruce to semi-natural pine woods and covered lowland and upland woods. This means that it is difficult to know to what extent the results were due to differences between sites, altitudes and tree species.

French *et al.* (1986) carried out a very similar study in a smaller geographical area, Deeside, but their results are similarly difficult to interpret and they appear not to have benefited from the shortcomings of the early trail-blazing study of Moss. Again, the Common Birds Census mapping techniques were used, and only a small number of woods were visited (14). No evidence is given that the woods were chosen randomly or that they were at all representative of the woods of the area. Between six and nine habitat measures were used in multiple regression models to explore the potential causes of the differences between the 14 woods. And the woods were of a wide variety of different types, tree species and altitudes. Although French *et al.* offer a long list of management guidelines (many of which seem very plausible) as a result of their work, the definitive study of the relationship between woodland structure and bird species diversity still remains to be done. Such a study would probably be based on a large number of point counts spread through a small number of woodland types (e.g. spruce plantations, semi-natural Birch woods, sessile oak woods) so that it could be tested whether relationships within woodland types also applied between such types.

We would suggest that the science on this subject has not yet caught up with what can be appreciated from a commonsense look at it. It seems highly plausible that some species of bird will be able to cope better than others with particular types of physical structure in the habitat. For example, foliage-gleaning species such as Goldcrests are more likely to be able to live in dense forests than flycatching species such as the Spotted Flycatcher. However, the scientific studies of woodland structure have not progressed far beyond this level of understanding. It remains necessary for the woodland manager to use his commonsense to create a varied habitat which is most likely to be suitable for a range of different species.

THE CONSERVATION VALUE OF BRITISH FORESTRY PLANTATIONS

It is often stated that the birds of conifer plantations are common species found in many other habitats in Britain. This is true of the species which are common in conifer plantations (e.g. Goldcrest, Coal Tit) but it is also true of the commonest species found in all other habitats. For example, the commonest birds of the British moorlands are Skylarks and Meadow Pipits, which are common in other habitats. It is not these species which give the British uplands their importance for birds. Their importance lies in the rarer species which inhabit them, and a similar argument is true for conifer plantations. It is thus extremely misleading to state that afforestation causes a change in the bird communities from interesting upland species such as Ravens and Golden Plovers to common woodland ones such as Coal Tits and Goldcrests. It is more often the case that afforestation will lead to a change from a bird community dominated by Skylarks and Meadow Pipits to one dominated by Coal Tits and Goldcrests, but if this completely described the changes in bird community then there would be little conservation interest in the effects of afforestation.

Here we argue that commercial conifer plantations hold important numbers of both common and rarer species of bird, and that all of these have to be taken into account when assessing the conservation gains and losses which are caused by afforestation. Here again we are at a disadvantage because of the neglect of studies of birds in coniferous plantations. There has been no properly organized representative study of British commercial coniferous woodlands, so it is difficult to speak in general terms. Thus for the rest of this chapter we have been forced to take a more advocatory role than we would ideally like. We spell out the reasons for thinking that commercial coniferous woodland can have high conservation value and present the strongest case we can to show that the birds of these forests are of conservation value. It is a case which has not seriously been made before, and has largely been ignored by conservationists. More detailed studies are necessary to establish its true strength but we would stress that the case which we make here cannot be refuted by mere contradiction; data are needed.

Before we discuss those rare or uncommon (a necessarily rather arbitrary division) British species which have benefited from afforestation, it is worth pointing out that the large populations of common species inhabiting commercial plantations should not be totally written off as of no interest. After all, much of the argument about the effects of farming on British birds revolves around the common species. An example is the large amount of attention given to the issue of the loss of hedgerows in the countryside. Although grubbing out hedges is unlikely to be favourable to many farmland birds (although some open–country species might possibly benefit), it is unlikely that the loss of every hedgerow in the British countryside would lead to the extinction in Britain of any breeding species (perhaps Cirl Buntings, Partridges and Red-legged Partridges) because all farmland species which inhabit hedgerows also live in woods and gardens. That does not mean that conservationists should treat such losses with equanimity, but by the same argument nor should the gains of common species in upland plantations be totally ignored. For example, Moss found that the density of songbirds in mature Sitka Spruce plantations in southern Scotland ranged between 351 and 549 pairs/km^2 whereas O'Connor and Shrubb (1986) found that on agricultural areas devoted to tillage, and scattered throughout Britain but with a southeasterly bias, the mean total bird densities ranged from 140 to 340 pairs/km^2 from year to year. In other words, mature Sitka Spruce plantations hold birds in greater densities than do farmland areas. It is interesting to use the data of O'Connor and Shrubb (1986) to compare (very approximately) the loss in passerine birds due to hedge removal with the gain from afforestation. O'Connor and Shrubb (1986) show that the total density of birds on farmland is approximately linearly related to the densities of hedges on a farm (their Figure 6.2) and that this relationship has an approximate slope of 2; every metre of hedge/ ha increases bird density by two pairs/km^2. In England in 1962–1970 about 44 000 km of hedgerows were removed in a total agricultural area of 14.4 million ha, resulting in a loss of hedgerow of on average 2.5 m/ha. This would result in a loss of five pairs of bird/km^2, or 720 000 pairs. This is thus 80 000 pairs each year over the whole of England. If we take 400 pairs of birds/km^2 to be the increase in bird densities due to afforestation then it is necessary to plant 20 000 ha of Sitka Spruce each year to compensate for the loss of hedgerows. This is a very rough estimate which almost certainly underestimates the value of the commercial plantation in purely numerical terms, since we would expect that bird densities in plantations further south would hold even higher densities of birds, and it is likely that the bird densities would be at least as high and probably higher in other crop species, particularly Scots Pine. In addition, Moss only counted the songbirds, whereas O'Connor and Shrubb (1986) include all species. Such a ballpark estimate obviously cannot be relied upon except to give a flavour of the effects of afforestation. However, it does indicate that in terms of common species Britain's new forests do hold nationally important proportions of birds. At the present rate of planting of 25 000 ha/ year each year's new forests, when mature, provide sufficient habitat to compensate for the number of birds lost in England due to the removal of hedges.

The common species which gain most from afforestation are difficult to pin down. One immediately thinks of Coal Tits, Goldcrests and Chaffinches but these are all birds of the mature commercial forest, and we have stressed in this chapter that the birds of earlier parts of the succession are just as much forest birds as those of later stages. Some interesting birds of the early stages of forest succession include Grasshopper Warblers, Stonechats, Whinchats, Tree Pipits and Whitethroats. For all of these species it is likely that conifer plantations hold important fractions of the British population which are almost completely outside the scope of the current national monitoring scheme, the BTO Common Birds Census. Certainly for a species like the Tree Pipit, and possibly for the Whitethroat too, monitoring of the British population cannot really be attempted without including the large upland forests which at the moment are missed.

Whinchats and Stonechats are often common in the early stages of afforestation (Thom, 1986; Sharrock, 1976). Phillips (1973) found that Whinchats reached 20 pairs/km^2 in Sitka plantations on grass but only 2.2 pairs/km^2 in plantations on heather. The situations with Stonechats were reversed, with the higher densities on heather (9 pairs/km^2) rather than grass (4 pairs/km^2). In some of these young upland forests the densities of Whinchats can apparently reach very impressive heights (100 pairs/km^2; A. D. Watson, quoted by Thom (1986)).

In the very large number of restocked sites studied by Bibby *et al.* (1985) in Wales, the average densities of Whinchats and Stonechats were much lower than those found in southern Scotland (5 pairs of Whinchat/km^2 and far fewer Stonechats) but the overall densities of birds were very high and averaged 1860 birds/km^2. Because Bibby *et al.* (1985) used the point count method, it is difficult to compare their figures with those of other authors who used intensive mapping methods, but if their figures are roughly equivalent to 930 pairs/km^2 then this is higher than the most densely populated Welsh oakwoods studied by Hope Jones (1972) which only reached 680 pairs/km^2. Comparisons of the mean densities of some of these species between Welsh farmland and Welsh restocks is interesting (we have simply divided the birds/ km^2 by two to give pairs/km^2 for Bibby *et al.* data for simplicity, which will probably underestimate the true densities). Willow Warblers provide the most striking and favourable comparison with densities of 300 pairs/km^2 on restocks compared with 19 pairs/km^2 on farmland, but the densities of Wrens, Robins, Dunnocks, Song Thrushes and Chaffinches are all higher on restocks than on farmland. Blackbirds, Skylarks, Blue Tits and Great Tits show the opposite trend. O'Connor and Shrubb (1986)(their Figure 2.7) suggest that total farmland bird density in Wales should be around 200 pairs/km^2; easily surpassed by Willow Warblers alone on restocks. Bibby *et al.* (1985) point out that there are reasons why Welsh restocks might be expected to be relatively rich in birds compared with restocks which are further north, those areas which were planted on poorer soils and sites where grazing by deer occurs, but the densities found on these sites are still impressive. In the past 10 years approximately 85 000 ha have been restocked throughout the country and have entered the age range of the sites studied by Bibby *et al*. If the densities of 930 pairs/km^2 approximate to average national values, then this would represent some 800 000 pairs of birds. We would not suggest that these impressive numbers of birds provide vindication for planting moorland with trees, but whenever we read that conifer plantations are birdless we think of these Welsh restocks.

Leslie *et al*. found that Tree Pipits were the commonest birds recorded in the Breckland restocks. This is probably a species which has gained considerably from afforestation. Whitethroats too were often found in the Brecklands, usually on those sites where the stumps from the felled trees had been cleared and piled into rows. These tangles of vegetation provided ideal habitats for Whitethroats to sing from and nest in.

For most of the common species found in conifer plantations we have little or no idea what sort of densities exist. We cannot say what proportion of British Blackbirds or Chaffinches nest in conifer plantations. As is often the case, the information on the rarer species is better. These species will also contribute most to the perceived conservation value of the new forests. Which uncommon British species have high proportions of their populations in plantations?

RARE SPECIES USING CONIFER PLANTATIONS

Hen Harrier

The Hen Harrier is a difficult example, so we will deal with it first. This species has often been quoted as one of those which benefits from afforestation, since it often nests in young plantations. Certainly its spread coincided roughly with the spread of upland conifer plantations but was also associated with a probable reduction in persecution by gamekeepers during the 1939–1945 war and a coincident reduction in heather burning, which increased the amount of old heather available for nesting (Campbell, 1957; Watson, 1977). Although the first nests on the Scottish mainland were found on heather moors in the early 1950s the habit of nesting in conifer plantations was becoming more and more noticeable. Watson (1977) pieced together the details of the species' spread through the Highlands and into southern Scotland; it is clear that in areas where grouse shooting was prevalent, the spread of Hen Harriers was slow or non-existent, and that young conifer plantations provided a reservoir of safe sites from which further colonization was possible. It seems very likely that the spread of this bird would not have happened were it not for the conditions provided by the foresters. Hen Harriers, of course, do not need trees around them to survive and prosper; in Wales, reached in 1961, few pairs have occupied conifer plantations and the small population manages to survive on keepered grouse moors. It is not known what proportion of the British population nests in young conifer plantations but the success of the birds appears to differ between afforested and non-afforested land (Watson, 1977). Watson studied birds in several areas and found that the number of young fledged from forest nests was nearly twice that achieved on the moorland. Interestingly though, the number of young fledged from successful nests was slightly higher on moorland than in forest sites, and the percentage of failed nests was around three times as high on the moorland sites as in the forest ones. These data suggest that although young plantations are not ideal sites for harriers to nest (since successful nests do less well here than on moorland), their drawbacks are more than compensated by the reduced chance of complete nest failure in afforested areas. Current opinions, and commonsense, dictate that the main cause of nest failure on moorlands is the activities of keepers. This suggests that the spread of the Hen Harrier has owed more to the activities of foresters than to the protection afforded by the Protection of Birds Act or the efforts of conservationists in general.

Although plantations may be occupied soon after planting, it seems that the optimal age of plantations for harriers is in the 6–15 years range (Watson, 1977). At this age the populations of small passerines in the forest are high, the trees are small enough to allow low level hunting, vegetation will have grown high enough to provide ample nest sites, and the amount of routine work carried out by foresters, resulting in disturbance, is low. Watson (1977) attributed the drop in numbers of Hen Harriers on a moorland area to the fact that the adjacent plantations had reached their optimal state. Watson (1977) found that many forest birds were taken by Hen Harriers in southern Scotland, including Black Grouse, Pheasants, Wood Pigeons, Robins, Grasshopper Warblers, Whitethroats, Bullfinches, Crossbills and Chaffinches. However, the commonest prey found at the two forest sites was Red Grouse, which suggests that the Hen Harriers were still catching much of their food outside the plantations (although Red Grouse do persist on newly planted areas). It would be interesting to know whether successful nesting by Hen Harriers in plantations is dependent on the birds having access to open moorland for hunting; if so this would suggest that the Hen Harrier would be a species favoured by mosaic rather than blanket afforestation.

As the forest grows it becomes unsuitable for Hen Harriers, although Watson points out that one nest occurred in a small open patch within an 18-year-old plantation, of which he would have known nothing had it not been found by a forester.

Montagu's Harrier

Just as Hen Harriers have avoided persecution by nesting in young plantations, so too have Montagu's Harriers; and to a sufficient extent that Parslow (1973) stated that the presence of young conifer plantations was a contributory factor to the recovery of the species' breeding numbers in Britain in the 1970s. Hope-Jones *et al.* give one example of a pair which nested in a conifer plantation on Anglesey for several years and would have prospered were it not for the depredations of egg collectors. In recent years most of the breeding Montagu's Harriers have been in cereal crops on agricultural land (J. Day, personal communication). John Day tells us that his historical research shows that in southwest England Montagu's Harriers moved into newly planted ground from the adjoining grassland until canopy closure and then the birds reverted to their former grassland sites.

Capercaillie

The Capercaillie became extinct in Britain in the 18th century, probably as a result of felling of trees. It was eventually successfully reintroduced into Scotland in the 1830s and then spread gradually from Perthshire into neighbouring areas; it is now distributed over much of northeast Scotland (Thom, 1986). The most favoured habitat of this species is the relict Caledonian pine forests of Speyside and Deeside but it has spread into commercial plantations in the surrounding areas. To be used by Capercaillie it appears that plantations must be mainly Scots Pine at least 20 years old, although firs, larch and spruce are also used. There is little doubt that although commercial plantations do not provide an ideal habitat for this species, the large acreages of planted Scots Pine in the areas around the Capercaillie's British strongholds have greatly contributed to the species' successful reintroduction and re-establishment. In recent years the move towards planting Sitka Spruce and Lodgepole Pine may mean that the more recent plantings will not grow into suitable habitats for Capercaillies. It is possible that restocking with species other than Scots Pine in areas used by Capercaillies could have harmful effects on this species' overall numbers.

Scottish Crossbill

It is somewhat ironic, considering the tension between conservationists and foresters over the conservation values of open moorland and commercial forests, that the Red Grouse, arguably the most characteristic bird of heather moorland, lost its status as Britain's only endemic bird species and was replaced in this position by the Scottish Crossbill. This is a bird of pine forests, most notably of the same natural pine forests as are inhabited by the

Capercaillie and Crested Tit, but also commercial forestry plantations if they are composed mainly of pines (Stroud *et al.*, 1987).

Scottish Crossbills feed almost exclusively on pine seeds. They are not easily able to cope with the cones of Sitka Spruce, which is now the most commonly planted tree in the British uplands. Nethersole-Thompson (1975) found that in areas where old Scots Pine and new plantations of Sitka Spruce existed close together the Scottish Crossbills only used the old pines. This suggests that, paradoxically, one of the British species most closely associated with coniferous trees will not benefit at all from the planting of the tens of thousands of hectares of new plantations even though many of these new forests will be adjacent to or within its present range. As the only endemic British species it is surprising that more research effort has not been put into discovering its needs and present status. We wonder whether this species is using the new forests so far unobserved, and whether Lodgepole Pine and Scots Pine plantations will, or could, prove satisfactory habitats for it in the future? The problems of actually identifying this species in the field are at present the major factors which bar progress in this area.

Common Crossbill

The Common Crossbill must once have been a very rare bird in Britain. It is clear that the vast majority of the present British breeding population nests in commercial forestry plantations. Not so long ago it was common to read that the British population depended on frequent top-ups from continental populations for its survival. Although almost no detailed research has been directed towards Common Crossbills (and it is strange that this situation exists), it is almost certain that there is now a very healthy British population which can hold its own without help from occasional immigrants. Because they live in a habitat which has been neglected by bird-watchers we are not in a position to say quite how much this species has benefited from afforestation. However, before the large-scale afforestation after the First World War, Crossbills must have been almost restricted to the natural Pine woods of northern Scotland. These native Scots Pine woods may not even have provided a particularly good habitat for this species which seems, from the little we know from Britain, to be better adapted to feeding on spruce rather than pine cones. The present remaining area of the formerly much larger native pine woods is around 10 000 ha (Goodier and Bunce, 1977), compared with a total of around 1 300 000 ha of coniferous high forest (Forestry Commission, 1983a,b,c, 1984). This suggests that the new forests have increased the potential nesting areas for Common Crossbills by about two orders of magnitude. Few bird species can have been so well treated by man in this country.

Comparing the BTO atlas (Sharrock, 1976) distribution with those of new county atlases, Crossbills appear to be continuing to expand their range. Apparent increases in range are shown in Gwent, Devon, the Sheffield area and Bedfordshire. In Norfolk the range appears to be stable.

British Crossbills certainly deserve more attention from ornithologists than they have yet received. Crossbills are irruptive species, moving whenever food becomes scarce and breeding in places where food is temporarily plentiful.

They have been recorded breeding in Britain in all months from August to April (Newton, 1972), and this aspect alone would make a difficult but fascinating study for someone who wanted to make their name studying this major beneficiary of afforestation.

Long-eared Owl

This is one of the many species about which we do not know enough. Long-eared Owls are forest species found over much of Britain, but nowhere are they very common. Owls' nocturnal habits do not make them easy to census and, as Sharrock (1976) pointed out, a special survey would probably be needed to reveal the species' true distribution. However, we would predict that if such a survey were to take place, it would show that a high proportion of Long-eared Owls are associated with the presence of commercial forestry plantations.

The Long-eared Owl benefits from afforestation in two ways; from increased nesting sites and from increased food supplies (Village, 1981). Interestingly it may be one of the species which benefits from patchy afforestation rather than blanket afforestation, because it apparently needs open areas for hunting. As more and more forests enter the restocking phase, it is likely that Long-eared Owls will gain even more from forestry than they have in the past. Tree shelter belts provide favoured nesting sites for the species, which can rarely be found nesting on the ground in open moorland (Thom, 1986). Long-eared Owls may lose in competition with Tawny Owls (Sharrock, 1976). Little information is available on numbers or detailed distribution of this species but it is very unlikely that increases in the area of afforested land will have done anything other than aid it

Goshawk

The Goshawk has been re-established in Britain by a combination of deliberate release of birds and escapes from falconers (Kenward *et al.*, 1981; Marquiss and Newton, 1980). It may be a commoner nesting species in Britain than is generally believed, since its nesting sites remain closely guarded secrets (although well-known to the egg-collectors and gamekeepers who are the main persecutors of the species). Published data on the numbers of Goshawks ringed in Britain appear to indicate that the population is growing rapidly (Table 2.1) although it is certainly possible that this is merely indicative of increased ringing effort. However, 144 pulli were ringed in 1988, and this seems to indicate an absolute minimum population of around fifty pairs in Britain so presumably there are in fact many more than this. The first Goshawk to be ringed in Britain was in 1967 and since then the annual ringing totals have increased in most years. Goshawks are ideally suited to living in upland plantations and will take a wide variety of different prey including Wood Pigeons, squirrels, etc. Other prey include Pheasants (which is the main reason for gamekeepers' dislike of the species) and the occasional Kestrel, Long-eared Owl, Short-eared Owl and Merlin (which will not endear the species to many conservationists).

The BTO atlas did not show Goshawks breeding in Gwent, Norfolk or Devon, but the recent county atlases suggest that breeding is now occurring in

TABLE 2.1: *Numbers of nestling Goshawks ringed in Britain and Ireland in recent years.*

Year	Goshawks
1976	4
1977	14
1978	18
1979	18
1980	36
1981	17
1982	30
1983	40
1984	56
1985	88
1986	57
1987	107
1988	144

all three counties. In the Sheffield area the BTO atlas did indicate that breeding took place (although at least some of the dots in the atlas are probably deliberately plotted inaccurately). Hornbuckle and Herringshaw (1985) describe the spread of the bird in the Sheffield area. The first proof of breeding came in 1965, and in the next 20 years the population increased to between 18 and 20 pairs, with at least 120 young Goshawks being successfully raised. Most pairs have nested in coniferous plantations along the moorland fringe.

There is little doubt that Goshawks will continue to expand in numbers and spread into more and more new areas. They have probably already overtaken the Osprey in terms of numbers, and the stage seems to be set for them to increase rapidly to a population of several hundred pairs, mostly in upland forests. The British population at the turn of the century may be as high as 200 pairs and there is little doubt that Goshawks will become increasingly familiar birds on the British scene.

Nightjar
The Nightjar is a declining British breeding species. The cause of its decline is not fully understood but there is no doubt that its range has contracted considerably. Although regarded as a heathland species, these days many British Nightjars nest on FC land. Leslie (1985) found that in the North York moors, churring Nightjars were to be found on restocked sites, often deep in the forest. There is no information on the comparative reproductive success of this bird in its heathland and woodland habitats but the evidence of a distinct preference for nesting in forests is clear enough. Perhaps a number of factors combine to make restocked sites attractive to Nightjars. First, there is the lack of disturbance from people. This is very difficult to quantify, both in terms of the actual levels of disturbance in different habitats, and in terms of the effects of disturbance on this species' success, but it may be a factor. Second, the

general lack of pesticides used in mature commercial plantations means that large insects are often abundant. Restocked sites, surrounded as they usually are by mature stands, may be very rich in food compared with heathland, particularly if that heathland adjoins agricultural land. But really it is not known why the Nightjars are on restocked sites. This is certainly a phenomenon which nobody predicted and so trying to explain it with hindsight is rather unconvincing. It provides a good example of where the commercial forest holds a much more important bird community than was ever expected.

In 1988 the RSPB started a three-year research project on Nightjars in the Breckland of Norfolk and Suffolk. This population, in contrast to the national trend, is growing in numbers. Now around 200 pairs, 10% of the British total, nest in the area. The majority of these nest on restocked sites in the middle of the forest, and the increase in numbers of Nightjars corresponds with the increase in available habitat. Restocked sites of up to at least 15 years are used; particularly areas where tree growth has been poor so that canopy closure is delayed. Although at present a success story, the planting programme for Thetford Forest indicates that the available suitable habitat for Nightjars is now at its peak in the area, and that smaller areas will be felled and replanted in the future. It is hoped that the research may indicate ways to assist this species, including perhaps suggesting management which will make the adjacent heaths, where numbers are declining, more attractive to Nightjars.

Most of the small Scottish population survives in areas of young conifer plantations (Thom, 1986).

Woodlark

Much of what has been said about the Nightjar applies equally well to the Woodlark. Most British Woodlarks breed on restocked conifer plantations, even though the heaths which have always been regarded as their core habitats are still in existence. This is another species which has declined throughout its British range during the past several decades, and now an increasingly high proportion of the total population is found in restocked conifer plantations. Like the Nightjar, this is another species which has been studied by the RSPB in the Breckland. Woodlarks appear to need areas of open ground such as those provided by forestry activities during felling and replanting. Lack saw few Woodlarks in the Breckland of the 1930s. We discuss this species in more detail in the Thetford case study (Chapter 7).

Firecrest

The Firecrest is a rare breeding species in Britain whose numbers have appeared to fluctuate since the first acceptable breeding record in 1961 (Batten, 1971, 1973). Norway Spruce plantations of at least eight metres in height appear to be the most favoured habitat but other conifers (Douglas Fir, Scots Pine) are sometimes used. A small number of predominantly broad-leaved sites have held Firecrests, and these sites have all had Holly associated with them. This species may still be overlooked and we would expect that it could be found in small numbers in conifer plantations anywhere in Southern England.

Black Grouse

This is a species of the moorland edge which has adapted well to the presence of young conifer plantations in the uplands. It has undergone changes in numbers at the local and national level, coincident with a cyclical pattern of change which makes it very difficult to assess overall population changes (Parslow, 1973; Thom, 1986). Certainly there have been declines in some areas of the country, while in other places colonization has taken place. Only 150 years ago this species occurred in counties in the English Midlands and southern and eastern England. It was still recorded in Hampshire, Wiltshire and Worcestershire at the beginning of this century, and only recently has the bird disappeared from Somerset and Devon (Cadbury, 1987). Colonization in other areas has been closely associated with new afforestation. In eastern Sutherland, Black Grouse are present in the plantations around Lairg and north to Altnaharra and Borgie, and so are poised to occupy the large areas of new plantations further east in the Flow Country.

RSPB research, funded by the FC, in Wales has shown that the ungrazed vegetation in young plantations provides food and cover for Black Grouse. After canopy closure, Black Grouse numbers drop in the plantations, but if some suitable areas are provided for them then they may remain in reduced numbers. Areas of poor growth and windthrow may help to support small numbers of grouse until the restocking stage, when it is possible that their numbers may rise again. The research in Wales may suggest ways in which the mature conifer crop and restocked sites can be made suitable for Black Grouse.

In the Sheffield area this species has declined considerably from its numbers at the end of the last century, when it was numerous. There appeared to be an increase in numbers in the 1930s which was associated with afforestation but as time passed, and the trees grew, numbers fell. Whether or not this was due to the reduction in moorland feeding areas due to afforestation is unclear. Lovenbury *et al.* (1979) and Yalden (1978) suggest that the maturing of conifer plantations has been one adverse factor and another has been increased grazing pressure by sheep. Whether or not either factor alone would have been sufficient to cause the decline is not known.

Grasshopper Warbler

This is one of those species which is little studied or understood. As an inhabitant of wet scrubby areas it is at threat from agricultural improvement of land. However, it is also likely to benefit from the early stages of afforestation up until the time when the canopy closes, and is a colonist of restocked sites. Thom (1986) suggests that new plantations are as important a habitat for this species in Scotland as the damper areas with which it is more commonly associated in peoples' minds. In Kirkudbright, young plantations near the altitudinal limit of planting have been colonized (A. D. Watson, quoted in Thom (1986)). This is exactly the sort of species which, because it is easily overlooked, could be lurking in Britain's upland forests in interesting and important numbers without being noticed.

Woodcock

The Woodcock in the past 150 years has greatly expanded its British range from England northwards into Scotland and westwards too. It has decreased in numbers in some counties due to felling of woodlands, but these losses have probably been more than compensated by the new areas of conifer plantations which Woodcocks use in the Breckland, New Forest, north Wales and southwest Scotland (Parslow, 1973). Although a woodland bird, this was a species which was originally thought to be one which would not benefit from afforestation (Thom, 1986). These days it is abundant in the conifer plantations of southwest Scotland and is found as far north as the north coast of Sutherland in plantations (personal observation). Thom (1986) suggests that afforestation may eventually lead to the colonization of Lewis.

Short-eared Owl

A result of the removal of grazers and the application of fertilizers prior to planting is an increase in the numbers of voles living in an area. Short-eared Owls often benefit because they prey on voles. In areas of high vole numbers the territories of the owls shrink to maybe a seventh of their original size, and clutch and brood sizes grow enormously (Lockie, 1955). Other predators such as Kestrels, Long-eared Owls and Foxes also benefit from this short-lived glut of food (Picozzi and Hewson, 1970). However, we know of no published data which demonstrate exactly how long this increase in food for predatory birds lasts.

Afforestation in Scotland since the early 1970s has resulted in changes in range since the atlas fieldwork (Thom, 1986). Where afforestation has occurred on predominantly grassland areas, Short-eared Owls have benefited greatly from the new plantations. As with the Black Grouse and Hen Harrier, the gain from afforestation is in the early stages only, and Short-eared Owls cannot survive in the mature forest. However, the increases in reproductive success during the short stage of forest establishment are so great that it seems highly likely that the benefit is truly a net benefit for this species. Sharrock (1976) states that the general trend in numbers has been upwards and that this is undoubtedly due to the areas of young forestry plantations which have become available. This has led to the Short-eared Owl now being more firmly established in Britain than at any time in the past century.

Siskin

Thom (1986) regarded this species as being potentially the one which has benefited most from afforestation, and Parslow (1973) remarked on the large numbers of counties in which breeding has been recorded for the first time since 1942 as a result of afforestation. Its original habitat in Britain was the Caledonian Scots Pine forests of Speyside and Deeside, to which areas it was formerly probably restricted. Since the 1950s there has been a massive expansion in range and Siskins are now found breeding in mature conifer plantations over most of Scotland (including some of the larger islands: Arran, Islay, Mull and Skye), northern England, Wales, East Anglia and the New

Forest. As further conifer plantations mature they will provide ever larger areas of suitable habitat, and there is no reason why the species should fail to spread to every English county with areas of suitable habitat. It appears that it is spreading in counties which in the BTO atlas years held few pairs. The Gwent, Norfolk, Sheffield and Devon atlases all show apparent increases in range. In the Sheffield area the first indication of possible nesting was in 1965 but now there are few mature conifer plantations which are without breeding Siskins.

The Siskin provides an example of a species which has readily adapted and benefited from the introduction of large areas of alien conifer species into the country. On the continent Siskins rely to a great extent on spruce seeds and in Britain they have rapidly adapted to the alien Sitka Spruce. In Ireland the introduction of Scots Pine may have led to Siskins becoming established there in the second half of the 19th century (Parslow, 1973; Deane, 1954). It is likely that its population has increased 1000-fold in Britain during the past 30 years and is still increasing in many areas (Thom, 1986). The attractive spectacle of Siskins feeding on peanuts in suburban gardens has probably been greatly helped by large-scale afforestation of the uplands.

Redpoll

The Redpoll cannot now be regarded as either uncommon or rare in Britain, because Redpolls have benefited from afforestation in only a slightly less spectacular way than the Siskin. Redpolls inhabit conifer plantations at an earlier stage than do Siskins; often in the pre-thicket stages. There has been a marked increase in range in England and Wales since about 1950 (Parslow, 1973), and in many Scottish counties the numbers are still increasing (Thom, 1986). Thom suggested that the increase in numbers in Scotland was likely to continue as more and more forests mature to suitable ages for Redpolls to use. Parslow regarded the Redpoll as being the species which had benefited more than any other in Wales from afforestation although this status is probably now challenged by the Siskin, which until recently did not even breed in Wales. Condry (1960) noted that Redpolls were scarce and local up until the 1920s but that by the late 1950s they were breeding in young conifer plantations across north and central Wales from sea level up to nearly 2000 feet. In Gwent the BTO atlas showed Redpolls as being nearly restricted to the upland parts of the county but the Gwent bird atlas suggests that, whilst the stronghold is still in the hilly west of the county, the species has spread eastwards into plantations in the Wye Valley. Redpolls bred on Lundy in 1952 and in mainland Devon they were first proved to breed in 1957 (Dare, 1958). The Devon atlas suggests that a small expansion in range has occurred since the BTO atlas years and that this may be due to the Redpoll's spread from the areas of forestry plantation, which it initially colonized, to deciduous woods. Only 6% of Devonian tetrads were recorded as holding Redpolls. In Bedfordshire all 10-km squares contain them, and 42% of tetrads. They were singled out in the Bedfordshire atlas as the species which more than any other has expanded in numbers in the county; only one nest was known before 1894. The

Norfolk atlas indicates a very patchy distribution, and one which appears slightly reduced from that of the BTO atlas.

Redwing

This is not a species which would spring to most peoples' minds as being a candidate for gaining from afforestation, but Thom (1986) mentions that as well as nesting in Birch scrub, oak-woods, gardens and pine forests (Sharrock, 1976; Williamson, 1973), Redwings also nest in young plantations which have Birches to act as song posts. In most areas of upland afforestation it is open moorland which is being afforested, areas which would be unsuitable for breeding Redwings, so this species does seem to be one which will gain from afforestation. Since it is apparently birds of Scandinavian origin that stay to breed in Britain, it is likely that plantations in northern Scotland would be the places to look for breeding Redwings and we know of Sutherland plantations where Redwings breed. At least 26 pairs of Redwings were proved to be breeding in Sutherland in 1984 and it is almost certain that many are overlooked. In 1984 a maximum of 78 pairs of Redwings were reported as nesting in Britain, and the report of the Rare Breeding Birds Panel noted that many of these were found in Sitka Spruce plantations at heights of up to 25 feet.

It seems likely that the large areas of new plantations in northern Scotland will eventually provide homes for breeding Redwings and that forestry will have been a very major influence in helping them to become established as a regular, and increasing, British breeding species.

Barn Owl

Barn Owls are able to benefit from the increase in vole numbers in the early stages of afforestation providing that there are suitable nest sites which they can use. Thom (1986) regarded the new forestry plantations as having created much valuable new habitat for the species. Although the amount of evidence is small, it was thought that the Barn Owl population in southwest Scotland was probably remaining stable whereas decline had occurred in the more intensively farmed eastern and central belt.

Considering the declines in Barn Owl numbers which have occurred widely in lowland Britain the opportunities for local and temporary increases in numbers in the uplands are very welcome. It seems that the provision of nesting sites is a critical factor in encouraging them to use new upland plantations, and that when nest boxes are provided they can become as characteristic of the early stages of afforestation as are Short-eared Owls in many upland forests.

Crested Tit

In Britain this is a species of northern Scottish pine forests, although on the continent of Europe it lives also in spruce forests. Why there should be this difference is not at all clear. It is likely that if continental Crested Tits ever found their way into southern England in sufficient numbers to form a small founder population they would spread, since their continental range encompasses many areas similar in habitat to those of southern England. Cook (1982) estimated that about 40% of British Crested Tits inhabit Scots Pine plantations which are over 20 years old. Plantations of this age may support more insects than younger ones and are more likely to provide dead wood suitable for excavation by nesting Crested Tits. Crested Tits have increased in range this century largely because they have adapted to the new plantations. Despite the fact that pine plantations appear to be less favourable habitats for Crested Tits than the remaining Caledonian pine forests, the enormous areas of this sub-optimal habitat actually hold a very high proportion of the total population. This suggests that new afforestation, providing it does not take place on sites of existing old natural pine forests, will provide more opportunities for the species to spread and increase. The current age structure of different species of tree in the North of Scotland Conservancy of the FC means that during the 1970s very large areas of Scots Pine reached the apparently critical age of 20 years, which would provide a possible mechanism for the continued expansion of the population from the position shown in the atlas which was commented upon by Cook (1982). Further large areas have matured during the 1980s but only relatively small areas of Scots Pine have matured during the 1990s because the planting in the 1970s came to be dominated by Sitka Spruce and Lodgepole Pine. It remains to be seen whether Crested Tits will regard Lodgepole Pine plantations as favourably as they do Scots Pines.

POTENTIAL COLONISTS

The new forests have already resulted in massive regional changes in bird populations. Crossbills, Siskins and Redpolls now breed in conifer plantations spread throughout Britain, including many southern counties which they would never have colonized were it not for the spread of conifer plantations. There is considerable scope for more species to benefit from the large areas of new plantations which have spread to the very north of Scotland. The plantations of Caithness and Sutherland provide large areas of continuous tree cover in areas where none has existed for centuries at least. Is there scope for Scandinavian woodland species to use these forests as a route to colonize Britain? Already colonizing Britain are species including Wood Sandpipers, Redwings and Fieldfares, birds which are characteristic of Birch and willow scrub mixed with open wetland, the 'forest bog' of the northern forest. Britain's best natural forest bogs are on Speyside but further north the trees are absent and solid Lodgepole Pine around a Caithness dubh lochan system is very different from scattered broadleaved scrub. Although breeding mainly in open country in Britain, the Greenshank is more characteristic of mixed wetland and woodland habitat in Scandinavia and Russia (Ratcliffe and Thompson, 1988; Knystautas, 1987). Greenshanks used to breed in the forest bogs of Speyside (Nethersole-Thompson, 1959) and now persist in the young Flow Country forests (personal observations; Nethersole-Thompson and Nethersole-Thompson, 1979, 1986). The Bluethroat nests in low shrubs around the edges of bogs, and the Brambling in larger Birch scrub. Both are exceptionally rare breeding birds in Britain. They might be commoner given the right habitat. Where ground has already been ploughed and trees planted close to dubh lochan systems in Sutherland and Caithness, it might be possible to create the right sort of tundra edge woodland by planting the first 50 m of ploughed ground with low-growing broadleaved scrub of Birch and willow.

The Scandinavian forest avifauna contains some species which we tend to think of as southern European. The Red-backed Shrike (and Nightjar) are found on Scandinavian restocks (Hansson, 1983). Red-backed Shrikes have bred in northern Scotland in the last decade but appear to have died out again. It has been suggested that climatic change triggered these arrivals from Scandinavia, and in contrast to some other cases where climate is used as a reason of last resort, in this case the number of Scandinavian species involved does tend to support the hypothesis. Perhaps climatic change is now particularly favourable for some Scandinavian species to take advantage of the new habitat of the forests. Birds that regularly pass through Britain on migration like Red-backed Shrikes and Wrynecks have the opportunity to colonize new regions.

Several species that seem best-suited ecologically, the boreal forest owls and woodpeckers, are unlikely to arrive here unassisted because they are not migratory. The Ural Owl, considerably larger than the Tawny Owl, is the medium-sized owl of northern Swedish conifer forests. Tawny Owls are confined to the southern third of Sweden and barely overlap in range with Ural Owls. Tengmalms Owl and the Pygmy Owl are smaller species characteristic of the boreal conifer forest, but both appear to suffer in competition with Tawny Owls, the latter showing a preference for slopes and valleys, and Tengmalms Owl being confined to flat plateaux. The even smaller Pygmy Owl comes at the bottom of the pecking order, in the Alps only occurring above the highest Tawny Owl territories. Both species occur in southern Sweden and Norway and in the mountains of western Europe. All three have a preference for spruce forests (Cramp and Simmons, 1985). At the very top of the tree is the Eagle Owl. It needs seclusion and a mixture of open land and forest and the remote forests of the far north might be large and quiet enough to support an Eagle Owl population as they move into the second rotation.

The Black Woodpecker could be the single most important absentee from British forests, because it is the only species large enough to create the natural nesting holes on which birds like Tengmalms Owl and Goldeneye depend. Black Woodpeckers were apparently introduced to Britain by the release of seven or eight birds near Brandon in 1897 (Dutt, 1906), but how long any survived is not clear; although there were by then many pine shelter belts in Breckland, the overall tree cover on the Suffolk–Norfolk border hardly resembled the species' natural habitat of extensive forest. The prospects of them colonizing naturally from the north are remote but Black Woodpeckers are extending their range in western Europe, are now close to the English Channel and have apparently spread to Dutch islands further from the mainland than the width of the English Channel. The delightful Three-toed

Woodpecker, again characteristic of conifer forest, could also colonize if it moved. A woodpecker of conifer forest, it feeds on spruce bark beetles.

Some irruptive species which breed further away than southern Norway may be more likely to breed in Britain than the woodpeckers and owls. They are opportunistic; often fleeing a food shortage, they may be ready to settle down and breed wherever a food supply occurs and they arrive in Britain in reasonable numbers. Waxwings have stayed into the summer and shown territorial behaviour after big irruptions (Sharrock, 1976). They are birds of open conifer and Birch forest, and although a real long shot, this is another possible bird to encourage imaginative habitat creation in the new forests. The Nutcracker, again very much a conifer forest species, occurs in southern Norway, but, although irruptive is far rarer in Britain than the Waxwing, and there has not been any suggestion of possible breeding. Usually seen singly, it is unlikely that a Nutcracker would find a mate, even if a suitable habitat were available. In contrast, the Siberian Jay's range extends almost as far south as the Nutcracker's but it has not been recorded in Britain at all. It too is a bird of dense, coniferous forest.

Nutcrackers and Waxwings are hard to misidentify: the opposite applies to the most likely irruptive colonists, the crossbills. They are always on the move and readily settle to breed in suitable habitats, as continuous recolonizations by Common Crossbills show. A celebrated pair of Parrot Crossbills bred successfully for several years in the Holkham pines on the north Norfolk coast, but this is the key to the crossbill problem: the Holkham pines are a well-known place for migrant rarities and are closely watched by expert birders. Crossbill species are hard to tell apart even with good views, which are rarer in large forests than in the smaller conifer woods in which irruptive parties are often spotted. In Yorkshire (Mather, 1986), most records of Parrot Crossbills and Two-barred Crossbills are from small, well-watched conifer woods in the southwest of the county. There are a few east coast records, but none from the forest. Both are easterly species; Parrot Crossbills do breed in Finland and occasionally in East Germany. Two-barred Crossbills prefer European Larch, a widespread genus in the Siberian Taiga, and are rare breeding birds in northwest Europe. It is quite possible that either species could have disappeared into the depths of the large conifer forests and already be breeding occasionally. Pine Grosbeak, also a finch of conifer forest, is another occasional vagrant in Britain; Fair Isle, where many such wanderers turn up is not that far from the Flow Country.

An unexpected species which has made an appearance in the new forests is the Greenish Warbler; its normal range extends into West Germany but a male was singing in a Scottish Sitka Spruce plantation in the mid-1980s. The prospects for the new forests are good and there are probably several more outsiders that the speculating armchair ornithologist would identify as possibles. The real teaser is that the huge area of new forest is so poorly covered by birders, especially in the far north, that it is quite possible we already have breeding species of which we are completely unaware.

CONCLUSION

In this chapter we think we have made out a good case that there are interesting birds to be found in Britain's commercial forests; in fact we have been slightly surprised ourselves while researching this book at the number of beneficiaries from afforestation. We feel that the real picture is probably much better than we have been able to portray, because little serious survey work has been done in these plantations; much remains to be found. Perhaps the first breeding records of Greenish Warblers in Britain will be in a Sitka Spruce plantation on a Scottish hillside, but almost certainly more Goshawks and Redwings are there to be found. With the limited amount of information at our disposal we are unable to convincingly show, one way or the other, what is the true conservation value of British commercial forests. However, the mere fact that the information is so meagre is enough to prove that the strongly held views about the low interest of birds inhabiting conifer plantations are based partly on prejudice rather than on evidence. Compared with the effort which has gone into surveying the moorlands which plantations will replace, practically no information has been collected on the birds which inhabit the forests themselves. The work of the RSPB in Wales (Bibby *et al.*, 1985, 1989b) stands as the first large-scale investigation of the birds of commercial plantations. In that study, surprisingly high densities of birds were found on restocked sites, which just supports our contention that much remains to be discovered in these exciting new forests.

As well as a surprisingly high number of rare species thinly distributed around the British commercial plantations, we have demonstrated that they also hold large numbers of common or relatively common species. As pressures on the countryside increase the new forests are likely to hold ever increasingly more important fractions of Britain's common birds.

The suggestion that commercial forests can hold important bird species in Britain should be greeted with enthusiasm by conservationists but we are rather resigned to the fact that it will not be. The stated position of the NCC is that however good these forests are, they will never be as good as the moorland they replace (Nature Conservancy Council, 1986). This is a claim which we will leave until later to examine in more detail, but it is worth pointing out that the findings of the present chapter render that extreme stance difficult to justify as a generalization. In assessing the potential effects on bird populations of commercial forests, we feel it is incumbent upon the conservationists to get out into the forests and assess their value rather than prejudge the issue. It may be that in many areas conifer plantations will hold relatively few interesting birds, but that remains to be conclusively demonstrated. We are sure that there are plenty of interesting birds out there to be found. In recent years there have been encouraging signs that conservationists are taking a more positive attitude to commercial conifer plantations. For example, the RSPB has current research projects on Woodlarks, Nightjars and Black Grouse which are funded by the FC. We believe that more research into the birds of commercial plantations would strengthen the conservationists' cause when dealing with foresters. At present there is an unfortunate tendency to state that plantations are of low conservation interest, which means that those (many) foresters who are good naturalists learn to doubt all of the conservationists' pronouncements. Ignoring the present value, and future potential, of British conifer plantations does the conservation cause no good.

The fact that commercial forests hold some nationally rare breeding species means that foresters have a responsibility to at least safeguard those species, but we would go further. It should be possible to design commercial forests which are even more attractive to birds. In the next chapter we describe how forests are managed for tree production (which is their main purpose) and in Chapter 4 we discuss conservation management within plantations.

CHAPTER 3

Managing the forest

Britain's new forests are a vast experiment which is only now completing its first full cycle. The story of the creation and management of these forests is one of rapid technical development, based on a cocktail of theory and empiricism. Big risks were taken and there have inevitably been mistakes and surprises, but the biggest decisions have turned out better than might have been predicted.

The most important thread running through the creation of the new forests is technological development in the establishment and growing of trees. When the FC was founded in 1919 it was faced with afforesting sites far poorer in terms of soil, elevation and exposure than most existing planted forests. Limited recent experience of planting on poor sites came from the work of Sir John Stirling Maxwell at Corrour (Stirling Maxwell, 1929) and the Crown Office of Woods at Inverliever and Hafod Fawr in Merioneth. It was clear that simply planting trees straight into unprepared ground would not work on infertile, wet soils, and the FC established a research branch in 1920 with the objective of finding out how to get trees to grow (Wood,1974). Since then every decade has seen at least one development that has transformed the approach to forest management. Species selection, cultivation and nutrition have played particularly important roles in this development.

TREE SPECIES

Lack of a post-glacial land bridge to Scandinavia left Britain with only one major commercial conifer species, Scots Pine. Although it produced high-

49

quality timber on dry sites, Scots Pine did not perform well on the wetter soils that foresters were faced with planting after 1919. The obvious answer was to select European species, and Norway Spruce featured prominently in the early plantings, especially on wetter ground. The concentration on the North American Pacific coast species, especially Sitka Spruce but also Douglas Fir, was arguably the greatest single risk taken in British forestry. The species had been present in Britain for over 100 years and had thrived in the British climate, but their use had been mainly as specimen trees on good soils in arboreta or around large houses. Planting for forestry had been largely with Scots Pine, or larch.

The choice of Sitka Spruce had profound effects on the subsequent shape of British forestry. Sitka Spruce has performed so well as to be both a boon and an embarrassment. It is a tough tree; easy to establish and hard to kill. It grows faster than most other species on wet sites, almost invariably outstripping Norway Spruce in the uplands. It is a valuable timber tree; its pale wood reduces the need for bleaching pulp. Straight, regular sawlogs are ideal for modern high-throughput mills, and its sawn timber is gaining wide acceptance for structural use. Although high-yielding, Sitka Spruce cannot compete with Russian or Scandinavian timber for joinery; the narrow ring width of imported wood is the result of growth rates perhaps 25% of those of British timber. At restocking, prickly Sitka Spruce is most resistant to deer damage and bounces back quickest after browsing. Other species outgrow Sitka Spruce under certain conditions; it needs moisture and is replaced by pines on dry east coast soils, and on better soils it is outgrown by the equally valuable Douglas Fir. Species that have not been extensively used in British forestry, like Grand and Noble Firs, may locally outyield it, but tend to be flawed by problems such as difficult establishment, vulnerability to deer and pests, and concern over timber quality (Aldhous and Low, 1974). Almost the biggest problem of Sitka Spruce for the forester is its outstanding success; as its versatility and reliability have become established over more and more site types, it becomes increasingly more difficult to justify the planting of alternative species.

CULTIVATION

Early plantings direct through the grass mat into waterlogged peaty gley soils (predominantly clay soils overlain by a layer of peat) at Kielder killed even Sitka Spruce and the trees that lived did not grow. Once established, trees can dry out a wet site by transpiration, but young trees straight from the nursery cannot get started with waterlogged roots. The need for some kind of site improvement to give the trees a chance was obvious, and the cultivation story, which continues today with new methods for felled areas, began. The first successful method was 'Belgian turf' planting. A large turf is cut with a spade and turned over, and the tree is planted in the middle of it. This gave local drainage, weed suppression and release of nutrients but was labour-intensive, did not greatly improve drainage and was not deep enough to break indurated pans on dry sites. Subsequent research showed a relationship between early tree growth and the amount of soil disturbed.

Up to 1939 the main obstacle to progress with cultivation was lack of sufficiently powerful tractors, but agricultural ploughs were too weak to give the required depth of cultivation. Development centred around draining wet soils (Zehetmayr, 1954; Neustein, 1976a). Heathland planting was very important in the early years, and the other major area of cultivation research was breaking the indurated pan (a hard layer of oxidized iron through which tree roots could not penetrate and which also impedes drainage) of dry heathland soils (Zehetmayr, 1960; Neustein, 1976b). Research continued through the 1930s, but it was not until the early 1940s that real progress was made through the introduction of powerful tracked tractors and the RLR plough for dry soils and Cuthbertson plough for wet soils, both larger and tougher than anything used in British agriculture up to that time. That foresters realized the benefits of ploughing is illustrated by 1930s experiments in which furrows were turned by hand to simulate tractor ploughing!

The Cuthbertson plough made the vast planting programmes of the 1950s and 1960s possible. Although it was of equal importance to dry heathland sites before 1945, the wet gley and peat soils dominated later planting programmes. Continuous-furrow ploughing has major advantages over the 'patch' cultivation of individual turves. Each furrow acts as a local drain, leading large quantities of water into a drainage system, itself created by ploughing. It makes a multitude of bare planting sites, and a far greater quantity of soil can be disturbed than by a person with a spade. It is mechanically easier and quicker to produce a regular continuous furrow than intermittent patches.

NUTRITION

Ploughing gave access to new sites, and with them came new problems, and the third major thread of technological development, tree nutrition. Like any plants, trees need nutrients to grow. On infertile sites some essential elements were in very short supply or locked up in the soil. Most important was lack of phosphate. Britain is a phosphate-deficient island, and application of phosphate as basic slag began before 1939 and in various forms has subsequently been used on most upland sites. Phosphate tends to be locked up in the soil and released only gradually. This is a disadvantage in agriculture, as rapid crop growth may be limited by lack of phosphate and further additions do not necessarily increase availability to the plant. This tends to be an advantage in forestry, because one application will provide benefit over an extended period. Frequent fertilizer treatments early in the rotation will reduce the overall return on investment because of the long time before they generate income. The quantity applied matters more than its precise form (Atterson and Davies, 1967), and cheap, ground phosphate from Gafsa in North Africa is the most frequently used material. The poorest soils, especially peats, are also deficient in potassium, which is the only other nutrient applied widely to forest crops.

Nitrogen has rarely been used as a forest fertilizer, despite a number of nitrogen deficiency problems because it is quickly leached from the soil so that any gain is short-term. However, developments other than direct nutrient inputs solved this problem. Nitrogen is more often locked up and unavailable rather than being absent from the soil altogether. This is especially the case with heather and spruce; pines can grow through heather, but spruce cannot obtain nitrogen and goes into check, turning yellow, stopping growing and eventually dying. The same occurs on infertile, deep peats. Large areas have been planted with pine that would otherwise have had Sitka Spruce. Two solutions have been found, the killing of heather with the herbicide 2,4-D (Mackenzie *et al.*, 1976), and crop mixtures in which a second species, either larch or pine, seems to release nitrogen to the spruce. Urea has also occasionally been used as a nitrogen source to release Sitka Spruce from check; once the canopy closes the heather is shaded out and the problem disappears. The action of mixtures is not understood but in some situations and with some combinations of trees they seem to work.

These threads of developing knowledge have run through the day-to-day work of establishing the new forests over the last 70 years. As the land available to forestry has become poorer and poorer, they have continually met the need for ever more sophisticated methods of making trees grow.

Ploughing deep peat. Caterpillar equipped with extra-wide 'apical' tracks for flotation on the softest ground and double mouldboard 'trailed' plough. The wings above the mouldboard push the ribbon of peat back from the furrow, allowing the tree to spread its roots more evenly over the drained ground between furrows.

AFFORESTATION

The process of new planting, the afforestation of previously bare land, begins with land acquisition. The FC has compulsory purchase powers; its only attempt to use them on any scale caused such a political storm that it was never tried again. The land on which the new forests are planted has all been sold to forestry voluntarily or has been planted by its existing owners; that much has changed hands because of hard times in agriculture has caused resentment, but hill farmers in many parts of the country still see the option of selling to forestry as much as a financial longstop as a threat.

Once an area has been acquired and planted, there is an incentive to make further purchases to build substantial forest blocks. Until first thinning (15–25 years), planting of new ground is the main work and a steady programme of planting is essential to maintain a labour force in productive work; this was particularly important to the FC up to the 1970s because of its social objective of providing employment in rural areas. A constant supply of plantable land remains equally important to private forestry companies today. In contrast to the FC, which is involved in everything from new planting to harvesting, private companies tend to specialize in either planting and management or harvesting. Use of contractors reduces the commitment of management companies to the long-term maintenance of a labour force, but contractors need continuity of work if they are investing in expensive equipment like ploughing tractors. The considerable expense of supervision is also best spread over a large area. Looking after thicket-stage crops that need little work but still need a labour force to put out fires and maintain fences is cheaper alongside a new planting programme. Planting over a number of years will eventually provide a more even timber supply.

Establishment of forestry in an area may mean that the existing economy is fragile. A very high proportion of land sales to forestry originate through personal contact, and once foresters became an accepted part of a community and were known to be interested in buying land, continuous sales over several years often resulted, and forests that started with a single, small acquisition built up into very large blocks. Very few of the large FC or private forests have resulted from a single land purchase.

Money invested in land must start producing a return as soon as possible, which means planting trees as soon as possible after the land is purchased. So large areas are planted quickly; the main limit to speed is the size of the labour force, which tends to be determined by future work prospects. Glen Orchy Forest is an example of the very large programmes undertaken by the FC (Forestry Commission, 1974a); in financial year 1972–1973, 849 ha were planted, leaving a plantable reserve of 3555 ha. The incentive to create large estates is less for private companies because the size and time-scale of investment tend to be more important for individual investors than maximum economy. Individual ownerships of over 1000 ha of new upland plantings are rare. However, some large private forests have been created as a mosaic of multiple ownerships, most notably the Economic Forestry Group's Eskdalemuir during the 1970s and Fountain Forestry's Sutherland and Caithness estate during the 1980s.

The use that land within each acquisition is put to has changed over the years; from the start of the FC into the 1960s, the best land for agriculture tended to be set aside as smallholdings for forest workers. Forest villages were also established. Transport for most workers was limited to a bicycle, so people had to live close to their work. Petrol was roughly three times more expensive in proportion to a forest craftsman's pay in 1939 than in 1980. People's habits have changed as a result: there has been an increasing move towards living in established villages and travelling to work. With the advent of harvesting and the passage of time a generation of real forest workers has evolved: if a harvesting worker wants to earn some extra money at the weekend, he buys a parcel of timber or contracts for a timber merchant rather than grazing some sheep or milking a few cows. The large communities which were expected to live in places like Kielder never materialized: half-way through the construction of Kielder village, chainsaws came into common use and the estimates of labour needed to fell the forest dropped dramatically. The village's huge community centre is a monument to the age of the axe and the cross-cut saw.

By the mid-1960s new smallholdings were rarely being set up, although many are still worked today. Where farmland was kept it was generally in larger units making up a whole hill farm, usually at the insistence of the agricultural departments rather than as forestry policy. At the same time, the quality of the land acquired had declined so there was a far lower proportion of the better land suitable for smallholdings. The policy of keeping smallholdings rather than planting all open ground has had a tremendous influence on the landscape and wildlife of the older forests. At Dalby, Grizedale or Coed Y Brenin, for example, a beautiful mix of open land and woodland with long, often broadleaved, edges has developed. Places like the Dalby Valley come close to the 1/3 open ground 2/3 forest suggested as an aesthetic ideal by Lucas (in preparation). Smallholdings, perhaps because they were a second job, are often unimproved grassland. Fields running down to the Dalby Beck are the feeding place of Moorhens, ducks and Kestrels, all of which find shelter and nesting places in the dense cover of the forest. Although these forests cannot be compared with later plantings because their soils and sites are far richer, they demonstrate the value of open ground within the forest. It is important that later forests should become more like them in time, and that the open ground of smallholdings is maintained even if its economic importance has declined.

From the 1950s to the end of the 1970s the main objective of afforestation was to plant as high a proportion of the land acquired as possible. Planting was pushed as far as technology and tree growth allowed, sometimes rather further than economic targets justified, as the low economic return on large areas of Lodgepole Pine in north Scotland show. Topography and soil, especially shallow rocky soils and very wet soils, determined the pattern of planting. Regular, easily plantable sites, the Border hills for example, were planted almost solidly, but steeper, rockier sites were of necessity more varied. Some Highland forests have considerably more unplantable than plantable land. Ground which is too steep, or too high, to plough, means that out of 39 617 ha

An example of provenance problems with Lodgepole Pine: although growing well, the bend in the stems of these trees, known as 'sweep', means they will never have value as saw timber and will present big problems at harvesting. The species does, however, show signs of living up to its reputation as a 'pioneer' for the most difficult sites. Its roots can grow into waterlogged soil and by transpiration dry it out.

acquired in the FC's Fort Augustus forest district 19 673 ha will not be planted (Forestry Commission 1986). By the mid-1970s this approach had been taken to an extreme as technological and silvicultural advances meant that almost any land could be ploughed and some sort of trees grown on it. More consideration started to be given to economic return on marginal sites, and also other factors such as provision for deer and the protection of fresh water. However, before ideas started to change, over 500 000 ha of forest were created (Locke, 1987).

The economic approach of the FC from the mid 1960s onwards had a profound effect on the way forests were planted. It underpinned the restructuring of the FC's accounts in 1972 (HM Treasury, 1972) and was the basis for assessment of performance against the 3% target real rate of return on investment set by the Treasury at the time. There was a real need for a more stringent approach to expenditure in the FC. Prior to the introduction of the concept of Nett Discounted Revenue (NDR) (Johnston *et al.*, 1967), the main financial control in the FC had been on year-to-year expenditure budgets, a patently inadequate approach to a very long-term investment which includes the need to compare the relative costs of different inputs at different times in the rotation. NDR enables costs today to be compared with later income through discounting and compounding to a common date; thus a cost at the beginning of the rotation is compounded up at a particular interest rate to the year at which an income is expected. The NDR concept allows investments and incomes at different points in the rotation to be weighed against each other. As such it is highly effective and as a planning tool represented an advance on the same level as the Cuthbertson plough did in tree establishment.

However, like many good things, it has proved easy to have too much of this particular concept. The NDR system was extended way beyond the bounds of validity to become almost a cult; for the numerate it is possible to reach almost any desired result by tweaking the many variables in the economic model. The long time-scale of forestry means that even small alterations early in the rotation can have a disproportionate effect on the final return. A failure to recognize this has also led to another major fault, ascribing a high degree of certainty to prescriptions. The alacrity with which the system was grasped and the failure to acknowledge the problems were probably connected with the uncertainty for the future felt by a forest industry cast adrift from the old mooring of strategic timber supply.

The less numerate just got out their economics tables and applied them with rather less understanding and discrimination than they would have applied to a thinning management table. Discounting and compounding are complicated concepts and many people were confused by them. However, this was the way things were going, and questioning NDR was liable to be met with a barrage of incomprehensible calculations and a suspicion of stupidity in the questioner. As a result many forest managers just stopped thinking and applied the rules.

The rigid application of this economic approach has probably led to more ecological harm than any other aspect of afforestation, because it aimed the big guns of the powerful new technology. It increased the pressure to plant as much of the land within each acquisition as possible, because by 1970 it was always a struggle to make the sort of land available to forestry show a projected return close to 3%. 'Efficiency' became the only thing that mattered, and with it a more rigid industrial 'tree farming' approach than ever before. Small features, wetlands for example, were eliminated wherever physically possible. However, economic targets were often not met. The worst sites often cost far more to establish than model costs and were most likely to fail because they were too wet, too high or were frost hollows. Even had the Lodgepole Pine plantations of the north lived up to expectations silviculturally, they would still never have made a 3% return.

An important area where great ecological harm was done but model costs and silvicultural targets were rarely achieved was underplanting of broadleaved woods. Again, the proportion of an acquisition made up by a Birch wood often had to be included to reach the target rate of return for what was probably a relatively expensive bit of better land. This sounds like dishonesty, which it was not really, at least not conscious dishonesty. The need to meet both economic and planting area targets created an irreconcilable tension for field practitioners. Theory and practice simply became detached from one another. Forestry is vulnerable to unchecked enthusiasms, as is conservation, because the end result is so far off and so many untestable assumptions have to be made. A farmer operating a flawed system would be bankrupt long before the end of the 15 years over which this hard economic approach was unquestionably the dominant force in forest planning.

ACCESS

The benefits of the NDR system for investment appraisal are not in doubt in planning the provision of access to the planting area. Access for people, plants and equipment is a preliminary to planting. The road and ride system of the forest is the main artificial division between blocks of productive trees. However, whilst the final lines for roads may be laid out at or before planting, lorry standard roads are not now normally built at the start of the rotation in FC forests. Fully metalled roads are an expensive part of the forestry operation, and to make the capital commitment required to construct a 30-tonne weight limit road at the start of the rotation is not good economics, because it is years before it is actually used. It will usually be cheaper to ferry workers and materials in by Land Rover or all-terrain vehicles.

The best time to build roads was rather different for the private investor when it was possible to claim tax relief against costs. A later institutional purchaser would not be able to do this, so it made sense to invest at the beginning at a discount of anything from 60% to 90% depending on the rate of tax. Although private companies use the NDR system to project long-term returns, they may not be as constrained in spending money early in the rotation as the FC, and this may affect their approach to other establishment operations.

The standard of forest roads varies greatly; they are not simple 'dirt tracks'. Although almost always gravel-surfaced, the top-quality harvesting roads are built on a full foundation and are really of equal constructional standard to minor public roads, but without the luxury of a tarmac finish. The extent and cost of roading depends on soil type and harvesting method. The high density of roads in the early forests was planned for horse extraction. Lower-specification roads may be designed for use by off-road harvesting equipment only, and many older roads which do not reach the standard for articulated timber lorries have been put into this category. At planting, the formation of the harvesting road may be designed and dug out to provide basic access for all-wheel-drive vehicles to bring in plants, fencing materials and workers and to provide access for fire fighting. On the worst terrain, especially deep peat or very steep ground, building roads may be spectacularly expensive.

Although the FC has opened its forests to walkers, with the exception of a few forest drives, roads are kept for the use of forestry traffic only, behind locked gates. The main reasons for this are obvious: these are single-track roads, and motorists unfamiliar with loose surfaces and articulated timber lorries do not mix. Access by car also allows arsonists the opportunity to light a series of fires over several miles of forest. Keeping barriers locked is also important for birds: Goshawks in particular have benefited from the seclusion of the core of large upland forests, especially in the Borders. In the Peak District of Derbyshire, where many nests are close to public roads, most eggs and young have been stolen, and the same has applied to Thetford Forest, where most of the forest is within two miles of a public road over easy terrain.

FENCING

Fences are also built before planting to protect the young trees from grazing. Domestic stock are almost always present on open land adjoining a new planting area. In the first rotation, sheep probably did more damage than all other pests put together; especially in Wales, foresters fought a continuous battle against 'hefted' sheep, animals that are drawn to the area on which they were born and lived their early life. Rabbits were the biggest wildlife problem prior to their near elimination by myxomatosis. In the Scottish Highlands the large numbers of Red Deer on the open hill means deer fences have to be put up, because numbers are too great to control by shooting and also because adjoining landowners, whose main income often derives from deer stalking would oppose such reduction. Other deer species, especially Roe Deer, have never been a significant problem on upland new planting areas, as they tend to spread out only from established forest.

With limited exceptions (Foxes around plantation edges in sheep-farming areas of Scotland, Wales and northern England, and in a few places Carrion or Hooded Crows) foresters only control wildlife that threatens the trees. Human persecution remains the main limiting factor for most diurnal raptors, including Buzzards, Hen Harriers, Golden Eagles, Goshawks and Red Kites, even after decades of legal protection (Newton, 1984). The reduction in persecution after the change of ownership in forestry has almost certainly been of far greater importance to date to most of these birds than habitat change; this has been particularly the case for Hen Harriers, which have always been ruthlessly eliminated from grouse moors.

PLOUGHING

The main physical limitation today is slope; sites too wet to plough are unlikely to be economically plantable. This was not always so. The inability to greatly modify sites led in the early years to careful species selection by soil type (Anderson, 1950). Larch and pine were used on drier sites, Douglas Fir on brown earths and spruces on wetter areas. Soil variation on quite a small scale was carefully followed, sometimes to an unnecessary degree of detail, creating some complicated and interesting woods. Up to the late 1950s a considerable area of broadleaves was also planted on the best soils, including some complete forest blocks of Beech and Oak in the lowlands in southern England. The earlier upland forests are, as a result, far more varied and interesting than later plantings, because of both site type and available

© PS.

technology and also through changing attitudes to how forests should be planted; even if technology had remained static, these forests would not have been planted with the small-scale variation in the 1960s or 1970s, because simplicity of management and cost saving had become more important.

As it was, improving ploughing methods ironed out minor site differences. The expanded programmes of the 1950s also meant that a larger-scale approach was essential. Better understanding of tree capability resulted in a more relaxed attitude to using a single species over a wider range of site types. A large scale of operation to some extent compensates for poorer growth on less fertile sites; ploughing costs in particular are greatly influenced by the frequency of manoeuvring at the end of rows or around obstacles, so long runs across regular ground cut costs. At the same time, increasing efforts were made to plant all available land.

Sitka Spruce became the mainstay species as earlier plantings proved its exceptional performance and an increasing proportion of very wet gley and peat sites were acquired. By the mid-1960s the other key species in the uplands was Lodgepole Pine. This tree is remarkable because it can provide air to its roots from the surface, avoiding the waterlogging which remained a problem for Sitka Spruce on the very wet sites which could by then be ploughed. Over the years it can actually dry a site through transpiration, shrinking and cracking deep peat, and its planting on some sites is planned to 'pioneer' the ground for subsequent higher-production Sitka Spruce rotations. Lodgepole Pine also replaced Scots Pine on heather sites because it was expected to be higher yielding, before the introduction of heather-control methods allowed Sitka Spruce to grow.

DRAINAGE

Drainage is not the same as ploughing. Plough furrows usually run up and down the slope and act as local drains, vital in peats and gleys where the drainage effect of cultivation extends less than 2 m from the furrow (Pyatt *et al.*,

1985). However, as water accelerates downhill its erosive power increases. Long, steep furrows turn to gully erosion so they are cut by a system of collecting drains, deep furrows cut by ploughs or dug by hydraulic excavator, and aligned at an angle of no more than 2 degrees across the slope. Serious gully erosion has occurred in both plough furrows and drains in Britain. The spacing ing of collecting drains depends on the expected quantity of runoff and the vulnerability of the soil to erosion. Cut-off drains, in contrast, intercept water from spring lines or water flowing onto the site from adjacent steeper ground and lead it away before it reaches the area being protected (Pyatt and Low, 1986).

Until the mid-1970s, drains were, like a plumbing system, dug to flow directly into natural watercourses, which meant that the erosion that inevitably occurred in even the best-designed drainage scheme flowed straight into the stream. Ploughing and draining right up to stream sides allowed planting even of wet flats next to the watercourse. Trees planted within 2-3 m of small watercourses start to shade them out by early thicket stage, and the habitat of riparian birds like the Common Sandpiper is lost. As crawler tractor tracks became wider, it was possible to push machines into wetter and wetter ground and all but the softest bogs were drained. The 1970s development of the apical track shoe, with an A-shaped track plate rather than the inverted-T of conventional tracks, which give grip without cutting the surface, allowed tractors to travel on ground as soft as the wetter bogs of Sutherland and Caithness.

FERTILIZER

Fertilizer is applied just before ploughing or just after planting. Initially, each tree's ration was thrown on by hand. Later, fixed-wing aircraft were used to cut costs, flying off specially constructed extra-wide, straight forest roads. Fixed-wing operating problems later led to the universal adoption of helicopter spreading. Virtually all upland sites receive one application of 450 kg ground mineral phosphate at the start of the rotation. Overall application of fertiliser should be made before ploughing, because after ploughing about 30% of the site is furrow bottoms linked into the drainage system and that proportion of fertilizer is liable to be lost by runoff.

There have been continual and serious problems with the accuracy of aerial fertilizer distribution which show up clearly in the tree growth of many younger forests; there tends to be a peak of fertilizer quantity underneath the aircraft. If the planned width between swathes is too close, too much fertilizer will be used; too far apart and the outer edges of the swathe will get a smaller dose. On severely phosphate-deficient sites, a clearly visible 'wave' pattern of tree growth results, accentuated by the yellower colouring of the deficient trees at the outer edges of the swathe. This can happen even where the application has been carried out carefully. Flying to plan requires great skill over a featureless area of new ploughing and often in marginal weather conditions. Inexperience, cutting corners to cut costs and other factors like wet material jamming the spinner that distributes the fertilizer from the hopper slung below the helicopter, mean that some areas get no fertilizer at all.

Inaccurate fertilizer application is a major cause of variable tree growth in the new forests, and on the most phosphate-deficient sites has resulted in total failure, leaving patches of open ground scattered throughout the crop which is of proven value to deer, and quite possibly to birds as well. Fertilizer has a significant effect on ground vegetation as well as trees, and contributes to the burst of growth after fencing and before canopy closure. Plants may also be introduced in fertilizer, most frequently Willow Herb but sometimes also tree or shrub seeds which survive to provide a seed source for colonization of the clearfells. Recent tests have shown just how difficult it is to achieve even distribution of a material of the variable size composition of the phosphate used at present, and there is an increasing move back to ground application by agricultural-type spreaders or by hand.

Later fertilizer applications are made to correct serious deficiencies that become apparent once the trees start growing and also to increase the growth of the trees by boosting the initial input. The economics of top dressing to increase growth are finely balanced: at first it appeared that a lasting increase in yield class could result, which made top dressing profitable. Its use was taken to an extreme in the FC's South Scotland conservancy in the 1970s, when a 'high tree farming' approach included regular fertilizer applications throughout the crop's life. However, research subsequently showed that additional applications resulted only in a temporary increase in top height rather than the sustained yield increase upon which the idea depended for its economic viability (McIntosh, 1984). Runoff of fertilizer applied after canopy closure will be minimal compared to establishment phase crops.

PROVENANCE

Young trees are reared from seed in nurseries (Aldhous, 1972), usually for two to three years, and are then lifted and sent out to the forest as bare-rooted plants. Where the seed comes from is important (Lines, 1987): the collection of seed for much early FC planting was poorly supervised, and the seed is now known to have been picked from the smallest trees, which were easiest to reach. Although not a guarantee of good genotype, picking from the best rather than worst phenotypes would have reduced the risk of poor performance. Sitka Spruce yet again came out ahead, proving to be a remarkably consistent tree despite some dubious seed lots. Lodgepole Pine was the exact opposite: it is a spectacularly variable species, and different regional 'provenances' planted in Britain have varied from fine, straight trees to shrub-like bushes. Lodgepole Pine provenance problems are likely in the long term to turn out to have been one of the most significant failures of the first rotation. Fast-growing provenances from coastal British Columbia were widely planted because of their fast growth. Over 100 million trees of the notably disastrous Lulu Island provenance from the mouth of the Fraser resulted from nearly

1000 kg of seed imported in the early 1950s (Lines, 1976). Trees from south coastal provenances are heavily branched and bushy. Some provenances suffer from swept butts, which are especially serious because it is the bottom of the tree that produces the most valuable product, sawlogs, which need to be straight. The worst problems are early instability and snow damage, and quite significant areas of Lodgepole Pine have already been flattened shortly after going into thicket stage. As soon as these problems became apparent, slower growing but less vulnerable and better-quality provenances from inland British Columbia were adopted, especially Skeena River.

PLANTING

Big is not beautiful for new planting trees; size gives an initial advantage over weed growth, but the shock of lifting from the nursery bed is worse for older trees. In the early days each forest had its own nursery, but better transport and the savings from mechanized transplanting of seedlings, the most labour-consuming nursery job, gradually led to fewer larger nurseries. The number of trees produced at local nurseries sometimes influenced planting practice; in a good year, when more trees germinated than expected, the area programme stayed the same but trees were planted closer: some stands were planted as close as 0.9 m square, compared to a standard spacing today of about 1.8 m. Initial spacing is obviously far closer than the ideal for a final crop of large trees; 1800-2000 trees are planted but less than 500 will be left at the end of a normal economic rotation and with standard thinning. The main reason for close spacing of planting is to eliminate competition from ground vegetation for nutrients and moisture as quickly as possible, but as discussed below, spacing also affects timber quality.

Planting is an operation where a person with a spade is better on most sites than any machine yet invented. The planter can travel across almost any ground and select just the right spot to plant the tree. Very steep slopes are a case in point: on some, soil had to be carried in for the trees, and at Strathyre Forest in west Scotland, St Kilda men worked for the FC following the evacuation of the island and planted some ground that is so steep it is unsafe to fell today. Young trees must be positioned firmly in the soil and it is crucial that their roots are not allowed to dry out at any time between lifting in the nursery and planting. A worker can plant 1500 trees a day on a ploughed new planting site, and where a regular ribbon of peat is produced on flat ground, the planter can work hardly breaking the rhythm of his stride and plant over 2000 trees. Trees less than 0.1 m tall can survive on less fertile ploughed sites because vegetation competition around the bare plough furrow on which the tree is planted is negligible.

Conifers are generally planted in the spring in the uplands; on lower ground and on the warmer west coast, autumn planting is also carried out, but in harsher areas autumn planting exposes young trees to the rigour of winter before they are established. Broadleaves, on the other hand, are often planted in autumn. The length of the planting season varies; on lower and sheltered ground, planting may start in January, and how long it can go on is limited only by the need to plant the trees while they are dormant. In most upland areas this allows planting to extend at least to mid-May. On the highest sites, however, it can be as little as six weeks between the snow clearing and trees breaking dormancy. Planting is the main work on establishment-stage forests, and the longer the workforce can spend on it the less time is wasted on less productive work. Now the planting season can be extended right into the summer by keeping trees dormant in cold stores, taking them to the forest in refrigerated lorries and quickly planting them.

A small proportion of trees are raised in small paper pots, and because their roots are never exposed they can be planted all the year round, even when growing actively. Although giving almost guaranteed initial survival, their use is limited by higher plant production costs and transport costs to species that are especially vulnerable to root disturbance. Most produced is Corsican Pine, but Birch and Douglas Fir are also being grown this way now. Moisture is the key for young trees, and the most common causes of death are drying out through bad handling between lifting and planting, and through dry spells during the first season after planting. Frost is another problem, both frost heave lifting newly planted trees and killing of newly flushed foliage by late spring frosts. It is impossible to establish spruce in frost hollows, and more resistant pine is substituted. The precise position in which the tree is planted is affected by these factors; on wet upland soils a furrow-top planting position gives maximum local drainage, but on very exposed sites the tree may be planted in a small step cut in the side of the furrow, to give it some shelter. On dry sand, trees are planted at the bottom of the furrow for maximum moisture. Early losses are replaced by 'beating up', although on new planting sites today establishment is so assured that it is rarely necessary.

WEEDS

The flush of ground vegetation that is so important to rodents and birds during the early establishment stage can also compete with the trees; to the forester, a plant that competes with the crop he is trying to grow is a weed. Weed control is costly, and can also lead to extra 'beating up' of suppressed trees and eventually to uneven establishment. Weed competition is greatest on more fertile soil, though heather check is a particular problem on poor soils. Grasses, heather, brambles and bracken can all suppress young trees. Colonizing broadleaved trees are rarely a problem on open new planting sites. Competition, for moisture and nutrients, is as important as shading and swamping, especially on drier sites (Davies, 1987).

Traditional hand weeding uses a weeding hook, a short-handled sickle. Two or three treatments a year may be needed to protect the trees; hand weeding is little different from mowing the lawn, and stimulates vegetation growth and competition with the tree for moisture (Davies, 1987). During the 1960s, pressure to complete large planting programmes with inadequate labour led to poor subsequent maintenance and trees being overwhelmed by weeds. Substantial failures sometimes resulted; on one Yorkshire site, spruce and larch failed on the more fertile slopes, out-competed by inadequately weeded bracken, but Lodgepole Pine on the less fertile heather plateau survived. Weeding is rare on upland planting today, because trees grow above competition before vegetation can reinvade the plough furrow. The major exception to this is the effect heather can have on species other than pine, depriving them of nitrogen and halting growth. Although heather check can now be prevented by use of 2,4-D herbicide or an application of nitrogen, there is a substantial backlog of checked spruce areas, and many small patches are not important enough to treat. Weeding failures are often the main open ground other than roads and rides in upland forests and may continue to provide habitats for birds like Nightjar and Black Grouse through the thicket stage.

Despite the introduction of a wide range of herbicides, over half of the FC's weeding was by hand in 1983 and has represented something of a failure to apply new technology (Neustein and Seal, 1984). Carrying the volume of liquid needed to apply herbicide across rough sites in constricting protective clothing is uncomfortable for workers and expensive. Early attempts to reduce volumes, especially the sophisticated Ultra Low Volume incremental system (Rogers, 1975), were effective, but complex to use and limited by wind speed. Recent simpler applicators, especially the drench gun and weedwiper (Sale *et al.*, 1986), are not only easier to use but reduce the need for protective clothing because of minimal contaminating drift. The quantity of chemical, most often Glyphosate which makes up over half the herbicide used by the FC at present, is minimized by accuracy of application and spot rather than overall treatment. A one-metre spot around trees at 1.8 m spacing means only 30% of the site receives chemical.

FIRE

Next to grazing, fire is the most serious threat to the young plantation. The worst fires are in establishment and thicket-stage stands; crown fires, where the fire burns in the tops of older trees, are thankfully rare in Britain. Big fires occur most often during strong, drying winds in spring. The autumn fires following the hot summer of 1976 were exceptional, burning 2191 ha of FC forest (Forestry Commission, 1978a). This was the worst year since 1942; on average, about 600 ha of FC forests burn each year, but 1986 was as wet as 1976 was dry, and only 122 ha were lost.

In spring the winter's dead vegetation can dry out rapidly before lush new growth emerges even though the ground below may be waterlogged, and wind-driven fire can spread at terrifying speed. The fastest-moving fires are in grass, especially *Molinia caerulea*, where dead leaves blow across the ground in bundles. Less frightening but more frustrating are fires in deep peat, which can creep through the soil, popping up many metres away from their last appearance. They can burn for weeks, only going out after heavy rain.

Fire has had a significant effect on forest design. Even now, breaks in tree cover are invaluable for stopping a fire, but before 1939, when the main fire-fighting method was men smothering the flames with beaters, (either Birch brooms or industrial belting nailed to a handle), breaks were crucial. Wide rides were left and belts of larch or broadleaves planted because their lusher, less resinous foliage burned less vigorously. Although fire-break rides are often unpleasant straight lines in the landscape, in some forests they are the only open ground or tree species variation over a large area; the wide rides that are still grazed in some 1950s forests in Galloway are good examples of this.

In the age of steam, trains caused more fires than anything else. Land was sometimes left unplanted and cultivated beside the track to limit fires, and it was on one of these areas that Woodlarks first nested in Thetford Forest. Today most fires are caused by people in the forest, many by intent rather than accident, and by out-of-control moor-burning on neighbouring land. Arson is particularly serious where forest adjoins urban areas, and the intimate mix of towns and forest of the South Wales valleys leads to by far the worst forest fire problems in Britain. Lightning is the main natural agency of fire, and the fires caused may be especially difficult to fight because lightning often strikes on high, remote ground. Water is now the key to fire fighting. After 1945 both FC and county fire services became better mechanized, the FC acquiring War Department four-wheel-drive tanker lorries, the 'green goddesses', and later Land-Rover-drawn 'mobile dam units', made up of a tank and demountable pump. Heathland forests such as Thetford, the North Yorks Moors and Cannock lacked accessible water, so concrete tanks were constructed. Too small to satisfy present-day appliances, they have been of value to wildlife. At Thetford the Mayday Farm fire-tank was a famous place to see Crossbills drinking.

Half the battle in fighting fires in upland forests is getting to them; if sufficient water can be applied quickly small fires never grow into big fires. The revolution of the 1980s has been the extensive use of helicopters, both for lifting people and pumping equipment, and for water bombing with specially designed buckets. False economies of the past that sometimes delayed the deployment of expensive equipment like ploughs to fight fires have been abandoned with the full realization that big fires burn up trees far more expensively than helicopters burn fuel. Fire-breaks still play an important role, but mainly round the outside of plantations. Wherever machine access is possible, a swathe is mown with a tractor-mounted brushcutter to check out of control or accidental muirburn spreading into the forest. The risk of fire was for many years the main reason for restricting public access to FC forests. Many accidental fires have been caused by visitors but it is likely that the increased numbers of people ready to dash to the telephone to report fires means that on balance no more trees have been lost.

The area of recently burned forest is a small habitat nationally, and it is unlikely that fire does much harm to birds. Some birds may favour fire sites: in North Yorkshire, Nightjars were found on several burned areas in Cropton forest. Fire has created ecologically valuable age–class diversity in some forests. There were many fires between 1939 and 1945 because forest areas were used for military training with live ammunition. Fire in these early plantings was a factor creating diversity before trees were big enough to be subject to windthrow.

PESTS

Catastrophic pest attack on the new forests is one threat that has not yet materialized to the extent of many dire predictions on the risks of concentrating planting on so few tree species. The main threats are from fungi and insects. Because of the length of time between initial investment and eventual return, foresters, in contrast to farmers, do not normally take action against pests that only reduce yield. Pesticides or other management techniques are only used to combat pests which either kill trees or cause significant timber degradation. A few tree species are actually ruled out from use in British forestry because of their susceptibility to pests, for example the aphid *Adelges nordmannianae* on Common Silver Fir, and European Larch which suffers worse from *A. laricis* and the bark canker fungus *Lachnellula willkommii* than Japanese or Hybrid Larch.

Large-scale fatal insect attack on the major commercial species is largely confined to pines. Over the years, Pine Looper Moth has been responsible for most problems, with occasional outbreaks throughout the main pine regions. However, the largest problem in terms of area, the Pine Beauty Moth, only struck recently; occurring naturally on Scots Pine, it was known as a pest in continental Europe (Leather *et al.*, 1987) but not in Britain until its large-scale fatal attacks on non-native Lodgepole Pine on deep peat began in 1976. The Pine Sawfly also attacks young pine sublethally. It is rarely serious enough to justify chemical treatment, and a nuclear polyhedrosis virus has been used successfully for biological control.

The possibilities of birds as biological insect controllers, following an idea originating in Germany, has also been tested. W.L. Taylor set up an FC bird box experiment in 1942 to see if increased numbers of small insectivores could reduce insect attack on trees. The Forest of Dean scheme went on to be run by Bruce Campbell with help from students at the Dean Forester School (Campbell, 1967), and on the school's closure the RSPB took over and now looks after the bird boxes on its Nagshead reserve, which is leased from the FC. The scheme has now been monitored continuously for 46 years, and is the longest-running bird box scheme in Britain. Another part of the original FC scheme in the Dalby Valley in Yorkshire, which subsequently fell into disrepair, is described in Chapter 7. Further trials aimed specifically at control of Pine Looper Moth were undertaken by Aberdeen University in Culbin Forest on the Moray Firth. Recent work monitoring the environmental effect of spraying Pine Beauty Moth has confirmed that birds have little effect on major insect pest outbreaks (Crick and Spray, 1987). However, although the boxes did not help the trees they were great for Pied Flycatchers in the Dean and other broadleaved locations, and at Culbin Crested Tits used them.

The bark beetle *Dendroctonus micans* is a recent arrival and a potentially serious pest of spruce; unlike many species of bark beetle, it can attack and kill healthy trees. Imported on inadequately de-barked timber, it shows how pests left behind when tree species were imported can catch up with them later. It kills spruce, and as such presents a particular threat to the new forests. Older trees are most vulnerable, and the worst threat to pre-felling-stage stands is where trees are already debilitated for another reason. The short rotation length of British conifer forests means that bark beetles are unlikely to have as much economic impact as in Scandinavia, from where they came. This pest poses a real problem waiting in the wings to confound attempts at retaining some crops well beyond normal rotation. Vigorous control measures will slow its spread, but it is now well-established and likely to spread throughout Britain eventually. The main hope for limiting the species is the use of a specific predator, *Rhizophagus grandis*, which is being introduced to infested stands. Living just below the surface of the tree's bark, bark beetles are an important woodpecker food source in Scandinavian forests. *Dendroctonus micans* illustrates the particularly problematical threat that exists from imported pests with most of our timber being imported.

Young adult Pine Shoot Beetles bore into the leading shoots of conifers, deforming the leader, reducing growth and causing bends in the tree that reduce its timber value. It is a pest that can be prevented because the beetle pupates in dead conifer wood and it will not survive if no dead wood, including felled timber left for more than six weeks, is available during the summer. Unfortunately, this cross-infection and damage to the living crop is an important reason why it is undesirable to keep dead conifer wood, either as standing trees or fallen material, in commercial forests, despite the benefit to wildlife.

Fungi also present considerable problems, both killing trees and devaluing timber. The greatest problem is with the root rot *Heterobasidium annosus* (formerly *Fomes annosus*), present throughout Britain but most serious on lighter soils and in drier areas (Gibbs and Greig, 1970). It has been the single biggest management problem at Thetford in recent years. Persisting in tree roots after felling, it invades the cut surface of felled tree stumps. All stumps are now treated with urea (dyed blue to show it has been applied) immediately after felling to prevent colonization by *Heterobasidium annosus* or, on pine only, the competitor fungus *Peniophora gigantea*. However, on the most susceptible agricultural or high pH sites at Thetford, this is not enough, and to success-fully stop reinfection and early death of the successor crop, the roots of the previous crop are removed and stacked in rows.

Also affecting Scots Pine at Thetford is resin top disease, in which the fungus *Peredermium pini* causes cankers that weep resin and kill the top. Some older stands are being depleted so quickly by this dual attack at the bottom and the top of the trees that over the compartment as a whole, timber volume is being lost to disease quicker than surviving trees are growing. Even when not fatal, *Heterobasidium annosus* weakens the roots and causes decay in the valuable sawlog timber of the lower part of the tree. Spruce and larch are particularly susceptible, but although Douglas Fir and *Abies* species may not suffer timber damage, root rot may increase susceptibility to windthrow. Along with the staining fungus *Stereum sanguinolentum*, *Heterobasidium annosus* can enter growing trees through wounds caused by machines hitting trees or cuts from wire winch ropes (Pawsey and Gladman, 1965).

Trees are more likely to suffer from pests and disease where they are already under physical stress. Brunchorstia dieback caused by the fungus *Gremmeniella abietina* effectively limits the range of Corsican Pine. Corsican Pine will grow close to the coast in the north of Scotland and inland in the south; but more than five miles inland in north Yorkshire it suffers severe Brunchorstia attack, and a high proportion of trees die in the pole stage. Recently, Brunchorstia has killed Scots Pine on wet and exposed soils in the Borders, where trees were already under stress on a site hostile to the species. Norway Spruce is also particularly susceptible to dieback which, as with Corsican Pine, is probably climate-related, but unlike the case with Brunchorstia, a precise pathogen is often hard to pin down. Many Norway Spruce stands stop growing and turn brown before they reach economic rotation age, an important reason why most Norway Spruce planted today is for Christmas trees, not timber.

BOGS

The treatment of diversifying features has altered over the years. The changing attitudes and abilities of the planting foresters have had profound effects on today's forests. The main variable has been just how intensively an area was planted, and how much effort went into covering every available bit of land with commercial conifer plantations. As we have shown, the technical ability to plant all sorts of land increased over the years and on top of this there was a rather more sympathetic attitude to existing features before 1950. A study by Mowle (in Peterken, 1986) looked at the distribution of land as wildlife habitat in two Scottish forests, Glen More and Inshriach. Glen More was a felled native Scots Pine wood, and Inshriach largely open ground.

Both the Scots Pine remaining when the FC acquired Glen More, totalling 222 ha out of 1718 ha, and the small bogs characteristic of native Scots Pine forests, were initially left alone. More bogs survived better in older plantations. However, after 1945 bogs were more likely to be planted, and 178 ha (80%) of the remaining native Scots Pine was underplanted. In the 1970s some of the bogs that had been left at the original planting were planted with Lodgepole Pine. This sort of thing happened in most forests. How far it progressed depended on the availability of other work: where there was a large new planting programme, for example Glenorchy Forest in Argyll, small-scale and expensive work was left, but new ground was not available at Glen More, so every area that could carry trees was planted. Peterken (1986) rightly points out that although failures and felling increase non-plantation habitats, including roads and rides, by as much as 25%, they are not of equal value to pre-afforestation habitats like the old Scots Pine. It is an important lesson for the future that devoting more energy to getting the main conifer crop area growing regularly and well and leaving environmentally valuable features alone may not give such impressive programme figures but will make as much money and maintain conservation value.

A similar approach was taken to other natural features. Ploughing and planting were taken right up to the banks of watercourses, and as close as possible to lakes. On flatter ground this resulted in the complete shading of smaller watercourses. Birds of the watercourse, like Dippers, Common Sandpipers and Grey Wagtails are lost. The ability to drain and plant wetlands increased over the years, and was generally used to the full. Bogs were often ploughed and planted well after the main planting, as at Glen More. In some places like the floristically important dune slacks of Newborough Forest on Anglesey, attitudes changed so quickly that the trees were removed again before they reached thicket stage. Very occasionally the birds won: a 2-ha patch of Newborough remains open today despite numerous plantings, because it was a gull colony and every summer the gulls trampled the trees planted in the spring: unfortunately the gulls are gone because Foxes reached Anglesey across the Menai Straights and gull colonies on mainland Anglesey quickly disappeared.

Some bogs were always too wet to drain, so remained as open space within the forest. Where sites of special conservation interest were identified, management did in many cases aim to maintain their value, or at least make some allowance for it. That less allowance was made than would be today is not surprising, and many sites that would today be SSSIs did not qualify for inclusion in the much smaller number of sites scheduled in the early days of the Nature Conservancy. Some sites of identified conservation value were afforested, and some undoubtedly with a degree of malice, against the protestations of conservationists. However, it is very difficult to pin down exactly how people saw the importance of these sites and the effect forestry work was having on them in an era of very different attitudes in conservation as well as forestry. People in general were sufficiently persuaded of the need to 'improve' existing woodland for the National Trusts to follow the FC lead in converting existing broadleaved woods to conifers. There is no doubt that some sites were changed before their value was realized by foresters, if not by conservationists: this was still a problem into the early 1980s, despite there being more full-time conservation staff on the ground than in the 1950s.

UNDERPLANTING OF WOODLANDS

The treatment of remnant native Scots Pine at Glen More was also typical of national practice. Most accessible existing woodland was planted with crop conifers. Exactly what was done depended on how difficult the site was to clear for conifer planting. Big, scattered trees were surrounded with planting, and left as beleaguered islands poking out from the thicket. This was the common fate of scattered native Scots Pine, as at Glen More. Because underplanting peaked in the 1960s, few old Scots Pine have yet been killed by underplanting. They remain of exceptional value to large, tree-nesting birds, especially Ospreys. Paradoxically, the very intensity of underplanting, which was taken right up to the trunk of the old trees creates a dense thicket that protects nesting trees far better than an open moorland setting. However, the open forest and ericaceous vegetation so important to the special Scots Pine forest birds, Capercaillies and Crested Tits, is eliminated. Conifers were also planted around ancient broadleaved woodland which was expensive to fell. Big oaks have such broad crowns that they often manage to keep a hole in the conifer canopy and survive.

Denser broadleaved woodland was clearfelled or thinned sufficiently heavily to plant with light-demanding conifers, initially Norway Spruce and Douglas Fir, and later Sitka Spruce. Although planted conifers did find their way into some quite extraordinary corners, those woodlands on steep slopes, poor soils or in gullies, and where existing tree cover was dense and there was no local market to offset the cost of felling, had a better chance of being left untouched. Some fascinating woods survived, sometimes with interesting results, like the Ash woodland 'trapped' within the fence of a new planting area on Skye. In contrast to the SSSI area, left outside the fence and still grazed, the fenced Ash has turned into a positive jungle of natural regeneration, and feels like a little fragment of rain forest, full of old Hazel coppice stools covered in ephiphytes.

Later, to cut costs and deal with well-stocked woods, shade-tolerant conifers, especially Western Hemlock, were increasingly planted after a minimal felling of the existing overstorey. Large areas of broadleaved wood were treated this way. The value of broadleaved trees, and of old trees, to birds is now appreciated, and present-day policies which are restoring damaged woods as well as creating new broadleaved areas are discussed in the next chapter.

The results of conifer planting under existing woodland have been variable: trees have done well planted through scattered woods, but in well-stocked stands too little of the overstorey tended to be removed. Dalavich Woods from which late 1950s Norway Spruce underplanting was removed in 1986 is a good example. Too few Oaks were felled; it is likely that the intention was to fell more after the initial planting. The Oak remained, but the conifers shaded out the ground vegetation on this ancient semi-natural site. The density of the Oak canopy was too high and started to suppress the conifers by shading. By 1986 there was minimal conifer growth and the site's conservation value as a broadleaved woodland was being lost. This is not an isolated case; Western Hemlock plantings often established poorly and the species would not have been planted in any other situation, as it is not one of the most highly valued timber producers in British forestry (Aldhous and Low, 1974).

BROADLEAVES

Considerable areas of broadleaves were planted in the 1920s and 1930s, and some forests, like Thetford, have over 10% broadleaves. The fate of these plantings varied; many pure crops were neglected up to the late 1970s, but they survived well. Replacement of planted broadleaves with conifers on economic grounds was largely in the lowlands. Mixtures fared less well; by the time people started thinking broadleaves again, the broadleaved content of many crops had been eliminated by the conifers. Oak mixed with Norway Spruce suffered worst; Beech mixed with larch survived better.

Whilst larger existing woods were usually converted to conifers, smaller woods and individual trees tended to be left; often it was not worth the effort of clearing them. The FC acquired many small but attractive policy woods as part of estates, and even some ornamental woods. These generally survived; planted in the middle to late 1800s, the big trees of a wide mixture of species were expensive to fell and transport and difficult to market. Today these old stands are of particular wildlife value almost everywhere they occur, particularly in counties with very few old trees like Sutherland. At higher elevations, conifer shelter belts planted before afforestation, usually Scots Pine but also Norway Spruce and larch, provided small pockets of cover of great importance to some birds. They were essential nesting places that allowed raptors like Kestrels and Long-eared Owls to benefit from the vole food boom that followed fencing. Although sometimes clearfelled, the poor quality of old, windswept and rotten trees meant that most belts survived, and provided valuable islands of old, open-grown woodland right through the first rotation.

BUILDINGS

The most important artificial feature of the new forests comprised buildings, often abandoned and deteriorating. Although afforestation sometimes coincided with the abandonment of houses, it was rarely the cause; in fact, during the agricultural depression years of the 1930s in particular, it was only forestry that kept people on the land at all. Places like the North Yorks Moors had seen a contraction of agriculture dating back to the older depression on the land that started in the 1870s as a result of large-scale grain imports from North America. Only demolished when they were clearly unsafe, many dwellings disappeared quietly into the trees to continue their decline to the great benefit of birds like Barn Owls, which did well feeding on the vole supply of the new planting. Other hole or ledge nesters from Stock Doves to Kestrels also used old buildings.

Natural ledges on crags and cliffs were affected where conifers were planted right up to the cliff bottom: raptors could not use ledges enclosed by dense trees. In the flatter hills, tiny crags less than 10 m high could be essential for species like Peregrines, Golden Eagles, Buzzards and Kestrels to nest at all.

CROP DENSITY

Once the crop reaches the thicket stage, the worst hazards of early life are past. The forester's main priority is manipulating the stand towards the desired timber production objective. The crop's potential depends on fixed constraints, especially species, yield and windthrow, but there are also elements over which the manager has control. The way this physical potential is balanced against timber production priorities is quite complex, but it is also very important in understanding the basis for British forestry over the past 60 years. The two factors that the manager can most easily control are the spacing of the stand and the time for which it is allowed to grow. Trees compete with each other for light, nutrients and moisture. At high densities, trees support each other, and though total timber volume may be as high as that of a stand at wider spacing, each individual tree will have a lower timber volume.

Today trees are normally planted at a spacing of 1.8–2.1 m square, about 2000/ha (Hamilton and Christie, 1974). Thinning reduces this to between a fifth and a tenth of the number of trees after 50 years, so why plant so close in the first place? The well-founded traditional reason is the suppression of ground vegetation by quick canopy closure; many of the difficulties of early life disappear once the canopy closes and the trees dominate the site. Recently the importance of spacing to wood strength has been discovered: 'juvenile' wood laid down before canopy closure is weaker than timber produced during the rest of the tree's life (Brazier and Hands, 1986). Wider spacings increase the amount of juvenile wood, because canopy closure is delayed, reducing overall timber quality and reducing the likelihood of meeting the grading requirements for high-value construction use. As a result, extra-wide spacings (up to 2.5 m) adopted in the 1970s for crops that would never be thinned because of windthrow have now been dropped.

Foresters aim to produce larger trees because they are cheaper to harvest, and yield more valuable products. Tree diameter increases more quickly at wider spacings, so as soon as close initial spacings have done their job it is desirable to reduce the number of trees in the stand so the remainder can grow bigger. However, the premium for bigger trees is not so great that total volume produced over a rotation can be ignored, and removal of too many trees means that those left cannot exploit all the space available. Thinning must juggle these conflicting aims; where there are no physical problems, especially windthrow, the manager's ability to produce a particular type of final crop is considerable (Hamilton, 1976). Heavy thinning will result in larger individual stems, with a higher proportion falling into the high value 18+ cm diameter sawlog category more quickly. However, total volume production over the rotation may be reduced. On the other hand, a lighter thinning regime may yield a higher total volume but distributed between more stems, with fewer sawlogs and more pulp-sized material. More heavily thinned stands tend to have greater taper; unthinned trees go 'straight up'.

Thinning reduces the number of trees in the stand gradually, so that the remaining trees can quickly fill the space made by the removal of their neighbours. Although little is known about the importance of thinning to birds, open stands seem to be more accessible to birds of prey like Goshawks, and Capercaillies certainly need open woodland, probably more open than standard rotations and thinning prescriptions can provide. Management tables give optimum times and volume to be removed for different species and yield classes. In practice it may be difficult to work precisely to these guidelines: the first thinning is often delayed because it may not be possible to make a profit working the very small trees, and each additional year makes a big difference to working costs.

Both method and intensity of thinning have been variable in the first rotation of the new forests. The earliest thinnings of the 1920s plantings tended to remove the smaller trees. Although producing some timber, this had little effect on the remaining trees: it was once described as having as much benefit to the rest of the crop as removing last year's birds' nests! At the other extreme was 'eclectic' thinning which was concerned only with dominants, and left many suppressed stems that had no effect on the bigger trees but were potentially valuable as timber. These methods were all selective, but selecting and extracting individual trees from stands planted at spacings as close as 4 feet square added greatly to cost.

Line thinning, removing whole rows of trees at regular intervals, most often one out of three, was introduced. Traditional foresters frowned on line thinning as cheap and nasty and it is still sneered at today. Criticisms ignore a number of facts. Selective thinning aims to choose the best trees for the final crop. At first thinning the ratio of trees in the crop to final crop numbers is so great that removing a few that might otherwise have been left does not significantly reduce the choice available for the final crop. Line thinning increases the chances of the crop getting thinned at all; the idea that before line thinning everything was ideal is a myth. The thinning record for the first rotation of the new forests has not been good, with fluctuations in markets and equipment availability frequently overriding pure silvicultural consider-ations. Line thinning increases the average pole size of the first thinning because it includes a cross-section of the crop, not just smaller trees, again making it more economic. It also avoids the risk of 'robbing' the crop— selecting the largest rather than smallest trees, a tactic that has been used to reduce harvesting cost. A benefit of line thinning is in providing access both at the first thinning and for later thinnings; it should reduce the extraction damage to the remaining crop, which has been a major penalty of selective thinning in the first rotation. Line thinning incidentally creates a continuous gap in the canopy which might allow access for larger birds rather earlier than the scattered gaps of selective thinning.

Normal thinning practice in FC forests is now the removal of one line in three at first thinning followed by selective thinnings. In lower-yield-class crops these tend to be rather further apart in time than recommended by the management tables, because the reduced cost of working the larger amount of timber that can be thinned after, say, seven rather than five years, offsets the slight loss in yield from the delay. The effect and importance of thinning varies between species: at one extreme, larch languish if left unthinned, producing poor, thin trees, but respond vigorously to thinning and can be thinned more heavily than average without the crop suffering. At the other extreme, spruces, especially Sitka Spruce, produce almost as much volume over a rotation unthinned as thinned: obviously, the average tree is much smaller than in thinned stands, but this characteristic is vital where thinning is impossible because of windthrow. It means that one of the forester's objectives, maximum volume production, can still be met, and from the tree species with the wood most valued for paper pulp production.

ROTATION LENGTH

The way trees grow is a less obvious but equally important factor in determining how long they should grow before felling, the length of the rotation. Wood production is an S-shaped curve through the life of a tree. During establishment little wood is produced, but after canopy closure there is a period of sustained, rapid growth. Well before the tree reaches biological maturity, and starts to die, its growth rate levels off. Annual production is described as 'current annual increment' (CAI), which represents the increase in volume over a particular year during the stand's life. 'Mean annual increment' (MAI) is the average rate of volume increase up to a particular year. Maximum MAI is achieved at the point where its rising curve crosses the CAI curve, which is by then falling. Felling the crop and replanting at maximum MAI will give the maximum average wood production for the tree species and site over time (Hamilton and Christie, 1971). Maximum MAI is central to the yield class system which is used to express relative site and species productivity, because that is what the yield class of a stand is—the maximum MAI in cubic metres of timber production per hectare per annum. So a hectare growing yield class 12 Sitka Spruce is in theory able to produce an average of 12 m^3 of timber per annum in perpetuity.

In the field, yield class is derived from a sample of the top heights of the 100 tallest trees per hectare: conversion between top height and timber production is based on intensive study of sample plots of all the main commercial species scattered across the whole country. After species, planting year and area, yield class is the manager's most important piece of information about a productive compartment. With this information and forest management tables, the manager can identify not only overall timber production, but also the optimum felling and thinning times.

'General yield class' (GYC) is the yield class of a crop of average tree taper and management and is the one normally quoted. However, local biological factors, for example higher than average exposure increasing tree taper, may mean that crops differ significantly from the standard relationship between timber volume and top height. A non-standard 'production class' may be applied, either raising or lowering the yield class. This level of precision is only necessary for detailed production planning.

The growth curve of a higher yield class stand is steeper than for lower yield classes, so it will reach the age of maximum MAI quicker. Species vary greatly in growth pattern. For a given yield class, larch completes the cycle quicker than spruce, and pine are slowest. Given an average upland site, with no particular physical constraint such as low rainfall, spruce will be the highest yielding, followed by larch, followed by pine. To maximize return on investment, stands should be in the part of the curve where growth is highest for as much of the time as possible. The interest accruing on the initial investment in planting the trees means that as soon as the rate of timber accumulation of the stand starts to fall off, the main return of the rotation, clearfelling of the final crop, should be realized.

The third factor in this is the price-size curve; to some extent it parallels the growth curve (Mitlin, 1987). Generally, small timber (below 14 cm diameter) is less valuable than larger timber, because it cannot be cheaply converted to higher-value products. It also costs more to harvest, because the time needed to carry out operations like felling, trimming off branches ('snedding') and extraction does not rise in direct proportion to the size of tree. Eighteen centimetre top diameter (i.e. the thin end of a log, whatever length it is) is the minimum for a 'sawlog', and represents a considerable step in price; larger logs will be more valuable, but the increase is not as steep within the sawlog sizes as between them and smaller wood. Thus, like the growth curve, the price–size curve starts to taper off, and beyond a certain point adding to the rotation length to produce larger timber will not be rewarded by a proportionate return in increased value. For the largest trees, practical limitations may come into play; whilst there is a good specialist market for large timber, it may be limited in a particular locality, and very large timber may be impossible to extract with local equipment; this is more important today with the increased use of forwarders than in the past when most wood was skidded.

So, in the absence of other factors (especially windthrow), three superimposed curves determine rotation length, and to a lesser extent, stand management. They are predictions, not certainties. Predicted growths are based on sample plots and the precision of the predictions increases as the plots age and more information is gathered. The price–size curve predicts future performance from past trends; as with all forest economics, the best possible information can only be viewed through a crystal ball. However, because the prediction is not exact, it is a common mistake of critics of forestry to assume it is without value. Even assuming high accuracy, economic loss five years either side of an optimum felling age is generally minimal, and is slight over 10 years for crops completing a full rotation (i.e. not prematurely felled due to windthrow). Predictions change over the years; the latest FC models show optimum rotation lengths extended compared to the previous five years due to re-analysis of sample plot data showing a higher growth rate later in the stand's life, and the price–size curve re-analysis showing a greater than previous advantage in larger timber. None of this may seem relevant to birds, but it is, because the criteria for conifer crop management determine the nature of our forests more than any other factor. Many people believe that growing and harvesting trees is a hit or miss affair and that rotation length or thinning regimes are arbitrarily decided. There is less certainty in forestry than agriculture but we hope we have illustrated the financial consequences of moving away from the economic optimum: this does not mean it cannot be done, but if the will to change stand management exists, a lower rate of return on investment must be accepted.

WINDTHROW

In an ideal world the trees would grow, be thinned and be harvested according to the economic and stand production models. In reality, modifying factors intervene. Windthrow, which prematurely terminates a rotation through the trees blowing down, has been by far the most significant unforeseen problem of British forestry, far more important than more commonly advanced dangers like insect attack and dependence on a small number of species. Windiness is a component of Britain's oceanic climate, which is otherwise much better for tree growth than the continental climate of western Europe and Scandinavia. Windthrow dominates forest management on the more exposed and wetter upland sites; trees blowing down, rather than economic or management decisions, decides when trees will be harvested. However fast a tree grows, if it blows down before it has reached a size at which it can be profitably harvested, the rotation will be an economic failure. Tree and stand stability is a function of exposure, rooting, tree height and canopy structure. Windthrow can be either catastrophic (caused by exceptionally severe winds) or endemic (caused by normal winter gales).

Catastrophic windthrow is caused by gales of a strength occurring only once in many years in a particular area. For example, the gale of January 1968 in central Scotland, with wind speeds of over 100 km/h, sustained for over seven hours in the worst-hit area, was estimated to be likely to recur on average every 75 years (Holtam, 1971). On average, such severe storms hit Britain every 15 years, the gale of 16 October 1987 in southeast England being the most recent example. Exceptionally severe damage occurs within the worst-affected area, and sites not normally vulnerable to windthrow may be devastated. In the 1968 gale it was estimated that even on the best soils where trees were firmly rooted, any over 12 m high were unlikely to survive. The distribution of damage in southern England, which had more to do with exposure and topography than tree height or rooting, confirmed previous experience of catastrophic incidents. There is no real defence against catastrophic gales other than maintaining age class diversity to ensure that the whole forest is not vulnerable at once. Although the devastation of the worst storms is horrifying, and the greater incidence of broken (rather than simply toppled) trees and quantity of fallen timber cause serious harvesting problems, in the long run endemic windthrow is of greater economic importance to the forestry business.

Endemic windthrow is caused by normal winter gales of 70 km/h, gusting to 100 km/h, which occur several times a year in most upland forests, acting on trees poorly rooted in wet soils. In contrast to catastrophic windthrow, criteria can be identified that determine a stand's susceptibility to endemic windthrow (Miller, 1985). Soil and tree height are most important; wetter soils result in less stable crops. Worst are peaty gleys; because rooting often stops at the gley layer, and the tree root and peat layer are prone to peel off the gley layer. Well-drained deep peats do not have this impediment and trees tend to root better in them. As trees grow they present a greater surface area to the wind and greater leverage on the root system. The local exposure of a site and the variability in windiness over Britain as a whole (Miller, 1985) are the next most important factors; the top of a hill on the Argyll coast is far more vulnerable than a sheltered glen at the same elevation in the same area. The western half of the country is windier than the east, and combined with generally better-draining soils the hazard of endemic windthrow is not a serious constraint on forest management in southeast England.

The tall thin trees of a high wind hazard plantation fall like a pack of cards as an exposed edge is opened up. But once blow has started no one can predict how fast it will move: this stand could be gone in 6 months or still be intact in two years time. Note the characteristic lifted root plates.

Endemic windthrow tends to spread from a focus, often on an exposed edge of the crop but also from points within the stand. Local waterlogging is often a cause (and a forceful silvicultural reason for not planting riverine flats). But the main factor is tree height, and once trees have reached the 'critical height' (see below) for the site, onset of windthrow is likely to be rapid. The rate of subsequent spread is variable. It may take five years from the onset of windthrow for 40% of a compartment to be blown. Small patches of wind-throw within younger crops may stabilize. However, once started, windthrow can spread across large even-aged blocks at an alarming rate, sheltered internal compartment boundaries providing minimal stability to check its progress. During the early years of dealing with endemic windthrow, blown trees along exposed faces were quickly cleared up and the standing trees behind them were promptly blown down. The 1000 ha of the Trapezium block in Harwood Forest, Northumberland, was felled from one end to the other in a decade, southwest to northeast, down the prevailing wind. Subsequently, it has been found that blown edges may stabilize for several years, allowing some age class variation to build up, which is important silviculturally in developing stability between adjacent compartments.

Silvicultural practice can influence stability at several points in the crop's life, and some cultural practices have exacerbated the problem in the first rotation. Cultivation and drainage greatly influence the trees' ability to root on wet sites. Spruce will not root into waterlogged soil and inadequate drainage causes very shallow rooting. Roots may also be restricted by ploughing; shallow single-furrow ploughing in the 1950s is a serious cause of instability now. Roots do not cross the plough trench and the root plate develops parallel to the direction of ploughing, making trees very vulnerable to winds at right angles to the direction of rooting.

Susceptibility to windthrow is influenced by harvesting practice. External trees develop a windfirmness not shared by sheltered trees within the crop. Wind behaviour over the stand varies with roughness, a canopy broken by thinning causing disruption of wind flow and turbulence which exerts further force on trees. Because of this, thinning increases the chance of windthrow, and serious windthrow may occur at 25 years in a thinned crop which would have survived unthinned to 45 years. To put this in perspective, the onset of windthrow in a windthrow hazard class 5 (see below) GYC 12 Sitka Spruce stand systematically thinned is 11.5 m, at 28 years old, only five years after thinning, but the same stand will continue unthinned to nearly 50, only 10 years short of optimum economic rotation age. This is why stands in hazard classes 5 and 6 are never thinned and it is also an important reason for planting Sitka Spruce on sites susceptible to windthrow, because Sitka Spruce can produce an exceptionally high proportion both of larger timber and of volume in unthinned stands. In the FC estate 17.7% is in hazard class 5 and 8.7% in hazard class 6, totalling 234 527 ha. Over the whole of this area, forest management is physically limited to short rotation, non-thin and clearfell.

The likelihood of windthrow can be predicted from a classification which allots points for different characters and divides sites into hazard classes from 1 (least risk) to 6 (greatest risk) (Miller, 1985). Hazard increases towards the north and west of the country because these areas are windier and exposed, a function of elevation and topography. Soil type is important and for the reasons explained above, wetter soils score higher than drier ones. The practical application of the hazard rating system is in establishing the critical tree height and whether it is safe to thin: the age (depending on yield class) and average size of the trees at critical height, the point at which windthrow starts, can be predicted and are vital parts of the economic prediction for the crop. It is important to distinguish between 'critical' and 'terminal' height. The crop is rarely felled at the onset of windthrow, the 'critical height'. Often a stand may continue growing for several years and the extra volume accrued is vital to the economics of harvesting small trees. Terminal height is when 40% of the crop has blown; it is reckoned to be the limit for holding onto the crop, after which loss from deteriorating dead trees offsets the gain from those still growing. Usually crops are cleared between critical and terminal heights.

Windthrow is the sting in the tail of Britain's otherwise ideal tree-growing climate. Silvicultural techniques and forest planning may be able to reduce its dominance over management, but it is likely to remain the major constraint in the more exposed upland forests.

HARVESTING

In all the arguments over the impact of afforestation, it is easy to forget the main objective of the new forests: timber production. The scale of the increasing timber harvest is not obvious; thinning hardly shows in the landscape and less than a third of the FC's plantings have been clearfelled. The development of harvesting has been the major change in forestry operations of the 1970s and early 1980s. Now far more people are employed harvesting trees than planting them, and even if planting of new ground stopped tomorrow, the forest industry would continue to grow into the next century as the 350 000 ha planted during the 1970s come on stream for timber production. The growth of timber production from the new forests has been spectacular: in 1950 the FC forests produced slightly less than 400 000 m^3 of timber, in 1987 production passed the 3 000 000 m^3 mark and forecast conifer production will be over 5 500 000 m^3 by the year 2000. Between 1982 and the end of the century, national production of conifer timber from state and private sectors is forecast to double from 4 470 000 m^3 to 8 760 000 m^3 (Forestry and British Timber, 1984).

Harvesting machinery and the inherent economic constraints of different methods may affect important aspects of forest structure, especially the shape and size of felling areas and the distribution of roads. There may also be smaller-scale effects that are still important. The type of machine used to extract the trees is important in deciding thinning pattern. The cost of thinning, which also depends on the harvesting method, may affect when thinning is carried out, or whether a crop is thinned at all. The way in which the branchwood trimmed off the trees (known as 'brash') is spread over the site may affect the pattern of vegetation regrowth.

Most trees are still felled by workers with chainsaws who also trim branches ('snedding'). They may measure and 'convert' the tree into products by cross-cutting at the point of felling, or timber may be extracted as whole poles and converted at the roadside. The main products are the smaller material for pulp, which is cut into 2–3 m lengths, and the lower and larger parts of the tree for sawlogs, which are cut into lengths of anything from 4 to 8 m. Assisted by a wide range of hand tools, the feller is able to direct all but the largest conifers to fall precisely as required on clearfells. Thinning is more difficult: 'takedown', the problem of the felled tree catching on the trees that are being left, increases the time and effort it takes to do the job and contributes to high thinning costs. Although on the better, more stable, soils, crops of high yield class Douglas Fir and Sitka Spruce can grow to average more than 1 m^3 timber volume per pole, the average size of conifers clearfelled by the FC nationally is very much smaller, nearer 0.25 m^3.

Felling a tree is spectacular, but it is also about the easiest part of the harvesting operation. Trimming the branches takes the chainsawyer far longer than felling the tree but the removal of the timber from the felling point to the roadside across steep or soft ground is the most difficult part of the work. Although a lorryload of sawlogs may be worth over £1000, softwood is still a relatively low-value product, and handling efficiency is vital to the viability of harvesting. Harvesting must be on an industrial scale to meet the massive demands of processing mills, to fully use expensive machines and to make practical day-to-day management possible.

The quantity of produce coming from a particular area is of great importance at every level: at a compartment level the density of timber and how far a machine must move to pick up a full load has a significant effect on cost. It is another factor that makes thinning expensive, as less than 50 m^3/ha may be felled for a thinning whilst a clearfelling is rarely less than 200 m^3 and sometimes over 400 m^3/ha. For a forest a large harvesting programme allows a larger, more cost-effective extraction machine to be introduced. At a regional level, long-term timber availability will influence the location of timber using industry. At an international level, Britain's new forests represent a new and uncommitted source of raw material which, against a background of shortage in Europe as a whole, has successfully attracted major new industrial investment during the 1980s.

In 1960 nearly 400 horses extracted timber from FC forests. Horses worked relatively steep slopes but pulled a small load slowly. Roads had to be close together, a major expense in mountainous country. Horses do little damage to the ground, extracting low densities of small poles from first thinnings, but can turn most soils into a quagmire if they travel over the same place several times. By 1969 the number of horses had declined to less than 200, but tractors had increased from 380 to 550. They were mainly 'skidders', dragging whole trees along the ground behind them. A few were specially built forestry machines, but adapted agricultural machines which were cheap to buy and run were commonest. Although they extended the economic range for pulling timber, maintenance costs could be high because of shocks fed from the dragged load to the tractor, and the load and wheels could also churn the ground up severely.

Cable crane systems, which work on the same principle as alpine chair lifts, were developed to work the steepest ground. Because the logs run attached to a carriage suspended from a fixed cable, there is little damage to the ground and the small cable cranes used to extract the relatively small timber harvested in Britain are cheap compared to giant American 'yarders'. However, the time taken to set up the fixed cable and anchor the tractor and the tower to which the bottom end is secured makes the system expensive to operate. It is also inflexible: the cable must run straight from the tractor's location and has a

Clearfelling Sitka Spruce. Most trees are still felled by a man with a chainsaw. Keeping a narrow 'hinge' of wood at right angles to the planned angle of fall gives a high degree of control over where the tree falls. For a fairly large tree like this one a wedge is used to start its fall. Note safety equipment - helmet, earmuffs, visor, gloves, boots, which combined with training makes safe use of a chainsaw possible.

limited range. High road densities, often on slopes where road building requires expensive blasting into hillsides, are required. Timber density is absolutely critical in cable crane operations because the cost for each set-up is fixed and must be spread over as much timber as possible. It is far more difficult to justify leaving part of an area for landscape or conservation reasons, as the cost of setting up again to harvest it later will be high. Extraction by some kind of tractor running on the harvesting site is almost always cheaper, and cable cranes are used only where it is impossible to get to the timber any other way.

Cable crane extraction from a steep slope in Argyll. The base unit is a small agricultural tractor but the cableway and anchoring arrangements are complicated. Note the accumulating pile of logs next to the tractor.

Cable cranes and skidders have one big problem in common: what happens to the wood when it reaches the road? They may be able to pull it out fast enough but there has to be somewhere to put it until lorries come to remove it. Space is at a premium. On soft ground every bit of hard standing has to be built up with imported stone and can cost pounds per square metre. Whole poles must be converted into products and that takes space too. Stacking timber makes space go further but cable cranes and skidders have limited stacking ability so a hydraulic crane has to be introduced to the system. In the end there may be several workers and machines involved, and if only one machine stops, everything soon stops because each element depends on all the others. Also, however cosy the machine's cabs, everything also stops when it rains because half the team work outside.

Timber prices were high in the late 1970s so the high costs of skidder systems were hidden, but the crash in world markets in 1979 left smaller trees and remoter areas in Britain making an operating loss on harvesting. The necessity to increase efficiency or go out of business led to a revolution in harvesting method that overcame the chronic technology transfer inertia that still besets most changes in forestry method today. At the centre of the change was the introduction of 'forwarders' and the flexibility of working they allowed. A forwarder is a tractor-trailer combination with a hydraulic grapple crane for self-loading. They may be purpose-built forestry machines or adapted from agricultural tractors; the former tend to be fully articulated and 'frame' steered by bending in the middle to turn and have all-wheel drive; the latter pull the trailer through a standard drawbar and only the tractor wheels are driven. Purpose-built machines were in use in Sweden by the mid-1960s (Holtam *et al.*, 1967). Mechanical development of cranes and drive increased speed of loading and terrain capability. The Volvos of the early 1970s were a breakthrough in reliability, speed and terrain capability, but they still could not travel over the softest British sites.

The introduction to Britain of the Mini-Brunnett in the late 1970s, with its eight balloon-tyred, hydrostatically driven wheels, finally cracked the problem, and now 62 of these machines operate in Britain. The Brunnett could 'float' across ground that other machines could not attempt, assisted not just by its low ground pressure but also by the smooth delivery of power to the wheels. Although still the best on the softest sites, the Brunnett set a standard that was followed by improvements in other Scandinavian machines, including the larger 10-tonne capacity Kockums, Valmets and Lokomos that are now in widespread use in Britain. One hundred and sixty purpose-built forwarders now extract the majority of the timber harvested in the new forests (Forestry Commission information, 1987).

Forwarders extract 'shortwood', the tree converted into its various components, because their trailer is generally less than 5 m long and logs over 8 m will fall out. As each lift of the crane takes roughly the same time, picking up a single piece of pulpwood is likely to be rather inefficient, and there is also the problem that small material gets lost in a sea of brash. For the first decade of forwarding in Britain, trees were felled flat on the ground, the sawlogs left where they lay, and pulp gathered together by sheer brute force into heaps. This was very hard work and expensive because it was time-consuming. The other big factor in the growth of forwarder extraction was bench or organized felling, which allows material to be gathered into heaps far more easily.

Organized felling uses balance and pivoting to allow the feller to move large timber around with little effort. In the simplest system, two trees are felled parallel and a third is felled across them at right angles (Husqvarna, undated). Further trees felled onto this 'working bench' are raised off the ground, allowing them to be rolled along the bench and giving the operator a more upright posture for snedding. A series of trees fall off the end of the bench together and it is then easy to gather the pulp from their tips into piles. Sawlogs, which are not moved once they have been converted after rolling off the bench, and pulp are separate and are laid out in a neat, concentrated pattern, making extraction easier.

A major economic gain of organized shortwood/forwarder systems is their simplicity and lack of interdependence. The feller presents timber on the ground. The forwarder can come in any time from the next day to the next month to pick it up. The machine is completely self-contained, with a single driver, and has the ability to stack already converted products high at the roadside using its own loading crane, so that large quantities of timber can if necessary be accumulated along a short length of road. If the forwarder breaks

Brunnett 578 extracting from organized premature clearfell in Kintyre. In this machine the crane is mounted on the cab. The driver's seat swivels, facing to the rear when loading. Produce density is high and as it travels the brash mat between rows of timber the forwarder may need to move only two or three times to pick up a full load.

down, only one machine and one worker stand idle. The operator can work in the wet in the comfort of an ergonomically designed cab. The forwarder normally extracts only one product at a time, and the organized felling layout means it is possible to select either pulp or sawlogs first to meet urgent orders. However, with a 10-tonne machine costing £100 000, the forwarder must be kept working all the time, so there must be a large harvesting programme within a reasonable distance of the forest centre.

Forwarder systems have a number of effects on forest conditions. One of the most obvious is the strip of concentrated brash created by organized felling. Forming under the bench where the trees are snedded, it provides a carpet for the machine to travel on, reducing direct soil damage and allowing travel on softer sites. Half the site has little brash on it at all, but the density of brash on the rest of the area may reduce vegetation regeneration and affect the success of restocking. Less obvious is the effect on forest planning of the greater flexibility of forwarders: although they are cheapest when travelling the shortest distance to extract, neither costs nor damage to machine and site rise as fast with increased extraction distance as for skidders. Road density can be lower, and size, shape and precise location of felling areas have a less critical effect on cost. The number of working sites is also less important when only one machine rather than a whole circus of workers and equipment have to be moved.

Organized felling in pine at Thetford. The neat presentation is achieved entirely by accurate felling and leverage and balance using simple hand tools. This is a 'strip' rather than bench system: the trees are too big, scattered and coarse for the sort of bench system, used in smaller, straighter spruce, to work well.

This contrasts with the common belief that bigger machines reduce flexibility. It is equally true for the machines increasingly used to fell, delimb and convert trees. Again, a sufficient volume of timber to keep a machine busy is the main constraint: otherwise, they can work wherever they can travel. Most are purpose-built Scandinavian machines. Harvesters do the whole job, including felling the tree; processors delimb and convert trees felled for them by chainsaw. The main disadvantage of harvesters is their greater complexity; it does not matter that the processor part of the machine is still working if the felling head has broken down.

The unique operating advantage of harvesters is their ability to deal with windblown trees as easily as standing trees. A major reason for felling before windthrow is the difficulty of disentangling the mess of fallen trees for men with chainsaws. Not only is the job time-consuming but it is also dangerous because fallen trees lying over each other have complex tensions, and careless cutting can result in trees whipping in all directions. Harvesters reduce the risk of hanging on to crops up to and beyond the point at which they should theoretically blow down. The potential value to birds of keeping some crops beyond theoretical felling age is discussed in the next chapter.

Kockums 88-65 clearfell harvester. The tree is felled by a chainsaw at the end of the hydraulic crane, laid on the processor bed between two large rubber tyres which drive it past the delimbing knives. A chainsaw connected to a sophisticated computer-controlled measuring system crosscuts to a pre-set programme.

The powerful hydraulic crane also solves the problem of takedown in thinnings. The width of the harvester is a problem for access to the crop, especially working close-planted crops: smaller machines are generally just under 3 m wide and can work down a single row removed at 2 m between the rows (i.e. 4 m total width), but obviously not between rows spaced at 1.5 m. However, once into the crop they can reach out either side to cut and pull down trees that would be inextricably stuck for a worker with a chainsaw.

The rate of mechanization has accelerated in the last few years: harvester numbers increased from 11 in 1986 to 23 in 1987, and processors increased from 44 to 60. Catastrophic windthrow has in the past been a particular spur to mechanization: it simply is not possible to go out and hire and train the extra workers needed to clear the fallen trees. The October 1987 gale has been no exception: the FC added two more large clearfell harvesters to the single machine it was then running. Mechanization also has a lot to do with labour supply; this has been the main spur in Scandinavia, where it is impossible to get sufficient chainsawyers in the north of the country. The same is likely to occur in Britain, especially on the west coast of Scotland: the small town life of foresters and the nature of chainsaw work means that unemployment in the towns cannot be turned into employment in the countryside. As programmes rocket, bigger machines will be introduced to take up the slack. There will be fewer jobs per unit timber harvested but this is not automatically a bad thing. Chainsaw felling, although far more skilled and safer for trained workers than is generally realized, remains an exceptionally arduous job. Mechanization means a better quality of job involving higher pay for higher output, new skills and greater responsibility and far less physical effort. The actual felling of the trees is only one part of the picture: other jobs such as road haulage must continue to increase unless there is a spectacular investment in improving public roads in the uplands.

RESTOCKING

Planning for the next forest operation, restocking the felled site with young trees, should start even before the trees are felled. A felled area must be a practical shape for restocking as well as felling, and the method of felling may influence subsequent treatment of the ground, for example breaking up the concentrated brash of organized felling.

Land is replanted rapidly after felling; it is important to get the next crop growing so that the capital invested in land is once again giving a return. Quick restocking also beats the surprisingly rapid regeneration of ground flora which competes with newly planted trees. Restocking is planting the second time around; planting succeeded once, so the lessons have been learned and everything is that much simpler, or so it seemed. The earliest extensive restocking was approached with a casualness only partly explained by the involvement with the new experience and pressure of large-scale harvesting. It quickly became obvious that restocking involved a completely new set of problems, and that the trees were not growing on many of the early sites (Low, 1985; Tabbush, 1988).

Part of the problem was overconfidence. By the early 1970s techniques for new afforestation had become almost completely reliable. There was minimal weed competition on the infertile peats and peaty gleys that dominated acquisitions, and very small trees survived well. Planting was almost exclusively with Sitka Spruce and Lodgepole Pine, which were able to survive handling that often owed more to economy than care. Low costs also dominated decisions on ground preparation and planting. A new range of problems quickly emerged. It was impossible to plough clearfell sites using new afforestation equipment. The plough had to get below root plates to cut through the roots rather than simply lifting them and turning the whole site into a battlefield. With a caterpillar tractor twice the size of the types used for afforestation towing a 1-m depth plough slowly over rough ground, it was not surprising that costs were generally four times higher than for ploughing fresh ground. Only sites with an ironpan were routinely ploughed.

Most sites were planted without cultivation, using the same size of tree as for ploughed afforestation sites. Dense brash cover made planting difficult, especially for people under pressure to meet piecework targets. Finding the ground between the branches can make it impossible to select an ideal planting site; sometimes trees were not planted in the ground at all. On fertile sites, rapid weed growth out-competed the small trees, whilst on gleys, waterlogging checked growth. Although larger trees were used for beating up, the weeds had grown too. It becomes increasingly difficult to establish the new crop after an initial failure.

Douglas Fir restocking suffered especially badly; a more fragile tree than Sitka Spruce, it was easily killed by bad handling and carefree planting. It proved particularly vulnerable to damage by the pesticide Gamma-HCH in which trees were dipped to protect them from the Large Pine Weevil and the Black Pine beetles. A new factor on restocks, these insects feed on the stumps and roots of the previous crop and emerge to damage the stems and roots of young conifers. The majority of trees will be killed (Scott and King, 1974) unless protected. Traditionally, sites were left for three or four years for the beetles to complete their life cycle, but weed growth becomes rank in this time and where clearfelling is extensive they can invade from adjoining areas.

But the factor which more than any other turned much upland restocking from a problem into a disaster was deer. Problems with Gamma-HCH, plant quality and ground preparation have proved soluble; deer follow only wind-throw as the second most intractable problem in upland forest management today. Our understanding of deer in upland forests and their impact, not just on the productive conifer crop but also other forest wildlife, is still growing. Deer proved painfully that the new forests do have a natural ecology and that it is something foresters ignore at their peril. A combination of deer and escalating restocking costs are bringing to an end the era of technology as a means of subduing the site and creating a new awareness of the need to manipulate site factors, including wild animals, rather than trying to suppress them.

All British deer are woodland animals. However, in Highland Scotland Red Deer successfully adapted to life on the open hill as forest cover declined. This was not the case for Roe Deer and Fallow Deer, the predominantly lowland range of which was limited by human pressure and lack of woodland cover. Large herds of Red Deer presented an obvious problem to foresters, and because they are preserved for stalking on adjoining open hill estates (known paradoxically as 'deer forests'), they could not all simply be shot, so the forest had to be protected by fencing. Fence lines were designed for economy and frequently cut across traditional lines of deer movement, including bad weather escape routes from high ground. Break-ins, sometimes of hundreds of deer, occurred but were easily dealt with by shooting or driving deer out while the trees were young and the deer could be seen.

Once trees were into thicket stage they became less vulnerable to browsing. Red Deer still broke in but now there was cover and they became established and began to breed. Once again they became woodland deer, seeking dense cover when threatened rather than running like hill deer. Roe Deer spread into the uplands as the new forests provided a suitable habitat. Most upland forests now have Roe Deer, except in Wales, which has only Fallow Deer. The colonization of upland forests took foresters by surprise for a number of reasons. Within the Red Deer range, deer managers were accustomed to seeing deer out on the open hill and did not realize how they could melt away in the forest. In Galloway, for example, hill counting methods were used right up to the early 1980s to assess the Red Deer population. A great deal of confusion resulted when severe damage started because it seemed to be caused by very few deer. The rapidity of the spread of Roe Deer into the dense cover of thicket-stage forests was similarly unexpected. Logically, the deer could not thrive in these conditions because the forest had shaded out the deer's food supply, the ground vegetation. In fact, the high-tech forest was never as regular as its proponents would like to have believed, and failed patches, hidden rides and unplanted bogs or rocky outcrops provided hundreds of secret deer lawns throughout the forest (Ratcliffe, 1985b).

Without natural predators the deer thrived and soon began to damage trees. Bark-stripping by Red Deer started before the first rotation was complete; the bark of trees anywhere from early thicket stage to pre-felling is stripped, leaving bare timber exposed to fungal attack, and smaller trees may be ring-barked and killed. Lodgepole Pine and Norway Spruce are most vulnerable, and at Fiunary Forest on the west coast of Scotland large areas of Norway Spruce were so badly damaged they had to be felled prematurely. The introduced Sika Deer, which is spreading rapidly in Argyll and Sutherland (Ratcliffe, 1987b) also strips trees and is damaging the young Lodgepole forests of Shin. However, the most serious problems have been the browsing of the young trees on restocking sites.

Both Red Deer and Roe Deer eat young trees. Damage to prickly young Sitka Spruce is a sign of overpopulation, leading to starvation. Damage is worst in hard weather, especially if the young trees are all that is above the snow, and in the early spring, when other food is exhausted. Unfortunately, this is the worst time for the trees; they have survived the rigours of winter, their own reserves are exhausted and spring growth is just starting. Small trees are only a few mouthfuls, and deer damage compounds other stresses, such as waterlogging, turning weakness into death. On better sites, for example much of Argyll, Sitka Spruce is remarkably resilient and in the end grows above the browse line. It even makes up for lost growth, so that by canopy closure there may be little difference between browsed and unbrowsed stands (Low, 1985). In some forests with rich alternative food supplies of vegetation, grass and bramble, such as the North Yorkshire Moors and Thetford, high populations of deer may not damage conifers very much at all. However, in many forests, including Kielder, restocking has been virtually wiped out by deer several times on some sites.

The main methods for protecting the trees are shooting and fencing. Both are expensive and they do not always work. Where damage was worst more and more deer were shot but they just seemed to keep coming. Application of continental census techniques and lessons learnt from familiar lowland Roe Deer habitats did not tally with practical experience in the new forests. The research commissioned to investigate this applied problem developed into the most important ecological study to date of the relationship between our native wildlife and the new forests (Ratcliffe, 1984b, 1985a; Staines and Welch, 1984;

Ratcliffe and Rowe, 1985; Ratcliffe *et al.*, 1986). It showed that the fecundity of both Red Deer and Roe Deer in upland forests was far higher than expected; the difference between hill and forest Red Deer is particularly striking (Ratcliffe, 1984b). It is not surprising that predictions based on hill populations vastly underestimated numbers in the forest and the number it would be necessary to shoot each year to check population growth. The importance of the age structure of the forest to practical management also became apparent, as did the need for a firm factual base before undertaking wildlife management. The lessons learned from the deer story are equally applicable to birds that do not damage the forest but may require positive management to survive there. The difference was that the damage to the trees meant that the deer could not be ignored.

In the early years of forest deer management attention focused on the method of shooting deer. The FC led in getting the rifle rather than the shotgun accepted as the humane weapon for deer control, and the substitution of deer drives by continental-style stalking. It was assumed that culling would be selective in the same way as hill stalking, where stags are chosen from a herd of varying quality. Once damage started, it was gradually realized that with so many deer there was little chance of affecting the balance of the population by selection and that the big problem was seeing the deer to shoot them. Within regular, thoroughly planted forests in the stage between canopy closure and the start of felling there is minimal open ground accessible to the stalker. Restocks are often far from ideal in shape for shooting and tend to back onto large areas of inaccessible high forest that the deer emerging from the forest edge to feed can melt back into.

The idea of the deer glade developed; a space kept for its attractiveness to deer, its accessibility to the forest ranger and a lie of the land suitable for shooting (Ratcliffe, 1985b). For the first time, planned open ground became something more than a waste of space to the forester. Now deer glades are incorporated in an increasing area of forest at new planting, or following felling, and their value has highlighted the potential importance of reserved open ground to other wildlife. It looks as if Red Deer can be controlled in extensive upland forests, and to achieve this glades must be created even in some of the large areas of late 1960s and 1970s thicket crops where open ground is exceptionally scarce.

In contrast, the fecundity of Roe Deer is so high that it seems unlikely that their population can ever be controlled in big upland forests. Shooting must be concentrated on and around the areas where there are vulnerable young trees. Where damage is heaviest, temporary fencing has been used for the first five years to allow trees to grow tall enough to escape lethal browsing. In contrast to external fencing round the outside of a block, which aims to separate the hill Red Deer from the forest population, the idea of fencing restocks is to keep all deer out and prevent the escape of any that get in so that they can be shot. Fencing is expensive but at Kielder it has proved a great deal cheaper than beating up the same sites year after year.

Fencing also showed the profound effect that deer may have on ground vegetation, and through it on birds. At Kielder, broadleaved trees, not seen on restocks before, regenerated inside fences, and ground vegetation was similarly richer. Ratcliffe (1988) estimates that less than 5 deer/km^2 (100 ha) depress plant diversity and that 6/km^2 will prevent most broadleaves growing. Heavy browsing of Sitka Spruce takes place at 15/km^2, but density is not regulated by starvation until there are over 20/km^2. Densities of around 15/km^2 are common in the new forests. Unprotected broadleaved trees cannot regenerate and the diversity of ground vegetation is limited in exactly the way that domestic stock or hill deer limited it before afforestation. Deer restrict the managers' options; for example, the sort of wide-scale management of rideside shrubs advocated for lowland forests (Anderson and Carter, 1988) is impossible in upland forests where everything is eaten.

The result of these difficulties has been large areas of very poor restocking, with wide-spaced trees or substantial areas of bare ground. There will be a very significant loss of timber volume as well as quality problems with excessive quantities of juvenile wood resulting from very wide spacings. In Wales some restocking sites have been overwhelmed by regenerating broadleaves and most have rather more broadleaves than might have been planned. Scattered throughout western Scotland, the areas blown down in the 1968 gale are particularly noteworthy (Holtam, 1971). Generally, only the area damaged was restocked and the often small pockets were virtually impossible to protect from deer; some have still to close canopy. Although a disaster for forestry, the poor quality of much early restocking is probably of value for birds, providing a mixture of trees, open space and non-crop trees.

Restocking is now recognized as quite different from new afforestation and new methods and machines are developing to solve the main problems. Larger plants do better on restocks, with more reserves to become established on the site (Tabbush, 1988). However, the practice of simply growing trees longer in the nursery produced lank, weak plants. Wider spacing in the nursery bed allowed plants to spread outwards as well as up, producing larger and sturdier trees. Expensive in the area of nursery land they took up, these trees soon proved themselves economical by surviving after planting. It even proved possible to establish Douglas Fir reliably if sufficient care was taken in lifting plants from the nursery, transporting them to the site and planting them. It became clear that the planting microsite is crucial: it does not matter how sturdy a Sitka Spruce plant is if it spends its first winter with waterlogged roots. Raising a plant a few centimetres above the general ground level can give sufficient local drainage to make the difference between death and survival, and the practice of planting close to the old stumps was successfully adopted on gley sites.

Cultivation, however, remains the best guarantee of survival and growth. On the wettest sites 'dollops' or mounds of soil dumped on the ground by a hydraulic digger provided the best sites with good local drainage and no restriction to root growth but were very expensive to make. On dry, fertile sites weed suppression and quick establishment meant that ploughing could save money despite its high initial cost. Attempts to reduce costs led to the introduction of Scandinavian scarifiers designed to roll over immovable obstacles rather than smash through them. Much quicker, moving less soil and requiring less power, dry ground scarifiers like the TTS Delta can give most of the benefits of ploughing for far less cost. They are also particularly suitable for breaking up the dense brash lanes created by organized felling which are virtually impossible to plant through (Tabbush, 1988). Scarifiers for wet sites have proved more problematical: the Sinkkila is designed as a high-speed mounder but it is hard to create ideal dollops on sites with heavy brash. The scoop of soil may be left perched on top of a heap of brash with air between it and ground level, or broken up by the brash before it can be deposited by the scarifier tine.

Technological development and experience gained from the first rotation resulted in changes in which trees are planted where. Now more is known about the geographical potential of different species; Corsican Pine is limited in the north and west, as is Sitka Spruce in the English Midlands, and although Douglas Fir grows well on the Thetford sands, it is almost impossible to establish because of frost damage in its early years. Increased knowledge of cultivation and fertilizers has also made a big difference; the ironpans of many heavily indurated heathland soils in north Yorkshire and northeast Scotland were not broken by first rotation ploughing and produced low-yielding Scots Pine. Breaking the pan with deep tine ploughs raises potential yield from GYC 6–8 Scots Pine to GYC 12–14 Sitka Spruce. Conifer mixtures, which have had

TTS Delta scarifier mounted on a caterpillar tractor. The rotating toothed discs create a shallow serrated furrow, rolling over rather than ploughing through obstructions like rocks or stumps and smashing aside dense brash. The Delta does the job of a plough in producing good planting sites for restocking – but without as much effort or disruption to the site.

a chequered past, also seem able to solve some cultural problems, for example larch-Sitka Spruce mixtures seem to offset the nitrogen deficiency suffered by pure Sitka Spruce on heathland soils.

As a result the forester is under economic pressure to plant a less, rather than more, diverse range of species; experimenting with a variety of species was justified in the first rotation because of uncertainty over how they would perform. Now we know the differences in biological and economic performance between first and second choices; these are often large. This is resulting in a higher and higher proportion of Sitka Spruce restocks: Sitka Spruce has been restocked on first rotation Douglas Fir sites because it is so much easier to establish, and it will not be possible to redress the balance later by planting Douglas Fir on Sitka Spruce sites, because the area of better soils is limited. Douglas Fir is the only species that can normally match Sitka's economic performance: Sitka outperforms larch and pine on most upland sites. On a Borders gley site at lowish elevation, where Sitka Spruce grows at GYC 14, larch and Scots Pine are likely to be no more than GYC 8. On a brown earth, Sitka Spruce might yield 16–18, Scots Pine or larch 12 and Douglas Fir 18–20.

Pine still has an important role on light soils, especially sand, and down the dry east coast, although Sitka Spruce is being planted on heathland soils in northeast Scotland and Yorkshire. The proportion of Sitka Spruce in the new forests will continue to climb as existing forests are restocked and new land is afforested. Despite the poor quality of early restocking, the second rotation should produce substantially more timber per unit area than the first rotation, because of advances in species selection and technology. These factors are vital to considering how wildlife can best be managed in the second rotation forest, which is discussed in the next chapter.

WILDLIFE

But what effect did wildlife have on management thinking in the first rotation forest? We have not said much about it because until very recently species that did not damage trees hardly touched management of the forest outside designated nature reserves. Wildlife management in the FC has had a varied history. It is difficult to identify all the successes and failures because much of the most valuable work was done at a regional or local level and usually failed to reach a wider audience.

General attitudes towards conservation were particularly important. Prior to 1945 it was a minority interest and from 1945 to 1981 site protection dominated thinking. The FC's record on the protection of legally designated sites of special scientific interest was, in hindsight, rather variable, and considerable damage was done to areas that would today have been ruthlessly protected. However, on many sites the requests of the Nature Conservancy were respected; whilst the Nature Conservancy might have wished to ask for more, they were very aware of their lack of real power and the risk of losing more by pushing for more stringent protection of wildlife interest. It is easy to pass judgement; the conditions of money, custom and attitude that people worked within were very different and there is little doubt that we are making equally serious, if different, mistakes today.

It has often been said that foresters are also conservationists; usually by foresters, of course. There is more than a grain of truth in the statement: throughout the FC's history, interest in wildlife has always been high amongst staff. However, it was not until very recently that this interest was seen to have wide practical application. The changes brought about by forestry were observed, special projects on limited areas were undertaken and small acts of kindness to birds, like not weeding around a Short-eared Owl's nest, were commonplace. But the changes as the forest developed, and the impacts of particular forestry methods, seem to have been seen as inevitable, almost beyond the control of man. The failure to recognize the importance of the developing natural ecology of the artificial forest contributed to a lack of awareness of the potential for altering management to help wildlife.

Another contributory factor has been the cost of conservation: as late as 1979 (Balfour and Steele, 1979) the idea of the proportion of a forest that needs to be 'set aside' for wildlife was still being discussed. Five per cent or 10% of land or expenditure does not sound a lot until you start applying it to over 1 000 000 ha and a budget of over £100 million. Some conservationists have persisted in the 'if it's going to do you good it's got to cost' school of thought, and this also did not encourage change amongst foresters. Very little consideration was given to the impact of forestry on open-land wildlife; again, changes were observed from a distance almost as if they were inevitable. By the time conservation pressure got through to foresters, they were in the worst possible state of mind to accept them, following the critical Treasury review of 1972, and subsequent pressure has led to increasing defensiveness rather than greater understanding of the problems (e.g. Garfitt, 1983).

New ideas on forest management, described in the next chapter, did not suddenly emerge fully formed in the 1980s. The FC's annual report for 1964 (Forestry Commission, 1965) makes interesting reading. In that year an officer was first appointed to co-ordinate wildlife policy at a national level, and quarterly meetings at a senior level were set up with the Nature Conservancy. Twenty years later there is still a wildlife officer, but the post is a grade lower than when it was established and national level meetings with the NCC are almost back to the originally conceived quarterly frequency after periods when years went by without any formal meetings between senior staff at all.

A lot has happened in between. The original promise of 1964 was not fulfilled for a number of reasons. Roe Deer did not behave the way they were expected to; deer did more and more damage to the trees and the wildlife officer's job became slanted increasingly towards protection of the trees rather than conservation of non-damaging wildlife. This only changed in 1985. By the early 1970s ideas on wildlife were developing rapidly. Amongst a plethora of largely descriptive FC wildlife publications, Steele (1972) and Campbell (1974) are outstanding in starting to see forest management as a process with the potential for manipulation to help birds and other wildlife. They remain limited in some respects; Steele, for example, still felt that there would be little that could be done for wildlife in some of the less interesting upland forests.

Eskdalemuir Forest. Generous open space follows the line of the river. Pools and deer glades add to the definitive wildlife design of the 1970s. Tree species diversity – especially broadleaves – is lacking by 1980s standards and insensitive ride layout, particularly obvious on these gentle, white hills is very poor compared to the thought given to wildlife management.

They did, however, provide a base which was not built on at a national level. The main reason for this seems to have been the aftermath of the 1972 Treasury Review (HM Treasury, 1972), which led to the total concentration on economic return within the FC at a national level.

Much of the current thinking on the management of wildlife in the new forests has been brought together from more local developments led by individual enthusiasm. Of particular importance was the influence of J. S. R. Chard in the FC's northwest England conservancy. Although special projects like the Grizedale Tarns (Granfield, 1971) are best known, an approach that came closer to unifying control of damaging wildlife and conserving harmless species through habitat management also developed. It led eventually to the publication of the Forestry Commission Wildlife Rangers Handbook (Springthorpe and Myhill, 1985). That these ideas were able to evolve as far as they did may have been due partly to the small new planting programme of this conservancy compared to Scottish and Welsh upland areas. However, there is no doubt that, with a few local exceptions, the FC delayed far longer than it should have done in applying these ideas more widely, and during the mid-1970s it was left to Rose, wildlife manager for the Economic Forestry Group, to develop forest design in a large, new upland forest, the Economic Forestry Group's Eskdalemuir in the Scottish Borders. Generous open space, especially along natural watercourses, was the key new feature of this forest. Although aimed primarily at deer management, that other wildlife, including birds, would also benefit is clear. Although it fell down on landscape design and by today's standards has a frightening lack of tree species diversity, it represented another important step forward.

By the end of the 1970s the base for more general change was being laid, and has subsequently been more widely applied as forestry has abandoned some of its most ruthless economic practices. By 1980, for example, all forests in the FC's North Wales Conservancy had formal conservation plans and ideas were developing on broader forest management for birds (Forestry Commission, 1982). In northeast England, the management of wildlife became one dimension of the comprehensive 'restructuring' approach discussed in the next chapter (Hibberd, 1985; Leslie, 1981).

©P.S.

CHAPTER 4

Conservation management

British commercial plantations contain many interesting bird species but they are also the workplaces of foresters whose main aims are to grow trees rather than to promote bird conservation. In this chapter we deal with how the conservation value of these conifer plantations can be increased in the future. Some progress is already being made towards incorporating conservation into the management of the trees in forests, but much remains to be done.

In the early 1960s foresters were under attack from a different group of environmentalists than today's conservationists. In those days it was land-scaping issues that dominated the headlines. The first sustained approach to long-term environmental improvement was landscape-based, starting with the appointment of Dame Sylvia Crowe as consultant to the FC in 1963. This was followed in 1975 by the training of a forester to become a full-time FC landscape architect. The integration of landscaping ideas into forestry prac-tice is relevant to conservation management in two ways. First, it shows that new ideas can be assimilated by the forestry industry. This augurs well for the acceptance of changes based on conservation arguments. Second, landscape criteria now act as constraints within which foresters work, just as much as soil characteristics, deer and windthrow.

The many unsympathetic shapes within and on the boundaries of the new forests were obvious. How to deal with them was not so obvious: commonsense had often suggested trying to hide them by doing everything on a small scale. Crowe pointed out that small felling areas in a large landscape produced a moth-eaten effect. In the extensive, sweeping landscapes of the new upland forests, design has to be on a grand scale and felling areas of as much as 50 ha may best fit the scenery, but in an intimate landscape of rolling lowland hills, 10 ha may be too large. Problems of shape are more obvious, especially where straight, external boundaries cut through rounded landforms. Many prominent forests are already being redesigned: Ennerdale in the Lake District is one of the more notable examples, and includes planting some new land and not restocking other areas to fill prominent open areas within the forest and remove sharp, external boundary lines.

Landscape design has to take account of practicalities such as harvesting extraction routes, and the project approach involves the landscape architect getting to know the area and balancing ideal landscape design with timber production and other factors such as deer control or recreation. The considerable time this takes has been the most important problem for landscape design, because although completed projects have worked well, limited resources have meant that too few forests could be redesigned. Unfortunately, much of the rest of the forest was left with no input whatsoever, and forest managers found detailed landscape principles hard to apply (Hibberd, 1985). The shortage of published guidance, to be corrected soon (Lucas, in preparation), exacerbated the problem.

RESTRUCTURING

In the late 1970s the concept of 'restructuring' developed (Hibberd 1985). In contrast to a pure landscape appraisal, restructuring starts by looking at, and questioning, operational and silvicultural methods. It stresses the need to reconsider forest design from first principles at felling, and advocates some redesign for all the forest. Although its approach to landscape can be criticized as being unnecessarily mechanistic, it is still important because forest management and environmental problems are considered together. For wildlife this is crucial because the scale on which simple ideas can be applied in upland forests is important in realizing conservation benefits. The implications of landscape design for wildlife had not really been considered; fortunately, their needs have a lot in common and future conservation management in the commercial forests will benefit from the large-scale approaches introduced by landscapers. The comprehensive restructuring approach has demonstrated the extent to which genuine multiple land use can be achieved; the value of a broadleaved tree to a bird is the same whether or not that tree has another function, for example as a feature in the landscape. The restructuring approach should allow conservation ideas to influence the whole forest; but only if conservation ideas become incorporated into routine management.

Up to this point, conservation in the forest had been seen largely in terms of special sites. There had been a tendency to suggest that a certain percentage of the forest should be 'set aside for wildlife', and forest managers have been rightly alarmed that after one area has been lost for landscape, another for wildlife, some more for recreation and something for archaeology, there will be hardly anything left for the trees. However, rather more critical consideration of the effects of forest management on birds was developing. Moss *et al.*, (1979a) pointed out that the maximum number of songbird species in a single growth stage of the upland forests they studied was only 12, but over the first rotation the forest supported up to 26 species. At the same time, ideas on forest structure, and especially the significance for both birds and mammals of restocking, were developing (Leslie, 1981; Ratcliffe, 1984). Two factors were obviously crucial for birds; the growth stage of the conifer crop, and features like broadleaves, open ground or water. Restructuring provided a vehicle for integrating management of these features for wildlife as forest managers started to realize the impracticality of single-purpose plans (e.g. for felling, landscape, recreation, conservation) as management became more complex. By 1982 such ideas had developed sufficiently for broad recommendations to be made (Buckham *et al.*, 1982).

This new approach is based on the idea that the entire forest has some value to wildlife (Forestry Commission, 1986a; Leslie, 1986). This is clearly true for rare birds such as the Goshawk, which may nest in a stand of suitable size anywhere in the forest, but also recognizes that there is wildlife everywhere in commercial forests. In practice, for example, this means that if a rare bird is found nesting in any part of the forest, forest operations such as felling will normally not be carried on nearby during the breeding season. At the other end of the scale there are far more complex issues, in particular the question of adjustment of felling age to create a variety of habitats. However, in contrast to the special conservation site, in the broader forest wildlife management is one of several priorities and generally subsidiary in importance to timber production.

COST

Practicality and cost-effectiveness are as important in wildlife management as in landscape design. There are potentially many different ways of benefiting wildlife through practical management, and their costs vary greatly. Perhaps appropriate to the age of the new Victorian morality some people take the line that conservation measures have to be expensive and inconvenient for foresters if they are to be worthwhile; if it doesn't hurt it can't be doing any good. In fact, relatively simple management, often on ground that cannot for one reason or another produce a timber income, may produce greater benefits to birds than an elaborate, costly project. The approach should be pragmatic in that costs incurred in changing forest management should result in clearly demonstrable improvements for wildlife. This attitude does not rule out spending quite large amounts of money where a significant benefit can be demonstrated, but what it calls for is well-researched information on the effects of different management options. It has to be recognized that at present, knowledge of the effects of different management techniques for birds is so sparse that the efficacy, let alone cost-effectiveness, of many potential management prescriptions is unknown.

CONSERVATION GOALS

How money is spent must be based on a judgement of the relative benefit which will result. In determining the best way to integrate conservation objectives into forest management there must be a goal. How should the new forests change to increase their value for birds? The answer to this question which has most often sprung from the lips of conservationists has been that the forests should be absent! Since this is not an option open to the forest manager once a forest exists, there has been little common ground between forest managers and conservationists. The existence of artificial forestry plantations, with great unrealized wildlife potential, poses a challenge which has not yet been met. One reason for this may be a lack of vision about what the new forests could be like.

So much effort has gone into fighting the expansion of forestry in upland Britain that it seems conservationists have found it difficult to imagine the potential of the new forests. Comparing the new forests to a conservation 'ideal', such as an oak woodland or a well-managed heather moorland, is likely to create the feeling that it is not worth doing anything to improve these forests for wildlife. In improving the conservation value of the new forests, the starting point is the forest as it is today, however poor that is, and any gain is a plus.

The objectives for bird conservation in the new forests have rarely been discussed (e.g. Williamson, 1972), just as they have rarely been discussed for other habitats. Since the new forests are totally unnatural, the aim cannot be to preserve an existing habitat. Instead the aim must be to create a new and exciting forest habitat. But what should this be like? To ask such a question is to get to the roots of what is meant by conservation value. What do we want from our environment?

That we need a different kind of forestry has become a truism, the conclusion to every article on the subject, but ideas of what this new forestry might be are usually an unimaginative and impractical mirror image of present dislikes. If big forest blocks are bad, then smaller ones must be better; if conifers are bad, then broadleaves must be better. Often, it is suggested that the new forests should be more 'natural', though what this means in the uplands is not terribly clear. The characteristics of the upland forests that are seen as unnatural are that they are mainly non-native conifer crops of short rotation which end with clearfelling. Regardless of other factors, the tree species used in commercial forestry alone mean that the new forests cannot masquerade as imitation native forests. Non-native trees are central to the pursuit of productive forestry in the uplands and their abandonment would lead to something so completely different from what we have now as to bear little relationship to it.

The obvious starting point in considering how to make British forestry plantations more natural should be existing natural British forests, but there are not any of these. The ancient woods we see today are the result of over 2000 years of management and are quite different from the original 'wildwood' that existed before man had a major impact on the landscape (Rackham, 1976, 1980, 1986; Peterken, 1981). What we know about the wildwood of the lowlands suggests that it was very different from the planted forests of the uplands: regeneration probably took place in small gaps following the deaths of individual trees or small groups. What followed depended on the species: under shade-tolerant trees like Beech, regeneration would already be underway by the time a hole appeared. Growth from coppice or suckers would compete with regeneration from seed. Light-demanding pioneers like Birch and Aspen were probably rare in the middle of the forest. However, this is no more than informed speculation. Some new forests have been planted on uplands and heaths that must once have carried dense, broadleaved wildwood but they could not grow it again today: soil structure and fertility deteriorated as removal of tree cover led to leaching.

Native Scots Pine forests more obviously resemble the new coniferous forests but although they seem more natural they too have all been touched by man (Callander, 1986; Carlisle, 1977). All have seen felling in the past. Some have been managed for timber for over 100 years. Others have been felled exploitatively with no attempt at regeneration. In contrast to lowland broadleaved woods, which do not burn easily, fire may have played a major role in the natural pine forest, killing mature trees and creating open land for regeneration. The resulting forest would have included areas of even-aged trees. Their density we cannot tell: it is unlikely that the open forest of large old trees typical of native pine forest today is anything like the original wildwood. In some managed, native stands, tree densities and spacing are similar to those in managed plantations, whilst unmanaged stands are more open because the better stems were removed without much thought for what was left. Regeneration without protection from deer is rare. The results of fencing in places like Glen Affric suggest that the natural forest would have had a wider range of tree species, including Birch and Rowan, than normally seen now (Innes and Seal, 1972; Booth, 1977).

The totally unnatural nature of the new forests is one of the major blocks to sensible thinking in this area. Can unnatural habitats have conservation value? Our answer to this question is that they can. One of the best examples of this is well-managed heather moorland. This is not, in any sense, a natural habitat. It is the management of moorlands by controlled burning and by predator control which makes moorland such a valuable habitat for birds in Britain (Cadbury, 1987). Coastal saline scrapes, created by the RSPB around the east coast of Britain, are also intensively managed, unnatural habitats which are usually perceived to be of high conservation value in terms of providing breeding sites for rare breeding birds such as Avocets and terns (e.g. Sills, 1988). Provision of bird-tables and nest boxes in gardens are both completely unnatural influences on bird numbers which are widely encouraged by conservation organizations and seen as positive influences in the wider countryside. Elsewhere it is clear that conservationists have not concentrated on natural habitats alone and manage many nature reserves in ways designed to keep natural forces at bay. There thus seems no good reason to hold back from managing Britain's commercial forests for birds as soon as they are planted.

But agreeing that conservation within the new forests is worthwhile does not provide agreement on what should be the goal of conservation management in these forests. What do we want from our commercial forests? The aims of conservation management within forests should be a subject of much greater thought and debate than has been the case so far. However, we would propose three principal aims for conservation management for birds in commercial plantations. First, that rare species which were present before planting should be encouraged to persist within the forest wherever possible. This aim will often be very difficult to achieve; it is unrealistic to expect that forests will hold moorland birds. However, there are now some good examples of species which can persist in young plantations and which should be encouraged, e.g. Woodlarks and Nightjars. Second, rare forest species should be protected and encouraged wherever possible. Species such as Goshawks and Firecrests would fall into this category. The third aim should be to create a varied and rich bird community in every forest district.

These aims should be more widely discussed but could form the basis for conservation management in plantations. They are not the only possible ones. For example, it is arguable that a worthwhile aim would be to maximize the size of the British population of Crossbills, and that all forest management should be directed to this aim. This could probably be achieved by planting all land with spruce, removing broadleaves from all areas, draining wetlands in order to plant more spruce, and abandoning any other conservation measures (e.g. nest box schemes) which would not help Crossbills. We do not propose this as a serious option, but it illustrates the fact that it is easier to see how to achieve a conservation aim when it is specific, rather than very general.

One way of producing realistic and specific conservation goals would be to do this on a regional basis, but within a national framework. Whereas in East Anglia the management of restocked sites for Woodlarks and Nightjars is a sensible conservation goal, this has no relevance for northern Scotland, and likewise Scottish Crossbills should figure largely in conservation plans in northern Scotland but not in southern England. But in forestry, as in all other areas, bird conservation can only be advanced when specific goals have been identified and steps taken to achieve them.

At present it seems reasonable to accept that creating more diverse forests will lead to more diverse and interesting bird communities and that this is a worthwhile aim everywhere in British forests. This general aim can be achieved by many different management techniques, many of which are already practised in commercial forests. In the rest of this chapter we describe management practices which, if used appropriately, might further the aim of increasing bird species richness within British forestry plantations. Many of these prescriptions are quite simply guesses, which need to be tried and evaluated before their utility can be assessed.

AGE CLASS DISTRIBUTION

Because the typical birds occupying a forest compartment change with tree age, it is clear that for the greatest diversity of species in any forest the greatest possible range of successional stages should be represented at any one time, and that if possible there should be no break in the availability of any particular stage. This means that the forests of the future should be more varied than those we see at present. It should be possible to walk through any forest area and see birds associated with every successional stage; from Skylarks and Tree Pipits in the early stages, through Willow Warblers and Robins to Siskins, Goldcrests, and Coal Tits in the mature forest.

The age structure of the first rotation of the new forests has been exactly the opposite of this ideal. For efficiency, large areas have been planted within a short time-span, forming large plantations of uniform age and growth stage. Because British forestry is still dominated by planting and the first tree crop, the variation in age within forests is very small compared with the national range of tree ages. As time goes on, and the forests enter the second and subsequent rotations, the variation within forests will increase markedly.

The FC's Glenorchy Forest is an extreme example of a first rotation forest where exceptionally large areas were planted over a short time; areas planted simultaneously move through each successional stage in large, uniform blocks. In 1980 less than 1500 ha had closed canopy and over 2500 ha were in establishment or pre-thicket stage. In 1995 there will be 4000 ha of closed canopy forest and hardly any establishment or pre-thicket. Although Glenorchy is an extreme case, all the new forests will have gone through a similar sequence; the degree of uniformity depends on how rapidly a particular forest was planted. Natural variation in tree growth and windthrow will start to reduce the uniformity but felling at the economically optimal age perpetuates the situation, with the whole forest returning to open ground over a few years. On the other hand, physical and economic reality make it impossible immediately to create what foresters call a 'normal' forest where age classes are equally represented by area. Windthrow limits the age to which trees can be kept standing, whilst every year below or above the economic optimum felling age costs an increasing amount of lost return on investment.

Creating a more diverse age class distribution seems to be the best way of improving the diversity of common birds in productive compartments, and there is a strong case for having as full a range of successional stages as possible in each large block of forest (1000–5000 ha). Moving towards this goal is the central concern of current second rotation management ideas (see Fig. 4.1).

The forester can adjust both the age at which trees are felled, and the size of the areas that are felled. At felling, boundaries can be effectively remodelled. The area felled, known as the 'felling coupe', is influenced most strongly by the need to find a windfirm edge. Otherwise, a combination of operational and environmental factors will shape the new management unit. Harvesting and extraction must be possible and can constrain options on steep or very soft ground. The new coupe must be the right scale for the landscape, and shaped in sympathy with the landform. In the uplands the scale of landscape-designed coupe size is generally large enough to provide a reasonable scale of working for harvesting operations. Coupe boundaries should, for landscape reasons, be related to natural features such as watercourses. Once felling and restocking are complete, the felling coupe becomes the new compartment.

Felling the first rotation at Glenbranter, one of the earliest of the new forests. The variation in successional stage in this heavily forested landscape is a key feature of habitat diversity.

Although simple in theory, in practice restructuring is a considerable challenge to the forest manager's skill if he is not to suffer economic costs: varying age class in particular is a demanding balancing act. The point at which it is most profitable to fell can be quite clearly defined and although economic penalties five years either side of the optimum are slight, beyond that they climb rapidly. This, of course, is not the only problem faced by the forester, who also has to meet production targets with existing labour and machinery. Timber has to be brought onto the market, to meet medium- or long-term commitments. For example, a sawlog contract cannot be satisfied with an equivalent volume of pulpwood just because a decision has been made to fell a stand well before rotation age on conservation grounds. Workers and machines also have to be employed constantly and, with a 10-tonne forwarder and operator costing nearly 50 pence a minute, a single day spent idle is a significant expense. With such constraints, incorporating structural diversity into uniform forests appears to demand the reconciling of the apparently irreconcilable. But it is not the impossible task it first appears if you take the opportunities that are there.

The growth of the forest is not nearly as regular as most foresters would like to think. In early fellings the difference in felling age, often only five years, of a small yield class difference was often eliminated by combining adjoining stands into one coupe. However, were the crop with the earlier optimum felling age advanced one period (five years) and the later crop delayed five years, a 15-year age difference could be created out of the original five years at minimal economic cost. With careful planning most stands can be felled within five years of the economic optimum and still leave significant age class differences. Species differences provide a major element of existing variation in optimum felling age, and again can be exploited: the difference in optimum felling age between equal yield class Scots Pine and Japanese Larch is so great that no move from optimum felling age is necessary to create a difference of 30 years.

Windthrow has been the most significant factor in creating accidental variation in age class distribution. For example, the effects of the catastrophic blow of 1968 (Holtam, 1971) can be seen in thicket-stage patches amongst mature forest and new clearfells right across west and central Scotland. Practice has changed, from felling large areas as soon as endemic blow starts, to trying to retain stands which show resistance as long as possible. The stage of wind damage at which a stand is felled can contribute to age class diversity: at the onset of windthrow in a large even-aged block, starting felling well before 40% of trees are blown and extending the coupe to the nearest windfirm edge may create an early area of restocking well ahead of the felling of the rest of the block. On the other hand, the last standing compartment in an area that has blown down over a few years should be kept as long as possible.

FELLING COUPE SIZE

The size of felling coupes determines the scale of the age class patchwork. It has generally been assumed that small areas will be of greater benefit to birds because there will be greater numbers of edge species. Tawny Owls nest in mature conifer stands but in upland forests their main food is voles, and breeding success fluctuates with the vole cycle (Petty, 1987b). Tawny Owls can often be seen perched on restocks at dusk and appear to benefit from the patchwork of open clearfells and older forest. In Kielder Forest territory density is highest where clearfells are smallest (Petty, 1989b). However, Tawny Owls are familiar birds that have adapted well to the forest edge habitat provided by suburban garden hedges and farm woodlands. Bibby *et al.* (1985) showed no effect of restock size on passerines on Welsh restocks. In Sweden, Hansson (1983) demonstrated an apparent avoidance of the first 100 metres of open clearfell by restock birds. Interesting species like Red-Backed Shrikes, which are reasonably common on Swedish clearfells, tended to be well away from the forest edge. As expected, there is a higher than average density of forest birds in the 100 metres of mature forest next to the clearfell.

An important role for large felling coupes may be in attempting to retain some of the original moorland species nesting within the forest. Species such as Hen Harriers, Short-eared Owls and Curlews will nest in young plantations, even amongst quite tall trees. This suggests that they might be species which could be attracted to restocked sites after felling. A single case of Hen Harriers nesting on a restocked site in Northumberland (Petty and Anderson, 1988) should not be overrated, but is an encouraging sign that some moorland birds

Tawny Owls move in to nest in maturing forests. In the absence of natural holes they nest successfully at the base of trees but quickly move to boxes where they are provided. In Kielder they were found to predate Sparrowhawk at the nest and predation between raptor species is likely to become an increasingly important factor in the ecology of upland forests in future.

will be able to make use of upland restocked sites. The chances of this happening may be improved if large felling coupes are used in areas which are most likely to attract moorland species, for example on the edges of upland forests which adjoin moorlands with high densities of birds. Large felling coupes may be of little wildlife value in lowland areas or in upland areas which lack those species of high conservation value which it would be exciting to be able to attract into second rotation forests. In practice, landscape design will normally limit the size of clearfells and create a diverse range of sizes somewhere between 5 and 50 ha; adjoining coupes felled within 10 years and plans disrupted by windthrow result in a proportion of larger restocks.

The future importance of restocks for birds is a matter of some speculation. In recent years the use of restocked forestry plantations by birds such as Woodlarks and Nightjars, which have traditionally been thought to be heathland or open-country species, has brought home to conservationists the fact that birds may not always share our preconceptions about whether artificial habitats are good for birds. Bibby (1988) regarded these two species as ones whose British status was much dependent on the management of their forest habitats by foresters. Research by the RSPB, funded by the FC, is beginning to discover what, if anything, foresters can do to make Breckland restocks even more suitable for these species than they already are. Only on the basis of similar research on other species will it be possible to advise foresters on management measures which will benefit particular target species. Rare species deserve special attention because general forest management, designed to increase bird species diversity, will not necessarily help them. Their particular needs should be taken into account.

A few forests lack any significant natural variation, and diversity greater than 10–15 years can only be created by adjusting felling age. Kielder is a case in point (McIntosh, 1988). Retention is economically preferable to premature felling and also benefits birds of the older age stages. In a majority of the large, even-aged forests, windthrow limits retention, so that some stands at least must be felled early. Setting an objective target is difficult with our limited ecological knowledge. At present a target of at least 10% of stands felled at least 10 years before or after optimum economic felling age looks likely to create a worthwhile degree of age class variation in the second rotation without losing an unjustifiably large amount of money. The restructuring of Kielder is being carried out broadly along these lines, and is likely to incur a cost of only 3% on the rate of return compared to felling stands at the economic optimum age (Hibberd, 1985). Figure 4.1 illustrates the way in which a single block planted over only 20 years has been restructured: the total variation over the block as a whole has been increased by only 10 years, but a whole rotation is only 45 years. What is important is that the very large blocks, especially the central block which was planted over only five years, will have a minimum of 20 years variation, windthrow permitting. It shows just how much can be achieved at modest cost.

OPEN SPACE

The ephemeral open space of restocks will always be the most significant area of early successional stages within the forest, but permanent open spaces like deer glades, areas kept open for landscape or specially for conservation in and around the forest can be important to birds. The scale of open space ranges from the narrow band of roads and rides, through the small-scale clearings of deer glades or small wetlands to much larger areas of retained farmland or unplantable hill ground.

Deer have been responsible for the acceptance by foresters of the need for planned open space in the forest, and through them many other species will benefit. When the canopy closes, deer can become established and thrive out of sight, damaging the trees. Deer are shot to limit the damage but you cannot shoot a deer if you cannot see it, and if its main feeding grounds are gaps in the middle of a thicket-stage crop then seeing a deer is impossible. Foresters attract deer to deer glades: planned permanent open space within the forest located where deer are most likely to want to feed (Ratcliffe, 1985b). The vegetation must be richer than average, and the good soils of the riparian zone are especially suitable. Alternating areas of broadleaves and open ground, a string of glades can be created along a watercourse, providing an intimate mixture of different habitats and varied shade conditions in the stream. Deer

Restructuring in practice

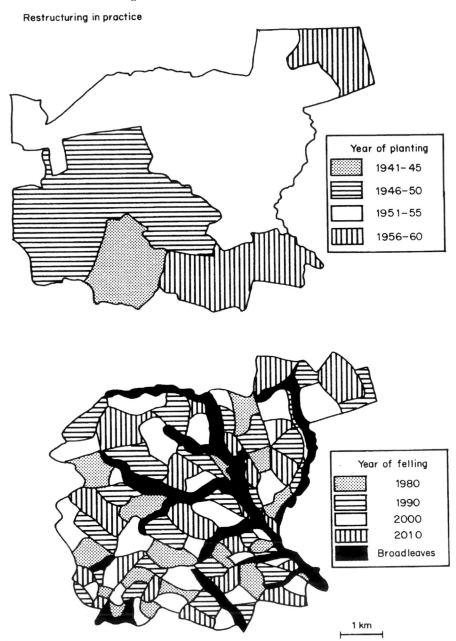

Figure 4.1   Age class diversity created in a typical large, even aged upland block (Kielder Forest).

glades away from watercourses will also be necessary and if deer become established early in the life of the forest they may pick their own spots, browsing down the young trees in their favoured feeding places. Fifteen years later, neatly mown Norway Spruce bushes may just have struggled to get leaders above the browsing level. It may be possible to combine deer glades with other open-land habitat requirements, although only if the primary need of attracting deer is met.

Roadsides in lowland forests are proving of high value for both plants and insects (Anderson and Carter, 1988) but their value to birds in upland forests is probably small. Rides in upland forests are designed as fire-breaks and administrative boundaries. They are often visually obtrusive and tend to be planted and lost at restocking. Narrow rides serve little purpose for wildlife: when the crop is young there is more than enough open space and when it grows up these rides become enclosed. Deer glades and other discrete areas of open space, for example around ancient monument sites, are likely to be more valuable for birds in the long run. They will not be much use to real open-land birds but may allow birds of the forest edge like Black Grouse to persist through the closed canopy stage.

Black Grouse are increasingly associated with forest, especially the early stages, but also require open land. Closed canopy forest is used for shelter and

Blackcock lekking on a forest road. One of the most dramatic forest birds, Black Grouse are likely to decrease in numbers as the intensively planted blocks grow into pole stage but providing glades and rides, and keeping unplanted bogs, could allow small numbers to survive through the rotation.

nesting, younger stages and unplanted land for feeding (Hope Jones, 1987b). In Wales, young forest planted on heather moor has the highest densities of them. Black Grouse occur on restocks (Leslie, 1981), and the broad structure of second rotation forests may suit the species well. Although living on heather and *Vaccinium myrtillus* for most of the year, Black Grouse chicks feed on insects and the high protein spring shoots of Bog Cotton may be important in bringing adult females up to breeding condition (Hope Jones, 1987b). Bogs, even very small ones, are an important component of the habitat and were often ploughed and planted in the past. Although a tiny percentage of total forest area, these special spots are now being protected for Black Grouse, and new open areas created in practical management experiments at five Welsh locations. This includes respacing parts of the conifer crop so that ericaceous shrub species survive longer before being shaded out, and also planting broadleaved trees for food (Hope Jones, 1987c). Heather is important in the species' diet and the maintenance of healthy heather within the forest can present problems. For example, in Wales it seems that many restocks which were heather at planting are reverting to grass. Black Grouse are not alone in the complexity of their habitat requirements but are one of the few species for which the research needed to generate positive management ideas is being done.

Larger open areas, which are sometimes retained primarily for forest design reasons, can also be good for birds. The entire grassy glen bottom at Strath Rory in Easter Ross was kept as a giant fire break and is large enough for waders like Oystercatchers and Lapwings to breed in what is an expanded riparian zone. However, such retentions remain hard for foresters to justify, as the land involved is often the best in the acquisition, and land is bought by the FC for forestry purposes only. Private companies have more flexibility but there is minimal overlap between the forestry and agriculture professions in Britain and both sides are nervous about moving into a business they do not understand. An important result of the 1988 Woodland Grant Scheme could be more planting by the present owners of land and a better balanced combination of trees and open ground (see Chapter 9).

Large open areas in or around the forest are usually either the best ground at lower elevation which has been retained in agriculture, or else the poorest ground above the plantable limit. Retained farmland still tends towards upland habitat types and may be valuable for hill farm species, especially waders like Lapwings and Curlews. The maintenance of traditional farming methods, without excessive use of fertilizer or pesticide, is likely to be the best management of such areas for birds.

Unplantable hill ground covers more than the area of plantations in some forests, amounting to thousands of hectares. It is usually in the higher altitudinal band, all the lower ground having been planted. Unplanted, often unplantable, high ground may hold significant numbers of the birds most affected by afforestation. It might be thought that the change of ownership of unplanted land would have no possible consequences for the birds of the area. However, purchase by foresters will almost certainly result in changes in land use and management which might have effects on birds. On the positive side, the removal of human persecution, which has always been an important factor in the distribution of raptors (Tubbs, 1974; Newton, 1984), may result when forestry acquires land which has been unsympathetically keepered. Whatever the longer-term effects on feeding habitat, forestry has also reduced the regular persecution of crag-nesting Golden Eagles, Buzzards and Peregrines.

Crags and buildings are raptor nesting habitats; both were lost unnecessarily in the past. Trees are now planted well back from crags to leave access for nesting birds. Considerable effort has also gone into the protection of rarer species from malicious or accidental disturbance. Petty (1988) gives some suggestions on how far forest operations like harvesting should be kept away from raptor breeding sites to avoid disturbance. For smaller common raptors like Tawny Owls and Kestrels within the shelter of the forest, which minimises visual disturbance, this may be as little as 100 m, rising to 400 m for larger, rarer birds like Red Kites, Goshawks or Ospreys. On the open moor, disturbance may occur at far larger distances, and 300 m is recommended even for Kestrels and Merlins. For Hen Harriers, Buzzards and Peregrines 600 m is suggested, and for Golden Eagles 1000 m. Although difficult for tree nesting birds that move around, these rules can be easily applied for regular crag nesting sites.

Moorland waders will continue to live on extensive, unplanted moorland areas, whoever owns them. Unplantable ground has traditionally had little management unless it can be fenced for grazing. Moorland vegetation is continually changing depending on grazing and burning pressure, and the short, young growth produced by burning or grazing is important for Red Grouse and Golden Plovers, for example. Large areas of open ground may be included within the forest deer fence if a shorter line can be designed to include several blocks within one ring rather than fencing them individually. Traditional burning regimes are now being used on several areas in forest ownership, including at Kielder and in the North Yorks Moors, to maintain suitable habitats for Red Grouse and waders. Management is also likely to be essential for strath habitats, and at Strath Rory grassland has been fenced for cattle grazing largely to maintain vegetation for waders.

Open space within the forest poses the forester some problems. Sometimes the ground may be of high quality for wildlife but often it is not, and then it is no good pretending it is a surrogate moor. As Peterken (1987) points out, it may well be better to plant large forests and maintain equally large, discrete areas as moorland than to try and mix the two by planting smaller, more scattered blocks of forest or incorporating a much higher proportion of open ground into forests.

OLDER TREES

Just as the existence and management of open space are important for both the diversity of common birds and the existence of rare birds in the forest, so too are the existence and management of old trees. To a forester an old tree is one which is still standing after the optimum felling age, because retention of part of the crop entails some economic penalties. On windfirm sites retention well beyond optimum felling age is possible, and in theory a far greater degree of age class diversity could be created in the second rotation. It has been possible to make quite significant retentions in terms of both area and age in some places, and some longer-term retention is physically possible in most forests. The smaller the area the more difficult it is to justify felling on the grounds of lost timber or economic return. Some areas of mature trees should be retained in most first rotation forests to bridge the gap to the earliest second rotation restocks reaching pole stage.

Whilst retentions of 10–20 years serve to create major variations in age class distribution, conifer stands of over 100 years old develop a particular value for birds related to their larger tree size, greater structural diversity within the stand (as opposed to between stands) and the presence of dead trees. In Wales, Currie and Bamford (1982) found birds more normally associated with broadleaved woods in 100-year-old conifer stands, including Spotted Flycatchers, Wood Warblers, and, using woodpecker holes, Pied Flycatchers, Redstarts and Starlings. The density of breeding birds in the 100-year-old stands was twice that in pre-felling stage conifers, Goldcrests, Blackbirds, Wrens, Chaffinches and Robins making up 60% of the population of the 20 breeding songbird species.

Very old conifer stands are extremely rare and are likely to stay that way unless some are set aside for retention now; whilst the principles of age–class diversity are widely accepted, the need for at least a small area of 'indefinite' retention is not yet being widely applied and causes concern to managers because of its high cost. It is because of this cost that long retentions can only be considered on a small scale so long as the pressure to maximize economic return is as intense as at present, but as Currie and Bamford (1982a) showed even tiny areas of very old conifers are worth retention. There are some places in most forests where continuous tree cover is desirable, for example, a key spot in the landscape or along a much-used forest walk, and such retentions will hold many smaller birds. For larger, shyer species, remoter retentions will be needed and birds may help the manager's choice in that there is a very clear reason for keeping a stand if it has an Osprey, Golden Eagle or Goshawk nesting in it. As Peterken (1987) points out, if we do not keep some conifers to biological maturity we will never know what can happen.

The way in which indefinite retentions are managed will affect both their cost and value to birds: very heavy thinning leading up to rotation age will not only open up the stand, allowing early development of a shrub layer but will also realize a higher proportion of timber value than 'by the book' thinning.

Economics, practical difficulties and the attitude of the forestry profession have all favoured clearfelling. The obsession with clearfelling has been so great that it has been applied to all tree species and site types, with disastrous consequences for lowland clay soils. Removal of tree cover led to an explosion of weeds which often swamped the planted successor crop. The government's broadleaves policy (Forestry Commission, 1985) acknowledged the appropriateness, both silviculturally and environmentally, of small-scale felling and regeneration in lowland broadleaved woods, the extreme opposite in Britain to windthrow hazard class 6 conifer forest.

Irregular management increases management costs; all harvesting is effectively thinning, with higher costs due to takedown (felled trees becoming hung up on standing trees), low produce density, and extraction machinery having to negotiate a route between standing trees. Keeping track of the timber stock is also far more difficult, as is selection of the trees to be felled, so the cost of both labour and supervision are increased. Against this must be set the larger size of trees on better sites, with lower harvesting costs and higher income, but in the FC at least, there has recently been pressure to cut costs to a minimum even on the most profitable sites which during the lean years of the early 1980s subsidized harvesting of lower yielding stands on poorer sites.

Management which assumes that new trees will be provided by regeneration rather than by planting also presents problems; non-native conifers do not always regenerate naturally in Britain. Norway Spruce hardly ever does. Douglas Fir regenerates well on most suitable sites. Sitka Spruce is very variable, regenerating vigorously in some places but normally not. Heavy regeneration (up to 3 million Sitka/ha have been recorded in Fernworthy Forest on Dartmoor) costs money to respace to something near normal stocking, and gaps may also have to be filled by planting. Irregular systems entirely dependent on planting would be very expensive because of the complexity and small scale of planting and subsequent maintenance. But deer are the biggest problem; at their present population level within forests, deer will graze off all broadleaved and most conifer regeneration. Holes left by the felling of half a dozen trees will provide ideal feeding for deer. The shelter of the dense forest close at hand also makes protection of the trees by shooting quite impossible. Where Roe Deer or Red Deer are present, fencing of sufficiently small areas for practical control, 20–30 ha perhaps, will almost certainly be necessary.

The lack of deer means that some Welsh sites give the best indication of the sort of bird habitat that irregular silvicultural systems might create. Where small groups of trees, usually Douglas Fir growing on forest brown earth soils, have blown to make gaps, a range of species are regenerating to form an understorey. Douglas Fir are mainly in the larger gaps, with shade-tolerant conifers under the canopy. Where there is a seed source there is Beech as well as Birch; although Birch does grow in these small openings, it is not so dense that it swamps other species. Irregular management is likely to be a far more practical way of growing broadleaves and conifers in intimate mixture than the row by row or group mixtures of conventional planting; with more than one species of both broadleaf and conifer to choose from and the advantage that their ability to regenerate naturally means that they are likely to be suited to the site, a mixed stand should be possible with minimum management intervention.

It is only possible to guess at the bird fauna of irregularly managed upland forest: several of its elements would complement measures already applied to even-aged stands. Retaining individual trees or groups of trees beyond rotation is likely to be particularly easy and worthwhile in irregular stands. In a Douglas Fir stand at Boulderwood in the New Forest, a small number of trees about 100 years old have been retained from the original planting. Around them are well-stocked stands of younger trees, ranging from about 50 years old to thicket stage. There appears to be little loss in timber production but enough of the huge, old stems have been retained to give an appearance of old woodland. Open-grown, without close competition, retained trees are likely to be suitable for large birds to nest, such as Buzzards and Ravens found by Currie and Bamford (1982) in old trees in Wales. Scattered through the forest, less obvious than blocks of even-aged retention, they should also help the spread of rarer birds like Goshawks, Ospreys and Red Kites.

Two elements set irregular stands apart from existing conservation measures applied to even-aged forestry. One is the continuity of forest cover: there is never the dramatic change of clearfelling, only the removal of some trees within what is always a fairly stable woodland habitat. The continuous availability of old trees will obviously provide habitats for birds like Crossbills, as well as nesting for larger species, but it would be wrong to jump to the conclusion that this would represent a nett benefit over providing an equal amount of habitat in the form of even-aged retentions. Birds like Crossbills are well enough able to follow their food supply for the indefinite retention of old trees in particular stands to be unimportant. In contrast, birds of open ground like Nightjars and Woodlarks would definitely suffer from universal adoption of irregular systems on better sites.

The second element is the actual structure of the stand; even single-species stands, like the Boulderwood Douglas Fir, are quite different from even-aged forest. Within the shelter of the tallest trees is a variety of tree heights and open spaces. In contrast to even-aged stands, irregular stands have both under-storey, of young forest trees or of shrubs, and big canopy trees. This type of habitat would suit Capercaillies if provided in Scots Pine plantations in the birds' range (Jones, 1982, 1984). Capercaillies have been found in high densities in old plantations, but young commercial plantations are poor for this species (Moss *et al.*, 1979b). Ericaceous ground vegetation provides much of the Capercaillies' food, and taller shrubs provide the best nest sites; young forest may be used for shelter but ground vegetation provides most of their food. Old 'granny' Scots Pine still exist in many central Highland forests, but were underplanted in the 1960s and 1970s, creating dense thickets and spoiling them for Capercaillies. Management experiments have started on 170 ha of the FC's Ardross Forest to see if the mosaic of thicket, open woodland and glades that Capercaillies prefer can be created artificially. Existing stands will be heavily thinned and small glades created. Ericaceous plants, juniper and broadleaved trees will be planted, and some new small wetlands created. In extending the idea more widely in future, a 'core' of a few hectares around the old pine might be intensively managed primarily for Capercaillies, a network of smaller glades spread throughout the block and the rotation of some crops be extended by up to 20 years to produce bigger trees and more open stands. The small clearings of early endemic windthrow may give some more idea of the birds of irregularly managed upland forests. Spotted Flycatchers, thrushes and even Redstarts occur in small patches of windthrow in Sitka Spruce (S.J.

Petty, personal communication). None of these birds is well served by the habitat of the even-aged forest.

The potential of the irregular forest for birds is probably much wider: based on what we know from the lowlands and Wales, this sort of forest structure, especially where there are some broadleaves, is likely to provide a habitat for *Sylvia* warblers like Blackcaps and Garden Warblers and also for Pied Flycatchers and Wood Warblers. Regardless of tree species, the structure of irregularly managed stands looks like providing a quite different habitat to anything presently available in the new upland forests. In this it is different from most other options for improving these forests for birds, which tend to be in the nature of 'more of the same'.

The greatest diversity of birds, and probably interest to people, will be achieved by a mixture of systems and methods. We believe that to more fully realize the potential of our new forests we need to get away from clearfelling-only systems. We would like to see a significant area of irregularly managed stands, providing a quite new habitat type, which may bring in birds rare or absent at present. However, just because irregular management may have benefits does not mean that clearfelling is bad for birds or that it should be replaced wherever irregular management is physically possible. For Wood-larks and Nightjars, living largely on clearfells, in forests where irregular forestry is technically possible, stopping clearfelling would be a disaster.

TREE SPECIES

The effect on wildlife of conifer tree species has attracted even more attention than coupe size. However, evidence for significant differences between conifer species for birds is limited (Chapter 2). The economic cost of using an alternative to the optimum conifer is generally very significant indeed; usually far more than the cost of delaying felling for 20 years (Chapter 3). If the the main objective of designing the forest for bird conservation is to achieve the greatest effect for the least cost, the variety that might be created through altering the conifer crop species looks like a poorer way to spend money for birds than varying age class. Certainly, conservationists can provide little evidence on which the economic costs of changing, for example, from pure spruce to a significant percentage of larch would be justified. Although greater effects may emerge with study, we are not aware that any such studies are in progress.

Scots Pine within its native area is an exception to this general rule: Crested Tits and Capercaillies use Scots Pine plantations even within the course of a normal rotation, and for all we know Scottish Crossbills may as well. Restocking with at least a proportion of Scots Pine where another conifer would be more economic may, in the central Highlands, compete in terms of conservation value with diversifying age class and planting broadleaves. There may be other good reasons for varying species, for example, limitation of pest attack, landscape or timber marketing, and such variety is often welcome for birds even if it cannot be justified on bird conservation grounds alone.

Native Scots Pine is now considered almost an 'honorary broadleaf'. Its treatment in upland forests over the last 30 years also parallels that of broadleaves. Although Scots Pine produces high-quality timber on most sites, it is very slow-growing and extensive areas were underplanted with spruce and pine. For once, planting Lodgepole Pine was not an outright disaster: because it is a different species it does not pollute the native Scots Pine gene pool in the way that cross-breeding with non-Scottish Scots Pine provenances would.

Native pine woods are of exceptional importance to some of Britain's rarest forest birds, especially Scottish Crossbills, Crested Tits and Capercaillies. Any management of the native pine woods should take account of their unique wildlife interest. Britain's only endemic species, Scottish Crossbill, occupies this habitat, and our knowledge of its ecology is at present meagre (Nethersole-Thompson, 1975). The increases in range of many British species which have been aided by afforestation, largely by chance, could perhaps deliberately be repeated for this truly unique group of species by management of Scots Pine plantations and choice of Scots Pine above any exotic species for further plantings in the region including Speyside and Deeside. Underplanting has been a common form of management which transforms open woodland into an enclosed habitat from which the ericaceous ground vegetation is excluded by shading. Clearing underplanting and encouraging regeneration of native Scots Pine is at least as important as doing the same for broadleaves. In the FC's Black Wood of Rannoch, work has started to remove over 100 ha of underplanted Lodgepole Pine, but for many of the fragments running to only 10–20 trees the work needed to maintain this valuable feature is on a much smaller scale.

Planting broadleaves is perhaps the most popular measure favoured by conservationists for forest management. Work in Welsh upland forests showed that broadleaved trees within essentially coniferous plantations brought in new species of birds; particularly interesting were Pied Flycatchers and Wood Warblers (Bibby *et al.*, 1989b). Welsh conifer forests which contained some broadleaves had a more diverse passerine fauna with a higher proportion of summer migrants than did pure conifer stands. Upland broadleaved forests are, however, poorer for birds than lowland woods. Many of the species which are restricted to, or favour, broadleaved woods are absent from Scotland, where most plantations exist and where most new plantations are likely to go. So it is unlikely that Nuthatches, Marsh Tits, Willow Tits, Nightingales, Green Woodpeckers, or Turtle Doves will ever become established in upland forests in areas such as Argyll or Perthshire, regardless of how many broad-leaves are planted there. The existence of a small proportion of broadleaves within commercial forests will never turn them into ancient woodlands, such as the New Forest, whatever we do.

It is important to be clear about the bird conservation reasons for planting broadleaves; if the aim of planting them is not to introduce some new species into the forest, but merely to increase the densities of existing common species, then one has to question whether it is really worth the considerable cost. In the areas where the scarcer colonists are more likely to occur, in England and Wales, and at lower altitudes, then on the grounds of increasing bird diversity

Caledonian Scots Pine in the Black Wood of Rannoch, a fine remnant of the native pine forest. Old trees at low density and an ericaceous ground flora are ideal for Capercaillie; Forestry Commission experimental management plans to reproduce this habitat where ancient trees survive in pine plantations

planting more broadleaves is most likely to be worthwhile. How they are introduced is also important: any broadleaved tree in a conifer forest is better than none, but a silvicultural plantation at close spacing produces a far from ideal habitat for birds. Even if we planted only broadleaves in the uplands, the resulting commercial plantations would, if they were managed primarily for profit, share many of the disadvantages of the present conifer plantations; dense monocultures which are felled long before biological maturity. With the increase in broadleaved planting, more thought is being given to how to make the best of new planting for wildlife and more sophisticated planting schemes that aim to create a complex habitat of varied species and structure. Semi-natural fragments of woodland may provide useful models of what to aim for, showing the species and woodland structure characteristic of the area.

Providing the expected results are not overrated, getting at least some broadleaves into all forest blocks is certainly the most obvious way of trying to improve the forest for birds. Most forests planted before 1945 had some broadleaves, or some were planted, but many 1950s and 1960s upland forests have no broadleaves at all: not only is there no broadleaved habitat today but the lack of seed trees and intense deer pressure mean that broadleaves will only spread through human intervention. That even very small groups can allow some birds to colonize predominantly conifer forests means that large areas of productive land need not be lost and that there is no real excuse for not introducing broadleaves everywhere; even in Scotland broadleaves may allow some species to colonize which would otherwise be absent.

The government's 1985 broadleaves policy (Forestry Commission, 1985) has consolidated the ideas developing through concepts like restructuring. Now some broadleaves must be planted in all FC and grant-aided private schemes; both new planting and restocking. A target of about 5% broadleaves for all forests, both FC and private, is suggested. The precise objective for which trees are planted is important: most FC broadleaved planting in the past has been with the expectation of producing sawable timber. Exotics like Red Oak were sometimes used both for their tolerance of poor soils and their appearance. Some recent plantings failed to identify clear objectives, usually at high cost. Silvicultural prescriptions for producing high-quality hardwood timber can be exacting and expensive (Evans, 1984). Planting Oak at the close spacing recommended for saw timber production at 300 m altitude in Scotland simply will not meet its planned object and is a waste of time and money.

It is now generally understood that most upland forest sites are incapable of producing high-quality hardwood timber because of their harsh climate and poor soils. Broadleaves are planted primarily for environmental improvement, and selection is based on ecological and landscape considerations rather than commercial value (Low, 1986). Birch, for example, is not at present commercially valuable. Birch used to be described by foresters as 'scrub' and underplanted with Douglas Fir, spruce and Western Hemlock. Birch is of value to birds, and in many parts of upland Britain is the only broadleaved tree species which is likely to regenerate from seed. Its aphids are of special importance to newly arrived spring migrants like Willow Warblers. Birch is one of the few broadleaves that can grow on peaty soils, so is a good choice for environmental improvement. Alder and willows are other widely distributed broadleaves of far higher wildlife than economic value.

Most forests can support trees like Oak, Ash and Wild Cherry on better soils at lower elevation (McIntosh, 1988). Sycamore and Beech, both introductions to most of the uplands, can produce timber on poorer sites. There is great confusion over their value for wildlife. Sycamore, which is not native to Britain, can be vigorously invasive, swamping the native trees in semi-natural woodlands. Beech seems to be poor for birds. That these species produce timber from poor sites may mean that it is economically possible to increase the area of broadleaves in a forest. If the option is pure Sitka Spruce or 90% Sitka Spruce and 10% Sycamore, then the Sycamore adds valuable diversity.

Attention always tends to concentrate on planting new broadleaves but the first priority should be to maintain and extend existing broadleaved woodland; existing woods may have established wildlife communities with an ecological value that cannot be created by planting. In flat, heavily grazed uplands, infrequent gorges may be the only places broadleaved trees and the flora of former woodland cover survive and from which natural regeneration is possible. Space should always be left around existing woods to allow expansion, though effective protection from deer is too expensive to be applied to every site. Planting should aim for some broadleaves in every forest block so that there is a seed source for the future. Since 1985 many underplanted broadleaved woods have had the conifers removed. The work of removing the conifers is expensive but the lost revenue may be small. The Dalavich oak wood in Argyll is a good example; it is an ancient semi-natural wood, last coppiced in about 1880, after 200 years of exploitation. This area was formerly important for iron smelting, and most broadleaved woods were managed on a coppice cycle for charcoal production. About 30 out of 55 ha were underplanted with Norway Spruce in 1958. Twenty-one hectares that had not been underplanted were later made an SSSI. By the time the conifers were removed, the canopy of the remaining Oak was suppressing them so badly that they were making little growth. As part of the improvement project the whole wood was deer-fenced, and is now safe from grazing, almost certainly for the first time in a century.

Where broadleaves are planted, the fact that quality timber will not be produced has important practical implications. Wider spacings not only save money but also produce bushier trees and more open woodland with better ground flora; poor for timber but good for wildlife. Planting patterns can be varied to give a complex of trees and open space rather than solid blocks of plantation broadleaves which may not be much richer in birds than conifer plantations. The development of understorey, possible under the lighter foliage of broadleaves (other than Beech), is important for birds. In Wales grazed woods favour species including Wood Warblers and Pied Flycatchers,

A Male Redstart; one of the most beautiful broadleaf-specialists. Associated with grazed western oak woods, it is a hole nester benefiting from boxes where natural holes are not available.

and ungrazed woods with a shrubby understorey tend to be better for tits (Stowe, 1987). Shrubs are now being introduced to planting schemes; Hazel forms an understorey characteristic of many native woodland types, whilst Holly provides evergreen cover in winter. Juniper is an important component of the native pine wood ground flora.

The proportion of the forest that should be broadleaved has been the subject of endless debate. In practice, the way in which broadleaved areas are distributed is as important as total area. There are few certainties: a group of species will be absent altogether from forests with no broadleaves, and a different group from forests with no conifers. Between the two extremes, the gain in conservation value will be relative: as long as a population of, say, Pied Flycatchers, is large enough to be firmly established, it really is not possible to say that twice as many will be twice as good. Achieving the clear benefit of extra bird species supports the idea that there should be some broadleaves in all parts of the forest. Bibby *et al.* (1989a) showed that as even very small groups of broadleaves could allow some species to live in conifer forests, the best theoretical distribution of a set area of broadleaves was in the smallest possible group, even individual trees. In practice there are other objectives than simply getting the maximum number of bird species as widely distributed through the forest as possible. However, the work of Bibby is of great practical importance and means that individual plantings do not have to occupy significant areas of productive ground to be valuable for birds.

Birches protected by tree shelters. Typical of the introduction of some broadleaves to each clearfell, here at wide spacing for environmental improvement rather than timber production. In the background there are older birches retained when the first conifer crop was felled.

Whilst small groups appear to give the greatest benefit for birds, keeping planting in scale with landscape should lead to some larger plantings running to several hectares. Protection methods will also dictate management approach. Very small groups are usually individually protected using plastic tree guards or tree shelters, but larger groups will be fenced, and it will be more cost-effective to create areas of 2–3 ha rather than $\frac{1}{2}$ ha; as with the size of clearfells, the greatest number of birds likely to benefit is from a variety of sizes of areas of broadleaves. An unquantified benefit of fencing against deer is the protection given to ground flora as well as the trees. A comparison of bird species lists for restocks in Wales (no deer) (Bibby *et al.*, 1985) and Northumberland (with deer) (Leslie, 1981) suggests that this is a far more important factor than is at present realized.

In treating broadleaved planting as a component of the overall forest design rather than something to be plumped down in unrelated patches, the restructuring concept may again solve a lot of apparent problems. Figure 4.1 shows how at Kielder, watercourses were identified as the best location for the main broadleaved component of the forest, and about 70% of the riparian zone varying between about 60 and 200 m wide will be planted (McIntosh, 1988). A benefit of this approach is that a quite large area of continuous broadleaved woodland is created and, because it is long and thin, a large area of the forest is also close to broadleaves. Smaller areas of broadleaves are included in compartments between watercourses and as scattered woodland in deer glades. Concentrating planting along watercourses is sensible because the soils tend to be better and a wider variety of both trees and associated vegetation can be developed.

There can be no natural invasion of broadleaves without seed trees but Birch, in particular, seeds prolifically and the bare soil of clearfell sites can be heavily colonized. On the best soils other broadleaved species regenerate naturally; shade-bearing Beech is becoming a component of some well-thinned Douglas Fir stands in Wales. In some places Birch regeneration has taken over conifer restocks. Foresters have tended to see Birch as a weed but the expense of traditional 'cleaning' with clearing saws just before canopy closure meant many areas were left untreated. Large areas planted on or near felled Birch woods are now as much Birch as conifer and present a severe harvesting problem if the Birch cannot be sold. Birch is very sensitive to the herbicide Glyphosate and it is now much easier to control, but at the same time rather more thought is being given to its environmental value.

Birch regeneration provides 'free' broadleaves of value to wildlife and there is no point in controlling it ruthlessly and at the same time spending money planting new broadleaves. The places where regeneration will be kept need to be identified and left, and Birch controlled in the other areas where growing conifers is the main choice. Already some forests have well over 10% Birch. In Wales, regeneration is very extensive. In North Yorkshire, the prolific Roe Deer probably have sufficient food to ignore the Birch and this is the case in Thetford too. At Kielder, protection by fencing results in natural regeneration on pure mineral (gley) soils but not on peaty gleys. Unplanned Birch regeneration tends to result in a dense thicket of mixed Birch and conifers. For practical reasons it is probably best to grow Birch in pure stands: Birch often does well to start with but the conifers later outgrow it. Birch trees do survive in the crop but they become drawn up with little bottle brush crowns. Their presence can greatly increase the cost of thinning, and when the conifers are clearfelled they tend to flop and cannot be retained. However, there is some evidence that mixtures of conifers and broadleaves may be better for birds than either on their own (Moss, 1978), and this is a subject on which more information is urgently needed for practical management.

DEAD WOOD, NESTING HOLES AND ARTIFICIAL NESTS

Dead wood is one of the scarcest features of the new forests. The value of dead wood to other wildlife groups is considerable; Steele (1972) estimates that as much as 20% of a wood's animal species are associated with dead wood, and it is especially important to beetles. Nothing is known about the importance of dead wood in supplying food for birds in the uplands: in lowland woods Great-spotted Woodpeckers depend on insect food from dead wood for the winter (Smith, 1987). Creating or retaining a stock of old trees and dead wood needs a positive will: traditionally, scattered broadleaves around felling sites were 'tidied up' as firewood and the growing value of this product has made old trees more vulnerable than ever. It is best to retain broadleaves which are generally natives with their own associated fauna and flora than dead conifer timber which harbours a variety of damaging fungi and insects transmissible to the living crop.

The absence of tree holes has not proved quite the problem for birds as was originally expected; two common forest species normally thought of as hole nesters, Coal Tits and Tawny Owls, both nest successfully on the ground. Great-spotted Woodpeckers feed on the dead wood on clearfells but finding nesting sites in upland forests could be a significant constraint on their distribution. For Pied Flycatchers, lack of tree holes is the single biggest limit on their distribution. Less is known about other birds that may be affected in the same way, Redstarts for example. The lack of nest holes has been tackled by a series of conservation projects, putting up nest boxes. The first big projects in the new forests were aimed at reducing insect pests of the forest. Most box projects were for small passerines. Nest boxes are clearly the icing on the forest conservation cake, and the greater importance of good management of habitat generally is now well understood. Although nest boxes are readily used by a wide range of species, for most they are a luxury rather than a necessity and do not actually increase bird breeding density (Currie and Bamford, 1982b). Recent projects have been more carefully directed to help species which really are constrained by lack of nest sites.

Probably the most imaginative nest box project has been for the Goldeneye duck (Dennis and Dow, 1984; Dennis, 1988), a bird of rivers and lakes in coniferous forests; the habitat of the central Highlands of Scotland seemed ideal but there were no tree holes. Boxes were put up in the late 1950s and early 1960s but it was in 1970 that Goldeneyes first bred. A slow start was followed by rapid growth from six breeding attempts in 1977 to 40 in 1981. By 1982 there was 49% occupancy of the 83 boxes then available (Dennis, 1988). Although some more boxes were put up, the biggest boost came when 800 were located at suitable waterside locations in FC forests in 1985 and 1986, concentrating around the original breeding area but also spreading as far as northern England. The missing link in the Goldeneyes' ecology in Britain is now available far beyond where they currently breed, and their expected expansion of range will make a fascinating study in biogeography. Tawny Owls and Goosanders also use boxes made for Goldeneyes. When Red Squirrels were discovered nesting in Goldeneye boxes, similar large boxes were put up specially for them. The Pine Marten has come back from the brink of extinction over the last 20 years, and, spreading back into Speyside, is now preying on Goldeneyes in boxes. As with rare raptors taken by Goshawks, this is an example of the conservation 'problem' of the rather more complete predator fauna that is now developing in Britain.

A declining species relatively new to tree boxes is the Barn Owl; boxes have been put into suitable buildings for many years, but in the Galloway forest, as abandoned cottages crumbled, an alternative for the healthy Barn Owl population had to be found (Shawyer, 1987). Wooden boxes and plastic 10-gallon drums have been put in trees and, after a gradual build-up, a mild winter and high vole population led to 30 being occupied in 1988 (G. Shaw, personal communication).

Thinking about what birds need rather than just banging up some boxes has also led to providing other sorts of artificial nest sites. Big, old trees are needed for Osprey eyries. Built on a triangular sawn wood frame, and lashed to the tree with wire and nails, artificial eyries have several advantages. They are less likely to fall out of the tree and a ready-made nest can advance breeding of young birds by a year, both important to a population still building up from extinction. Sadly, however, the most important reason is to persuade Ospreys to move to quiet corners of the forest, away from the prying human eyes that still represent a threat to this superb bird in Britain (Dennis, 1987b). A number of Ospreys have now nested successfully in artificial eyries in Scotland and these new nests are particularly effective at attracting young, inexperienced pairs.

Boxes may not be just for the birds: the willingness of many species to use boxes facilitates research. Large populations of breeding owls have been studied using boxed birds in Scandinavia. In Britain populations of Tawny Owls are being studied by Petty (1983, 1987a) which, whilst nesting successfully at the base of mature conifers, will generally use a box if provided. Crick and Spray (1987) used boxes for small passerines to create an observable breeding population for monitoring the effect of Fenitrothion spraying against Pine Beauty Moth on birds in Scotland.

Both Merlins and Long-eared Owls will nest in former Carrion Crows' nests around the edge of the forest. In Northumberland, Newton *et al.* (1981b, 1986a) showed that tree-nesting Merlins had higher breeding success than ground-nesting pairs. Both isolated broadleaves out on the moor and conifers around the forest edge were used. The crows' nests these birds need may be in short supply and Merlins will use artificial nests, woven from small branches and heather on a wire netting base. Village (1981) used the technique to study Long-eared Owls on new afforestation at Eskdalemuir in southern Scotland, and more recently the idea has been extended to put up over a hundred nests along forest/moorland edges in the Border forests. This is another example of a positive conservation project carefully targeted to the known needs of two rare species.

In the natural conifer forests of the northwest USA, dead wood habitats, known as 'snags', are the most important planned provision for wildlife, on a par with broadleaves in Britain. A great deal of detailed information is available on use by different species leading to predictive models of woodpecker numbers at different densities of snags (Brown, 1985). The Americans have largely given up bird boxes; the scale of their forests means that schemes would have to be enormous to have real impact and anyway there is an adequate supply of old and potentially dead trees; it is just a question of leaving them at harvesting. Where too few dead or dying trees are available, the tops may be blown off living trees with explosive. Enough dead wood to dispense with nest boxes is a worthwhile target for us in Britain, but however carefully our existing resource is conserved, the youth of the new forests means it will be a long time before we can achieve it.

WATERCOURSES AND WETLAND

Streams and rivers that are a focus for broadleaved planting are an important habitat in their own right. Watercourses are the main natural physical division in most upland forests. Because forest operations disrupt soil and cause massive changes to vegetation cover, they can affect fresh waters, their fish, invertebrates and birds, well beyond the forest boundary. Some of the processes involved, especially acidification, are complex (see Chapter 6). Most important are siltation from ploughing runoff and shading, both symptoms of the high-technology forestry of the post-war era.

Mills (1980) suggested standards for watercourse management which were widely adopted over the next five years. He recommended a minimum hydrological protection zone for ploughing, drainage and planting conifers and standards for the design of drainage systems. Correct angles and frequency of cut-off drains are critical in reducing the quantity and speed of water in any one drain, and thus the amount of erosion. Drains must not be more than 2 degrees in angle and must stop at least 20 m from streamsides to allow water to fan out and drop its silt load. Conifers must not be planted within 10–30 m of streams, depending on the width of the stream, to minimize shading and maintain bank vegetation to prevent erosion.

In practice these are minimum standards. Applied rigidly they would maintain much of the existing value of the actual watercourse for birds but would not improve it. Nor would there be sufficient open space for many of the species of the banks adjacent to the watercourse, like Oystercatchers, as opposed to birds of the stream itself, like Dippers. The riparian zone may be far wider. Often a natural break in the ground indicates the right point to stop drains and conifer planting. In Kielder the gently rolling hills tend to break into a steep slope near watercourses. These were planted in the past, and at the end of the rotation cable cranes had to be brought in specially to harvest them. The profit on this operation was far lower than for the surrounding flatter ground and the extra production was not worth the management complication it caused. Wet flats are also well worth leaving open: in the first rotation the poorly rooted trees on the waterlogged ground were a focus for windthrow. Initiating windthrow that spreads prematurely to other inherently more stable crops can easily offset the gain of planting extra land. These wet flats are ideal for receiving drainage water off the forest and may be large enough for some of the waders and wildfowl of the wider riparian habitat.

Draining the wettest ground at planting often proves disproportionately expensive; equipment becomes bogged and elaborate drains are dug to lead away large quantities of water. It should not be difficult for a manager to persuade himself to leave this sort of ground open but bogs and wetlands were extensively planted to achieve the maximum use of ground (Peterken, 1986). Very little seems to be known about the fate of bogs within plantations; the 'winter lochs' or 'dune slacks' of Culbin and Newborough Forests have dried out since planting, but these are rather unusual wetlands. Invasion by conifer regeneration is an obvious problem, leading to accelerated drying out and succession towards forest. Even small retained wetlands may be of value to species like Snipe and ducks. Greenshanks are still nesting in young Flow Country plantations. There does seem a good prospect that they can continue to use these sites in the immediate future but what may happen to these wetlands in the long term is less certain. Greenshanks used to be numerous in the small wetlands of the Spey Valley Scots Pine woods, but have declined greatly. The draining and planting of many of these bogs by modern equipment is possibly the cause.

Open water, lochs and pools within the forest can also provide valuable bird habitats. Within two years of construction a small alkaline pool in Dalby Forest had attracted Mallards, Teal, Pied Wagtails, Grey Wagtails, Kingfishers, Grey Herons, Moorhens and Little Grebes. However, many acid pools are disappointingly poor for birds.

Open ground around open water is likely to be essential to divers; divers are weak flyers and cannot climb over trees planted close to lochs. Polystyrene rafts have been floated on selected lochs to provide nesting sites safe from people and which will not be flooded by changing water levels. These were used by Black-throated Divers in 1988, although other species had used them previously.

Even tiny pools and puddles are important to birds in dry forests; the small concrete fire tanks at Thetford were famous among birdwatchers for attracting Crossbills. Along with them anything from Hawfinches and Great-spotted Woodpeckers to Golden Pheasants can be seen around the tanks or at any small puddle that survives in a dry spell. With their steep, concrete banks these tanks have always been a safety hazard. They are now too small to be useful for filling modern fire engines and some have been filled in with fly ash.

Staindale lake, Dalby forest, a man-made waterbody, designed primarily for recreational use which limits its value to shyer birds. However, with its associated broadleaves and open space it illustrates an attempt to create a varied and valuable habitat.

Making new pools in the forest has been a popular conservation project in the FC over the years. Most famous are the Grizedale tarns (Granfield, 1971). Twenty years after construction, the oldest pools fit naturally into the landscape and one has a Black-headed Gull colony on its islands. New pools are most valuable in forests without much natural standing water, so a high proportion have been built in the drier heathland areas. Large pools can be expensive to construct, and the poor productivity of acid, upland waters means that the wildlife return on the investment may not be especially good. The emphasis of forest conservation has also moved away from the development of special sites, of which a pool was often the focus, towards the broad habitat management described in this chapter. Increasingly, smaller, cheap pools are being dug as an extra to routine drainage work where water will stand without the need for a large dam. However, one of the largest new pools so far built is nearing completion in the FC's Newborough Forest on Anglesey; designed as a shallow scrape, it is only a few hundred yards from the sea. Its islands are planned to attract roosting birds at high tide, with less disturbance than alternative sites near the public road. In such a westerly location, rare migrants might well appear with the commoner estuary birds.

Open water is an ideal focus for imaginative development of broadleaved planting. Broadleaves planted near water can benefit the aquatic as well as the terrestrial habitat. In contrast to conifers, they do not cast dense shade, and insect and vegetable material that falls into the stream is more easily digested by aquatic animals than are conifer needles. Broadleaved thickets are nesting places for riparian birds like duck and Moorhens. Fish-rich rivers in the lower elevation forests are used by Kingfishers and Grey Herons nest and feed in many upland forests. Birds like Dippers and wagtails will live on streams almost entirely shaded by broadleaves, often Alder. However, open ground is also necessary to attract the greatest diversity of riparian birds.

PEOPLE

Although management for wildlife is valuable in its own right, it becomes even more so if people can see and appreciate the wildlife of the new forests. Public interest in wildlife may help to persuade foresters that wildlife management is an important part of their role in the countryside. Sometimes human interest may harm birds, for example through egg collecting. However, today hundreds of thousands of people in Britain care enough about wildlife to subscribe to voluntary organizations to help protect birds and the wider countryside. It is a powerful groundswell of support that should be changing the way all land managers see their jobs. Wildlife has always had a part in FC recreation, but actually seeing birds in the forest habitat is difficult. The majority of rarer species are nesting during the main visitor season and are vulnerable to disturbance. The contrast between the drama of wildlife on the television screen and the reality of the back end of a Roe Deer slipping away into the forest may be too great to sustain an interest in seeing wildlife in the wild.

There are some opportunities for wildlife spectaculars in the forests; places where, for one reason or another, it is easy to watch a rare bird without harming it. The classic example of what can be achieved in Britain is the Osprey watch on the RSPB's Loch Garten reserve. Over 1 000 000 people have now been through the viewing hide. Similar opportunities exist on FC land.

The best-known forest site is at Symonds Yat in the FC-managed Forest of Dean. Here Peregrines returned to a traditional cliff-nesting site above the River Wye in 1982, abandoned years before as the species retreated under the effects of DDT. The nest was clearly visible from a famous viewpoint and the RSPB worked with the FC to set up telescopes. In 1987, 70 000 leaflets were handed out and at least 100 000 people heard the story of the Peregrines' return and most saw them at the nest. This was the real thing jumping off the screen into life. Possibilities exist to introduce similar schemes for other dramatic species such as the Red Kite in Wales. Projects like this must always be carefully planned; the right conditions are unusual and the commitment needed to protect the birds and meet and inform visitors is high. But the benefits in the support generated for birds and in the pleasure given to many people are also great. Large, rare species, particularly raptors, are the obvious species for this type of approach, because their nest sites are large, traditional and occupied for several months of the year.

Bird or nature trails have long been a part of forest design. However, many of these have been poorly advertised or failed to live up to expectations. There can be few things as disappointing to a visitor than to miss all the species which are billed on the trail literature as being present. One of the problems is that the common birds of most forests are thinly and evenly distributed throughout the forest, so that there is little point in having special trails, with particular locations for seeing particular species of birds. Such an approach, however, does work with rare birds, and the Thetford Bird Trail (Forestry Commission, 1987) breaks new ground in showing birds to people in the new forests. A simple nature trail in concept, it is original in that it guides visitors around sites used by rare species including Woodlarks, Red-backed Shrikes, Goshawks and Golden Pheasants. The trail relies on trusting visitors: in helping people find the birds they are after, it also enlists their help in not wandering onto ground where nesting birds might be disturbed. Around the Goshawk nest it is especially important that birders do not wander into the nesting area, confusing watchers on the lookout for nest robbers and disturbing the birds.

CONCLUSION

People are also the key to the success of the ideas described in this chapter. Applied throughout the new forests they would represent nothing short of a revolution in management for wildlife. However, we have mentioned just how slow the application of simple new ideas that give clear cash benefits has been in the past (Neustein, 1976ab; Neustein and Seal, 1984). Parts of this new thinking are being applied in most forests, especially basic water management standards and broadleaved planting. Relatively few have developed the comprehensive planning necessary to carry through the full range of work. Variation of felling ages is the most difficult area because it affects the core harvesting operation and many forests are still felling to unmodified economic rotation age.

Environmental development of this sort faces two great difficulties: it tends to be the first thing to suffer when cash is short and it is far more complex than most technological change, making considerable demands on the skills of practical managers. Professional foresters have to face this challenge because only they have the practical ability to carry it through; good intentions alone are not enough in this complex task. It might be easier to have timber production, landscape, recreation and conservation neatly divided up on their own bits of land, but it would lose many opportunities for birds and for the establishment of the sort of approach to land management that really integrates and balances conflicting objectives to achieve both environmental and socio-economic aims.

Conservationists too have a role to play in making sure that conservation aims are enhanced in the new forests. Without encouragement it is unlikely that foresters will switch to less profitable practices just because these may result in environmental benefits for wildlife. To date, rather little attention has been paid to the conservation gains which can be made within the current area of plantations and so conservationists have been unable to make firm suggestions as to how the forests should be managed. Where suggestions have been made they have often not paid attention to the real world in which the forester has to make his living. Too often conservationists have told foresters what they should be doing in their forests on the flimsiest of evidence. Changes in forestry practice which will cost the taxpayer money can only be introduced on the basis of sound research. Too little such research has so far been done for conservationists to be confident that they have very much to teach foresters about wildlife conservation.

Restructuring gives conservationists a good opportunity to make sure that foresters really appreciate the fact that wildlife is important, and is seen as important by the public who, in the FC's case, pay the wages. Wildlife interest should be encouraged and monitored throughout the forests, not sidelined to special conservation areas. Perhaps it would be possible to build in wildlife monitoring as a specific part of the monitoring of crop growth and value that occurs throughout the forest. Once foresters become used to dealing with wildlife issues it will be much more difficult for them to ignore the effects of afforestation on birds of conservation value than it is at the moment. While the wildlife within the existing forest remains important, the greater issue in the relationship between forestry and birds for the next few decades is likely to be the effects of afforestation, and that subject will be dealt with in the rest of this book.

© PS

CHAPTER 5

The birds of the uplands

So far in this book we have been almost exclusively concerned with existing forests; their management and their value for birds. The existing forests are surprisingly rich in birds and are improving due to more sensitive management and better knowledge. They offer many opportunities for foresters and conservationists to work together. However, one of the reasons why this co-operation within the forests is likely to be slow and hesitant is that the biggest conservation issue in the uplands at present is that of afforestation. The threat of large areas of new plantations blanketing areas which formerly were of great wildlife interest has meant that there have been many cases where foresters and conservationists have been in public conflict.

The conservationists' case against upland afforestation has to rest largely on the strength of the evidence that the British uplands are of high conservation value. This statement may seem self-evident but it is easily ignored in the wide range of arguments which are produced for and against forestry interests. Arguments which relate to interests other than those of bird conservation, such as whether forestry is economically sound, whether it is scenically acceptable, or whether it plays an important role in the continuance of rural communities, are important but they are not conservation arguments; nor are they areas where the views of conservationists need be taken more seriously than those of other lobbies until they demonstrate their expertise and mastery of these subjects. When conservationists use such arguments then they are using them as convenient tactical weapons in order to achieve a specific goal. This is a fair enough ploy but here we will deal solely with the conservation issues. We do this for two main reasons.

First, we believe that the conservation arguments are the important ones for conservationists to highlight; rarely will anyone else bring up these arguments and even then they will not deal with the issues as well as conservationists should. There is a danger that if conservationists become too engrossed in, for example, the socio-economics of the situations they are dealing with, they will lose sight of those features of the natural environment which they are trying to conserve. By choosing to oppose some developments on economic grounds, there is a danger that conservationists give weight to just the sort of arguments which they normally are saying tell only part of the story. After arguing the case against forestry on the grounds of non-profitability in one area, it will be difficult to say that the high profitability of forestry in another area is irrelevant. This is why we believe that the arguments about loss of valuable bird habitats must be pressed home. This should involve a rational explanation of what the conservation arguments are and an educational process which stresses their importance. Conservationists should be confident enough to insist that in some cases conservation arguments are so overwhelmingly strong that they should take precedence over other interests.

Our second reason for concentrating on the conservation arguments is that we have not seen a cogent and comprehensive national analysis of the effects of upland afforestation on upland bird communities. One of the consequences of the arguments over afforestation in Caithness and Sutherland has been a tendency to regard the fate of that exceptional area as a good example of the effects of forestry in all other parts of the country, and this is not true. By providing the bones of a national analysis of the conservation value of different upland areas in Britain, we aim to identify the areas of the British uplands which are most valuable for birds. These are the areas which should have a high priority for protection through either voluntary or statutory measures.

How does one measure the value of a moorland? For many types of potential crop which might be produced in the uplands it is relatively easy to do this. For a forester the profitability of acquiring land and planting it can be calculated from the expected growth rates of the different potential crop species, given informed guesses about the future economic situation. Similarly for the farmer, if he makes a guess about the price of sheep he can make an accurate assessment of the profit he should expect to get from an area of land, depending on factors such as its altitude, average rainfall, and winter temperatures. Although it would be very difficult to put a monetary value on upland bird communities (Red Grouse moors would be an exception to this rule), this does not mean that their relative conservation values cannot be assessed. Just as a farmer or forester could lean over a gate and make a good estimation of the commercial value of a piece of land, so could an experienced bird-watcher lean on the same gate and predict the species of bird which would be likely to be found in the area. However, whereas it would be easy to compare the commercial value of land in different parts of the country in purely monetary terms, it would be much more difficult for a conservationist to say which of two moorlands was 'better' for birds if they held different bird species. However, to attempt to do this is a worthwhile exercise because it concentrates the mind on what one is actually interested in conserving. We would claim that there is plenty of scope for the conservation arguments against forestry to be honed and refined, and that one of the essential steps in doing this is to reach a much more rational and objective way of defining the conservation value of any area. In general, conservationists have found it easy to persuade themselves that the conservation arguments against forestry are strong, but very difficult to persuade either foresters or more disinterested parties that this is the case. This has been due in part to the emotive and vague arguments used in trying to explain the value of upland bird communities. Rarely is it possible to say, when singing the praises of a particular piece of moorland, that this is the best, twentieth or hundredth most important moorland bird community in Britain, only that it is good or interesting; and there has been a tendency for conservationists to say that all uplands are interesting, whereas foresters have been unwilling to accept that any of the uplands are of great conservation value. In this chapter we develop a method for ranking the importance of different moorland areas in Britain as a first step to defining the areas where continued afforestation would be most damaging and also those where it would be least damaging to upland birds. This exercise is worth pursuing because we believe that those areas of Britain which are the very best in conservation terms should be safeguarded from the threat of afforestation regardless of the economics of the forestry that might occur on those sites. The FC has an obligation to take into account conservation interests, and we believe that conservationists have been lax in neglecting to provide a national assessment of where forestry would be least and most damaging to conservation interests. It is relatively easy to define which are the best areas to grow trees in Britain but more difficult to define the best areas in conservation terms. However, a comparison of national maps of present conservation value and anticipated forestry value would be a first step in reaching any compromise between the two interests; at least both sides would have to have their cards on the table.

How good are the uplands for birds? In order to answer this question we need two main pieces of information; which birds live in the unafforested uplands and what is their conservation importance?

What are upland birds? A walk across any heather moorland in Britain, in mid-summer, will probably be dominated by sightings of Meadow Pipits and Skylarks. Are these upland birds? Well, of course, in a way they are; in some ways they are the most typical of upland birds in Britain, but they are found in many non-upland habitats too. These are species whose fate is dependent on far more factors than just those which operate in upland Britain. We shall first concentrate on those species which appear to be most dependent on the uplands, since these are the ones whose fate may be most closely tied to the future course of upland afforestation. In order to differentiate between those species which are restricted to the uplands, those which occur both in the uplands and outside them, and those which are absent from the uplands, we have used an objective method based on the breeding atlas of British birds (Sharrock, 1976).

The Atlas is the result of five years' effort by thousands of amateur birdwatchers. In each of the years 1968-1972 records were sought of breeding birds

The Meadow Pipit is possibly the most numerous bird on British moors. It is an important prey species for Merlin.

in 10-km squares based on the national grid. The findings, summed over all five years, were then mapped under three categories of proof of breeding (confirmed, probable and possible) for each species. The Atlas therefore provides information on the geographical range of every British breeding species of bird, although some rare species are deliberately and explicitly mapped inaccurately in order to protect their nest sites. The methods used to produce the Atlas have some disadvantages in that no attempt was made to standardize the amount of effort put into all squares, so that there are unknown amounts of bias introduced by uneven coverage, but it has never been suggested that the picture it gives is wildly inaccurate. The Atlas is now rather dated, and fieldwork is under way to collect the information to produce another version, but the picture it gives of breeding ranges is almost certainly still a reasonably accurate one. Overlays of habitat information are available with the Atlas and two of these have been used in our analysis. These show the distribution of land over 1000-feet and of heather moorland. Together we regard these two features as being good indicators of the extent of the uplands.

Our aim has been to isolate the group of British birds which is most dependent on the moorlands, since these are the species most at risk from upland afforestation. Ideally, one would know exactly where all the birds in Britain were and then it would be a trivial matter to see which species were most dependent on the moorlands. The closest we can get to this is to know the breeding distribution of all British species from the Atlas. There is no information on the numbers of birds in a particular 10-km square, so that a dot in the atlas can mean one pair or 1000 pairs. The new atlas will take us much closer to the ideal situation where relative numbers can be ascribed to each 10-km square, or tetrad.

Knowing the distribution of all bird species, we then need to know the distribution of the uplands. Here we were faced with a dilemma; to use the overlay maps showing altitude or those showing moorland? We chose to use the moorland overlay because quite large areas of northern Scotland, the Outer Isles, central Ireland, Orkney and Shetland are below 1000 feet in elevation but are obviously moorlands. It seemed to us to be perverse to leave these areas out of our analysis so we chose to use the moorland overlay. A consequence of this is that some upland areas, particularly south Wales, which are mainly grassland, have been excluded from our analysis. The effects of this decision are probably minor, and lead to a concentration on the best uplands for birds, since no upland bird species are predominantly associated with grassland.

To identify moorland species, those for which more than 50% of their Atlas records were in moorland squares were chosen. Fifty per cent is obviously an arbitrary choice. The resulting species are given below. These are the species which we concentrate on in our discussions of the uplands. This list agrees considerably with Fuller's (1982) analysis of the upland bird community (which, of course, is not surprising but is reassuring). Only Black-throated Divers, Red-breasted Mergansers, Goosanders and Black Grouse are present in our list and missing from Fuller's much larger list, which contains many species which occur in the uplands but are not restricted to them. Ratcliffe (1977a) lists 75 upland species, which include all of the moorland species which we deal with except the Eider. Thus there is a fair degree of agreement about the identity of upland birds.

The following 31 species are the ones which our analysis has identified as those most dependent on the presence of moorland vegetation during the breeding season in Britain. They are arranged in order of the percentage of their distribution in Britain and Ireland which is restricted to moorland squares as defined by the moorland overlay for the breeding Atlas. Each account gives information on the ecology of the species and the possible effects of afforestation.

1. *Ptarmigan.* This species has so far probably been completely unaffected by afforestation and is unlikely to be greatly affected in the future, since it lives high above the tree-line, rarely being found below 2500 feet. Ptarmigan are unusual in remaining in the uplands throughout the year, and must see more snow than any other British breeding species. The species occupies 195 10-km squares, of which 97% also contain moorland.

2. *Dotterel.* This species is also found only on the high tops in Britain and is therefore beyond the reach of afforestation. Unlike the Ptarmigan, Dotterels are summer visitors to the highest mountains, arriving in May and leaving in July. Their short stay is similar to that of many of the breeding birds of the

uplands. The estimated numbers of breeding Dotterels submitted to the Rare Breeding Birds Panel in recent years have been woeful underestimates of the true numbers, which must be of the order of at least 600 pairs in Scotland alone (Watson and Rae, 1987). According to the Atlas, the species occupies 34 10-km squares, of which 97% also contain moorland.

3. *Wood Sandpiper.* As the name suggests, this species nests in woods in its Scandinavian range but in Britain most pairs have been located on open moorland. It is a very rare British species (although more probably exist to be found) whose known numbers in any one year have never reached double figures. Wood Sandpipers have not been recorded as occupying conifer plantations in Britain (as far as we know) but there is little reason to think that they could not do so in small numbers, since in Scandinavia they are one of the species which inhabit forest bogs (Nethersole-Thompson, 1986). They sometimes nest in old Mistle Thrush nests (Ferguson-Lees, 1971). The Atlas map for this species, which will probably be an exaggeration of the picture in any one year, shows the species as occupying 18 10-km squares, of which 94% also contain moorland.

4. *Greenshank.* This is a rare British breeding species restricted to the northwest Highlands of Scotland, where they arrive, often weeks before the Swallows, in March and April. For a few weeks the moors are filled with their loud songs as males fly fast and high, but then during incubation the birds are hard to encounter. A few birds will be seen feeding by the sides of lochs or large rivers. Once the chicks hatch, Greenshanks revert to outspoken noisiness as both parents call in alarm whenever danger, or a fieldworker, approaches. Greenshanks have been studied by the Nethersole-Thompson family for decades, first in the forest bogs of the Spey Valley and then in the more dramatic hills and glens of Sutherland (Nethersole-Thompson, 1951; Nethersole-Thompson and Nethersole-Thompson, 1979, 1986). They have become a rather symbolic bird in the battle over the Flow Country of Caithness and Sutherland. This is in a sense ironic, since this species is one of the few which is known to live in forest bogs in part of its range and persist in planted areas for several years until the trees grow too high (Nethersole-Thompson and Nethersole-Thompson, 1986), but the Greenshank is one of the species whose British status is most threatened by afforestation. The population which Nethersole-Thompson studied on Speyside in the 1930s and 1940s was almost certainly extinguished by forestry filling the clearings and forest bogs that the species used, although Ratcliffe (1979) points out that plenty of good nesting habitat remains. The British population probably numbers between 850 and 900 pairs (Nethersole-Thompson and Nethersole-Thompson, 1979; Piersma, 1986). Greenshanks occupy 254 10-km squares, of which 92% also contain moorland.

5. *Red-throated Diver.* These are much commoner in Britain than Black-throated Divers, with a population of about 1100 pairs; the majority of these are in Shetland (about 700 pairs (Gomersall *et al.*, 1984)) and Orkney (about

90 pairs (Booth *et al.*, 1984)). Their ecology is slightly different from that of Black-throated Divers in that they will nest on tiny lochans and commute to the sea or to large lochs to find food. They are summer visitors to the remote lochans which they use for nesting, and lay only one or two eggs in a simple nest beside the water. Their young are usually taken to the richer feeding grounds of the sea by their parents. Afforestation could affect them in several ways. Simply surrounding their small nesting lochans with trees could prevent them from taking off unless the trees were kept well back from the water's edge. However, the effects of disturbance of the birds and changes in water levels or quality are other possible harmful effects. The species occupies 212 10-km squares, of which 92% also contain moorland.

6. *Black-throated Diver.* These are restricted, in Britain, largely to the northern and western Highlands. Their numbers have been monitored in recent years by the RSPB and it is thought that the population is stable at about 150 pairs (Campbell and Talbot, 1987). They are mainly found on large lochs with islands and nest within feet of the water's edge. Their traditional nesting sites make them easy prey to egg collectors; in 1986 two egg collectors were arrested with five clutches of Black-throated Diver eggs in their car, about 3.5% of the British population. So far, there appears to be no evidence to suggest that Black-throated Divers have been greatly affected by afforestation, but this may partly be because those areas of the species' range which have been afforested still have only young trees, and monitoring is being continued by the RSPB. The potential mechanisms of the effects of afforestation are similar to those for Red-throated Divers, although this species' preference for larger lochs means that problems of blocked flight paths should be less important. The species occupies 319 10-km squares, of which 84% also contain moorland.

7. *Goosander.* These have been nesting in Britain for the past century after colonizing Perth in 1871 (Baxter and Rintoul, 1953; Meek and Little, 1977). Goosanders nest in holes in trees and will take to nest boxes. Conifer plantations are unlikely to have trees old enough to provide nesting sites for Goosanders unless special arrangements are made to leave suitable trees within easy reach of the water's edge. Goosanders are piscivorous and so the potential acidifying effects of afforestation on water bodies might have deleterious effects on them, though these have not been demonstrated. The British population has risen steadily during the past few decades, during a period of rapid growth in afforestation, and probably numbers about 1150 pairs (Thom, 1986). The species occupies 412 10-km squares, of which 84% also contain moorland.

8. *Peregrine.* This species represents one of the conservation success stories of the post-war years (Ratcliffe, 1980, 1984a). After declining due to organochlorines in the 1950s and 1960s, the numbers of Peregrines nesting in Britain are probably higher now than at any time for 50 years. We know of no evidence to suggest that Peregrines are likely to be greatly affected by further affores-

tation. The species has expanded in numbers during the period of maximum growth of afforestation, so that there is little evidence to suggest that afforestation has had a harmful effect on them. In inland Galloway, where afforestation has covered a higher proportion of the uplands than in any other area in Britain, Peregrines are commoner now than at any time in living memory (Ratcliffe, 1984a). Mature forests provide nesting sites for many potential prey species for Peregrines such as Wood Pigeons. Provided that trees are not planted right up to cliff faces, it is unlikely that nesting sites will be lost due to afforestation. In afforested areas, persecution from gamekeepers would be expected to drop. Thefts of birds or eggs might either be simplified or made more difficult, depending on how the ease of access to sites has been changed. The species occupies 598 10-km squares, of which 80% also contain moorland.

9. *Golden Plover*. These are found in most of the moorland areas of Britain, although only a few pairs cling on in the most southern moors of the country. This is one of the characteristic birds of the heather moorlands. In early summer the males fly high, singing their melodious, liquid song, but later in the summer the incessant peeping alarm calls of either adult with chicks haunts the fieldworker. A direct loss of habitat is the most likely effect of afforestation on this species. Most of the species' British range is on ground which is feasibly afforestable (Ratcliffe, 1976). The British population has been estimated at 22 600 pairs (Piersma, 1986). The species occupies 915 10-km squares, of which 79% also contain moorland.

A Golden Plover on its nest. Its haunting call is often the moorland surveyor's first evidence of a good patch for waders. Golden Plovers are especially attached to heather moor and one of the species most clearly threatened by forestry expansion.

10. *Snow Bunting*. This is a bird of the high tops which is found alongside Ptarmigans and Dotterels in the Cairngorms (Nethersole-Thompson, 1966, 1971). This species is rarely found breeding below 3000 feet in Britain and so can be regarded as being fairly safe from having its breeding sites afforested; particularly since it often nests in precipitous rocky corries. The species occupies 14 10-km squares, of which 79% also contain moorland.

11. *Common Gull*. Common Gulls are colonial and nest in smallish colonies, usually besides pools and lochs on the open moorland. The species occupies 1054 10-km squares, of which 79% also contain moorland.

12. *Black Grouse*. Although birds of moorland, Black Grouse are found in forests during their early stages and can be sufficiently numerous to cause some economic damage to trees. They use young plantations until the trees are about 20 years old (see Chapter 2). They are hardy enough to remain in the uplands throughout the year. Black Grouse leks are sometimes on forest roads. The species occupies 603 10-km squares, of which 77% also contain moorland.

13. *Common Scoter*. In winter, Common Scoters are often seen merely as black dots offshore, but in summer they make brief visits to the moors of Scotland to nest. They arrive, in Caithness and Sutherland, during May, and their ducklings hatch in July, so proving breeding for this species requires special visits to the area, since by July the northern Scottish moors are very birdless. In Scotland they nest on open moors close to lochs and lochans but in Ireland and in central Scotland they often nest on wooded islands in large lakes. Under these circumstances it is not clear how badly the species would be affected by afforestation of parts of its Caithness breeding range. The most worrying possibility is that the growth of the forests may enable new species or higher densities of predators to colonize these areas. Changes to water quality might be important for this species but no effects have yet been demonstrated. If afforestation does have harmful affects on Common Scoters then this may be the species whose British population is most affected by afforestation, since much of its Scottish range lies within the Caithness Flow Country, where afforestation has spread rapidly in the past few years. In 1988 the RSPB surveyed the Flow Country Common Scoter population and found an estimated 45 pairs. At present, groups of Common Scoters can be seen in summer on lochs surrounded by large new plantations, but their future in this area is uncertain and monitoring of numbers and breeding success will be necessary. The species occupies 57 10-km squares, of which 77% also contain moorland.

14. *Ring Ousel*. This is the thrush of the open moorlands, particularly where there are rocky outcrops. It is a summer migrant which is one of the first birds to arrive back in the uplands in the spring. It has been little studied in Britain, although it has probably declined in numbers more or less throughout this century (Baxter and Rintoul, 1953). Afforestation of its nesting areas will certainly have contributed to this decline but is unlikely to have been the main

factor. The species has declined in Ireland too. Much of its range overlaps with land which is potentially afforestable although its preference for steep rocky slopes may put some of its favoured haunts out of the class of land which is subject to afforestation. Poxton (1986, 1987) found that Ring Ousels were closely associated with heather moorland and were commonest on areas which were managed for Red Grouse. Although two pairs of Ring Ousels nested in conifers at the edges of plantations, the rest of the 22 nests found in this study were on the ground under clumps of heather, and usually near streams. The species occupies 780 10-km squares, of which 77% also contain moorland.

15. *Raven.* This is a typical bird of the uplands, although many pairs are coastal. In some areas the croak of the Raven is one of the few bird sounds to be heard, and in the winter the Raven is one of the few species to remain in the inhospitable and bleak uplands. Afforestation of its home range has the effect of reducing the amount of sheep carrion for it to eat, although flocks of Ravens have been seen in young forestry plantations, presumably feeding on inverte-brates. In Galloway and Northumbria a study of Ravens showed that afforestation is one of the main causes of the decline in numbers (Marquiss *et al.*, 1979), yet an equally detailed study in Wales (Newton *et al.*, 1982) demonstrated that similar amounts of afforestation had had no effect on Ravens there. These studies are discussed in more detail in the next chapter. Recent observations indicate that Ravens can find food in young conifer plantations. The species occupies 1697 10-km squares, of which 74% also contain moorland.

16. *Twite.* This species must have been the prototype for the original little brown bird. Easily overlooked or misidentified, the Twite has a claim to be one of the most interesting of British birds, since it is a representative of the Himalayan bird community. Twite from the isolated Pennine population breed on heather moorland but feed on areas of upland pasture and reseeded areas (Orford, 1973). Perhaps they should be thought of as birds of the moorland edge. Little work has been done on this species, despite a long-term and continuing contraction in range and numbers in Scotland (Thom, 1986). Outside the Pennines most British Twite nest close to the coast, but perhaps this is a reflection of their moorland edge status. Ringing studies have shown that the Pennine birds winter on the saltings of the Wash, one of the few cases where we know both the breeding and wintering grounds of a substantial part of a British moorland breeding species (Davies, 1988). Afforestation will presumably reduce the amount of both nesting and feeding habitat for Twite, although we are aware of no studies which have investigated the effects of afforestation on this species. It is not known, for example, whether Twite can persist in the early years of a plantation immediately after planting. In Caithness, where Twite are almost absent from the blanket peatlands, one of us has seen small numbers of Twite feeding on the roads of new conifer plantations. The species occupies 785 10-km squares, of which 72% also contain moorland.

17. *Dunlin.* Britain holds the most southerly Dunlin populations in the world. That title goes to the few pairs which survive on Dartmoor. This species can nest in many different habitats, from the blanket bog of Sutherland and Caithness, to the machair grasslands of the Outer Isles and the saltmarshes of estuaries. Now much of its range within the uplands lies within areas which are feasibly subject to afforestation, although only a few years ago the technology would have been lacking to tackle the waterlogged sites which in the north of Scotland hold important populations of Dunlins. In Caithness and Sutherland the species is associated with the wettest parts of the Flow Country, where the ground is broken into systems of tiny pools. These areas are often the only ones to defeat the technology of today's forester. We have found Dunlins nesting within 20 m of ploughed and planted land on several occasions, but the loss of habitat to afforestation in Caithness and Sutherland has probably affected some of their best habitat in the area (Stroud *et al.*, 1987). The species occupies 537 10-km squares, of which 72% also contain moorland.

18. *Red Grouse.* With the Golden Plover, this is arguably the most characteristic bird of the British uplands. Until recently it was regarded as the only endemic British bird species. At the touch of a taxonomist's pen it has lost its specific status and been relegated to the role of a subspecies. The Red Grouse is dependent on man for much of its habitat. The wish of landowners for a few days shooting of this bird is why much of the British uplands are still under moorland rather than grazed by sheep. The management of the grouse moor is a complex job involving a rotation of burning to keep the mix of different ages of heather which the birds favour. Young heather is best to eat because of its low tannin content, high digestibility and high energy content, but high, old heather provides the best cover from predators for nesting and roosting. Poor heather management has probably contributed to the decline of Red Grouse in many parts of Scotland since the Second World War (Hudson, 1986; Barnes, 1987). Predators, or potential predators, such as Foxes, Stoats and Crows, are killed by gamekeepers whose jobs may depend on the quality of the day's sport. Not surprisingly, but illegally, numbers of other predatory birds such as Hen Harriers may also be 'controlled' if they come within shotgun range. The species has declined considerably in all regions since 1940, and the rate of decline accelerated in the 1970s (Thom, 1983; Barnes, 1988). Just about any area which holds Red Grouse could support trees, so almost the whole of the British range of this species is at risk from afforestation. Probably thousands of pairs have been lost to afforestation. The species occupies 1503 10-km squares, of which 70% also contain moorland.

19. *Wigeon.* The large flocks of Wigeon to be seen throughout Britain in winter are mostly composed of foreign birds. The breeding population numbers about 500 pairs (Sharrock, 1976). Many pairs nest on moorland where large lakes or lochs are available, but some nest in the lowlands. This species has a patchy distribution, and nowhere is it very numerous. The species occupies 283 10-km squares, of which 69% also contain moorland.

20. *Merlin.* In contrast to most other British breeding raptors, the Merlin is declining in Britain. Nowhere is its hold other than tenuous and in many parts of the country it is slipping rapidly away. A recent estimate of the population based on surveys of part of the range showed that the breeding population is now about 600 pairs (Bibby and Natrass, 1986). In four areas where there have been long-term studies (Wales, Northumberland, Pennines and Orkney) the population is declining. Recent work in Shetland suggests that a decline in numbers may be about to occur there too, as breeding success has become very low in recent years. It is not clear exactly what has caused the national decline in numbers but the Merlin now holds the dubious distinction of being the only British raptor whose numbers are falling. Afforestation of sites has probably contributed to the decline in Merlin numbers in some areas (e.g. Northumberland) but cannot be held responsible for the decline in areas where afforestation has not occurred on a large scale, e.g. Orkney and the Pennines. In Wales, Bibby suggested that improvement of moorland edges by agriculture was probably a more significant harmful factor for Merlins than was forestry. In fact, Welsh Merlins took a variety of bird prey, including species which are typically found in conifer plantations, such as Goldcrests. Practically any site which supports Merlins could grow trees. The species occupies 843 10-km squares, of which 69% also contain moorland.

21. *Whimbrel.* This species is a high-latitude version of the Curlew. Most of the British population are found in Shetland (Berry and Johnston, 1980). Occasional pairs nest on the mainland of northern Scotland but the numbers in any one year are always very small. There has been no recent indication that this species might spread back onto the mainland of Scotland, where it might more often come into contact with afforestation. The species occupies 59 10-km squares, of which 69% also contain moorland.

22. *Hen Harrier.* We have already dealt with this species in Chapter 2, as it appears to be able to gain from the early stages of afforestation. Harriers hunt for small birds, including young Red Grouse; this brings them into conflict with grouse-shooting interests, which unfortunately in some places still illegally become Hen Harrier-shooting interests. Grouse moors, were it not for the increased chance of an early death, provide what are probably ideal conditions for Hen Harriers to breed in (Hudson, 1986; Cadbury, 1987). Stands of old heather provide cover for nesting, and the patchwork of younger stands encourages high densities and species richness of moorland birds. Because Hen Harriers have changed range so dramatically, and because their use of young plantations means that numbers rarely remain stable in particular areas for more than a few years, it is very difficult to know whether their numbers are stable or changing. Certainly the numbers breeding in England and Wales appear to have declined since the late 1970s (Cadbury, 1987). The RSPB began, in 1988, a two-year study of Hen Harrier populations in Scotland which should provide useful information on both numbers and habitat use. The species occupies 529 10-km squares, of which 64% also contain moorland.

23. *Arctic Skua.* The Arctic Skua more commonly nests inland than does the Great Skua. Most of Britain's Arctic Skuas are found in Shetland (Thom, 1986), and relatively few nest on the mainland where they might be affected by afforestation. However, in Caithness several colonies of Arctic Skuas persist in the Flow Country near areas of growing afforestation. Although Arctic Skuas, like other skuas, feed on voles to some extent, there is no evidence to suggest that this species benefits from the flush of rodent life which characterizes the early stages of the new plantations. In some ways this is strange, but the Caithness birds do not seem to feed on the moors at all often (personal observation). The species occupies 139 10-km squares, of which 64% also contain moorland.

24. *Golden Eagle.* This is restricted to the Scottish Highlands, apart from one pair under RSPB protection in the Lake District. Golden Eagles occupy large upland territories throughout the year. They prey on a wide variety of large- to medium-sized mammals and birds, and also eat much carrion. Although a high-profile species, this is not Britain's rarest moorland bird; with a population of 500 pairs it is commoner than the Hen Harrier, Black-throated Diver, Common Scoter, Whimbrel, Wood Sandpiper and Snow Bunting (Cadbury 1987). It would be most affected by the loss of hunting territory when land is

Britain's Golden Eagle population has become internationally important as the species declined elsewhere in Europe. Almost certainly harmed by high levels of afforestation, real evidence to demonstrate the loss of Golden Eagles in Scotland has been hard to come by. Wide-ranging birds, of which Golden eagle is the best example, present real problems in developing land use patterns.

afforested, although it might be one of the species to gain some compensation from reduced persecution on afforested land. The species occupies 108 10-km squares, of which 63% also contain moorland.

25. *Short-eared Owl.* This species is the third, along with the Hen Harrier and Black Grouse, which may gain to some extent from afforestation in the early stages of the forest cycle (see Chapter 2) but which cannot persist in the forest as the canopy closes. Short-eared Owls are thinly spread throughout the British uplands and their population numbers are difficult to estimate. The species occupies 802 10-km squares, of which 62% also contain moorland.

26. *Red-breasted Merganser.* Unlike their relative the Goosander, Red-breasted Mergansers do not nest in tree holes. They are otherwise similar ecologically and are thus prone to the same sorts of influences as that species. Like the Goosander, Mergansers have spread through Scotland and into England and Wales this century. The species occupies 915 10-km squares, of which 61% also contain moorland.

27. *Eider.* The inclusion of this species is probably slightly unwarranted or spurious. The Eider is certainly not a species that many would consider to be typical of the moorland scene, although Fuller (1982) lists it as a species of paramaritime moorland or bog. It probably manages to scrape into this list because many of the coastal Scottish squares are likely to contain moorland. The species occupies 498 10-km squares, of which 60% also contain moorland.

28. *Black-headed Gull.* Some Black-headed Gull colonies are very large; some of the largest are on the coast at sites such as Ravenglass and Scolt Head. It is difficult to estimate what proportion of the population nests on moorlands. In recent historical time the species has increased greatly in numbers at some coastal sites, but it is not known whether the same trend, or an opposite one, is true in inland sites. The species occupies 1736 10-km squares, of which 59% also contain moorland.

29. *Great Skua.* The Bonxie, to use its Shetland name, nests on moorlands close to the sea so that it can steal food from incoming seabirds returning to feed their chicks. Outside the breeding season Great Skuas migrate southwards to warmer waters where they have a more marine existence. A few pairs nest on the coast of northern Scotland, and pairs have recently prospected inland in Caithness. However, because of its coastal habits, and restriction to Shetland and St Kilda, this species is largely beyond the reach of afforestation. The species occupies 85 10-km squares, of which 56% also contain moorland.

30. *Buzzard.* In Wales there is no evidence to suggest that Buzzards have suffered as a result of afforestation (Newton *et al.*, 1982). This may be because Welsh agriculture is relatively unintensive and has high stocking levels of sheep so that there is plenty of food available at all times of the year. In this case nest sites may be a limiting factor in population increase, and the growth

of mature conifers allows Buzzards to spread into areas which they otherwise might not be able to infiltrate. It is not known whether the results of the Welsh study will be typical of the effects of afforestation on Buzzards in other parts of Britain. In areas where prey are naturally (or unnaturally) scarce but nest sites are common it may well be that Buzzard populations decline when afforestation occurs. This species occupies 1451 10-km squares, of which 51% also contain moorland.

31. *Greylag Goose*. The existence of large feral populations of this species in parts of Britain tends to devalue the wild populations in the eyes of many people. It would be a great pity if the only Greylag Geese nesting in Britain were semi-tame introduced stock. It is a very different experience to come across an isolated pair of Greylag Geese in May when they nest on the open moors amongst small dubhlochans, than it is to see the species walking around your feet at an urban lake. Later in the season you might come across pairs leading their broods of bright yellow goslings through thick heather down to the lake shores. The British population, excluding feral birds, numbers around 700 pairs (Stroud *et al.*, 1987). The species occupies 208 10-km squares, of which 50% also contain moorland.

How satisfactory is this list as a register of Britain's moorland birds? Simply looking at the list there are a number of species which are missing which might be expected to be included. These are; Dippers, Common Sandpipers, Chough, Red Kites, Ospreys and Curlews. Of these, Curlews, Dippers and Common Sandpipers, while found in most upland areas, are also found too frequently in the lowlands to qualify under our definition. Chough is primarily a coastal species, particularly in Ireland. Red Kites and Ospreys would probably have qualified if their distributions had been accurately plotted in the Atlas, but we could not assess their status on the information available.

Another factor which will have influenced the composition of the list we have compiled is the definition of moorland vegetation. The Atlas overlay showing moorland uses a variety of typical moorland species. However, this list does not include heather itself, accounting for the fact that the Red Grouse distribution is only 70% on areas classed by us as moorland. This problem of defining the limits of the uplands or moorlands is one which could usefully be addressed in future, but we are content that our method defines moorlands in a way that is objective and broadly sensible.

These are Britain's moorland birds. But where do they live? We chose a random selection of moorland squares to examine their moorland bird communities. This was done by stratified random sampling so that squares were selected from representative parts of the country. Forty Atlas squares were chosen out of the 1000 British moorland squares for this analysis and their moorland bird communities assessed. The species richness of these 40 sample squares (Figure 5.1) shows that in general the species richness of moorlands in Britain is higher in the north than the south and that this trend persists throughout the whole of mainland Britain, so that northern Scotland is richer in species than southern Scotland. Interestingly, the trend does not continue onto Shetland or Orkney which although they have important moorland bird communities, lack some species which are present in Caithness and Sutherland, such as Golden Eagles, Black-throated Divers, and Greenshanks. Of the 40 squares which we chose to examine, it was remarkable that no two had exactly the same moorland bird community. These results, which are clearly just a way of looking at several species distribution maps at once, illustrate a very simple but important point; the consequences of afforestation will be very different in different parts of Britain.

The latitudinal trend in species richness provides information for assessing the conservation value of the moorland bird communities in different parts of Britain. However, this brings us back to the problem that conservation value is an undefined attribute. Is it fair to use species richness as a measure of conservation value? Surely some species are worth more than others? This is a very thorny subject and one which has received some attention in recent years (e.g. Fuller, 1980, 1982; Fuller and Langslow, 1986; Ratcliffe, 1986). How can one measure conservation value? We do not have the definitive answer to this question, but can here describe the drawbacks of previous attempts and suggest a method which appears to be useful for assessing the conservation value of moorland bird communities.

Before presenting our own method for assessing conservation value, we should discuss the most influential previous attempt which was that of the NCC in their *Nature Conservation Review* (Ratcliffe, 1977b). Using the criteria established by the National Reserves Investigation Committee (Huxley Committee, 1947; Moore, 1987) the NCC attempted to identify the most important examples of each of a series of habitats, two of which were peatland and upland grassland. The aims of this exercise were different from, and more difficult than, what we have tried to do, because the NCC was aiming to choose specific sites rather than identify general trends, but the overall aims of comparing conservation value were otherwise very similar.

The criteria which the NCC used were size, diversity, naturalness, rarity, fragility, typicalness, recorded history, position in an ecological/geographical

Species richness of
moorlands in Britain

Figure 5.1

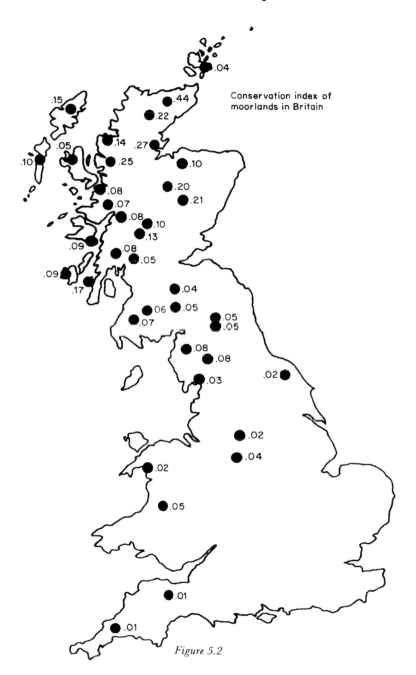

Conservation index of
moorlands in Britain

Figure 5.2

unit, potential value and intrinsic appeal. Whilst each of these attributes of a site is explained and justified in the *Nature Conservation Review* we would suggest that the mere length of the list and the vagueness of the terms are enough to seriously hamper the usefulness of this analysis. How, for example, does one compare the value of a large moorland which has been maintained by a well-recorded history of human management and supports a bird community of low diversity, but many rare species, which is not typical of the area in which it exists, with another interesting area different in all respects? The problem here is that the attributes which have been used are all things which some conservationists, but not all, would consider important, but they are not attributes about which any two conservationists are likely to agree over the relative importance of. These problems were recognized and mentioned in the *Nature Conservation Review* but not solved. Ratcliffe (1977b) commented that questions relating to the comparative assessment of conservation value 'cannot be answered with any precision or according to clear-cut rules. The problems involved are a matter for careful thought and considered judgement, having regard to the need for a fair balance between the claims of the numerous and sometimes conflicting requirements involved in the assessment and selection of sites. It is possible to find a consensus of opinion on these matters amongst informed people working in this field who between them bring to bear an enormous body of knowledge, experience and wisdom.' This cannot be a good basis for assessing conservation priorities, since it seems to mean that conservation value is whatever the small number of professional conservationists say it is at any particular time. Without some form of objective justification for the assessment of conservation value, it could appear to the outsider that conservation value is defined as the sum of conservationists' prejudices, which is hardly a situation likely to inspire respect for conservationists' arguments. Ratcliffe comes close to confirming this impression when he states 'Since the purpose of the *Review* is ultimately to satisfy a spectrum of human interests and activities, and is therefore heavily dependent on a series of value judgements, it is, in any case, difficult to see how this process can ever become completely objective.' and 'The ... review is an attempt to prescribe the contemporaneous requirement for conservation of special sites according to current values, views, information and constraints.' If this is all that conservation value adds up to then one wonders whether professional conservationists are the only people who should be allowed to have a say in whether an area is of high or low value. Ordinary people, including foresters, are quite capable of participating in questions which involve 'a spectrum of human interests and activities' and which are 'heavily dependent on value judgements'.

The problem of measuring conservation value in the *Nature Conservation Review* has arisen, we claim, for two main reasons. First, the list of attributes includes a mixture of terms which are descriptive of what the site is like and can be measured (e.g. size, rarity, diversity) with other measures which are judgements of how the site should be valued (naturalness, typicalness, intrinsic appeal, potential value) and cannot be measured. These two types of attribute should be separated. The first are the types of attribute which all parties should be able to agree upon. For example, the rarity of the species on a site should be establishable by reference to known facts; it is not a matter of opinion. Such firmly grounded attributes are the ones which should form the basis for assessing sites. It is important to arrive at a definition of the conservation value of a site which is based on what is there and can be measured rather than on intangibles. The neglect of rigorous definitions of conservation value has probably hampered the more widespread acceptance of the importance of conservation by decision-makers at all levels.

The second, but related, disadvantage of the approach adopted in the *Nature Conservation Review* is that the different attributes which were used cannot be converted into a single measure of importance. Any assessment of conservation value which is going to be useful must be able to rank different sites in terms of their value, which requires them to be measured in the same currency. This currency might be a combination of many different attributes but in order to do this each would have to be a measured one. Whereas it is relatively easy to compare the size of two different sites, and their species diversity, it is usually very difficult to compare the naturalness or typicalness of two sites. The relationship between some of the criteria used in the *Nature Conservation Review* is most unclear. How does a site with a low bird diversity of a number of rare species compare with a site with a high diversity of commoner species? And how does the low typicalness of the only British site for one species affect the site's conservation value, which must score highly for rarity? Such awkward questions are not merely academic, they will recur with every new site that is encountered. To the cynic the list of criteria used in the *Nature Conservation Review* is a recipe for describing everything as of high conservation value since the most apparently uninteresting site may be highly rated either for being so typical or for being so unusually boring! At a more serious level, we would say that a lack of clear methods for measuring conservation value has handicapped nature conservation in Britain when it has dealt, as it usually does when sites are threatened, with more hard-nosed interests who have clear ideas of the value of the sites which are under dispute in their own terms.

So in an attempt to provide a more useful and widely accepted basis for judging conservation value in the British uplands, we should look for a measurement which can be unambiguously defined and by which different areas can be unambiguously ranked in order. Having defined the moorland bird community in an objective way, we are in a position to know all the species which might be present on a piece of moorland in Britain. Using the Atlas, we are also in a position to know the geographical distribution of each of these species. For any species a particular site can be regarded as having an importance to the species in proportion to the percentage of the species range which the site occupies. Thus if two species are present in a 10-km square, the site will be more important for that species which is the rarer. If the square contains half the range of one species but one fiftieth of the range of another species, it seems reasonable to give the square 25 times the value for the rarer species compared with the commoner one. This seems intuitively reasonable; any particular site is less crucial for a widespread species which lives in it than it is for a rare species which lives in it. If we treat all species as being equally important, and there is no logical reason to do otherwise, then a conservation index can be made from simply combining the importance of the square for each species. To do this we have used the reciprocal of the total number of occupied atlas squares for each of our moorland species as the measure of species importance in Britain. Thus a square which contains a Ptarmigan has an importance for Ptarmigans of 1/195, because there are 195 Atlas squares in which Ptarmigans were recorded, whereas a square which contains a Snow Bunting merits an importance score of 1/14 for Snow Buntings because there are only 14 squares in which Snow Buntings were recorded during the Atlas years.

This index is then best expressed as a percentage of the maximum possible score (the sum of the reciprocals of the atlas square numbers for all species) since a score of 0.345 means little to anyone. This index can then be used as a measure of how close a particular area comes to containing all the British moorland species; and the amount by which an area falls short of the maximum possible score is determined both by the number of species which it contains and by how rare they are.

We have calculated this index for the 40 sample moorland squares with which we have dealt earlier (Figure 5.2). Not surprisingly, the trend in species richness is reflected in the trend in conservation value, but the north of Scotland now emerges as being even more important than in terms of simple species richness. In essence, this means that areas of high species richness are ones which contain most rare species. Although this is not a surprising result, it would have been possible for some areas to have had very low species richness but very high conservation value if they had contained communities of a small number of very rare species.

As far as we are aware, our index has not been proposed by other workers so it requires some assessment here. It has the following advantages. First, it is objective; it therefore would be calculated in exactly the same way by a dedicated conservationist, a keen forester and a politician. This is a very great advantage because it removes the possibility that it is possible to say that anywhere is important for birds. Every site has to be evaluated by fixed rules, which cannot vary with political expediency. Second, it is responsive to change in status; as a species becomes less widely distributed, the contribution of the occupied remaining sites becomes greater, and as a species becomes commoner, then that species will contribute extra importance to new sites but less and less to its former (and still occupied sites) as it spreads in numbers. Third, it provides a national perspective; species with restricted national ranges contribute most to the index and every site is measured against an achieved national maximum score. An alternative which we consider less acceptable would be to assess the conservation importance as a percentage of the maximum score actually achieved by any site in the sample. This seems less useful, as the maximum attained would change with time and thus make comparisons more difficult to make. It is important in assessing the importance of anything to put the arguments in context. Fourth, all species would be treated equally if they were equally widespread; thus there have been no value judgements made about whether Greenshanks are worth more than Golden

Eagles, or any other similar imponderables. Fifth, the index is related to sites, in this case 10-km squares. It is therefore a useful means of quantifying conservation value over the whole country in any assessment of where upland afforestation will have the most harmful impacts. Any approach which was species-based would be less suitable for this. Sixth, to calculate the index requires complete knowledge of the whole country. We regard this as a useful feature of the index, since it forces one really to know about the species one is dealing with. Any measure of the value of a site must be set in a wider context, not necessarily national as we have done here, but possibly European or regional, but in any case the information to do this must be available, otherwise rational assessment of the site is impossible.

The main disadvantage which we see with our index is that by treating all occupied sites as equal the index does not take account of the differences in densities or breeding successes in different areas. It will almost always be impossible to have very accurate or reliable data on breeding success for all species for all years, so this cannot be seen as a great disadvantage; however, it would easily be possible to adapt the index to take account of the different numbers of individuals of each species which occupied a given area. A second disadvantage is that this index takes no account of whether Britain holds a large or small proportion of the European, Palaearctic or world range of the species. Our index therefore tends to treat species which are rare in Britain (but which might or might not be widespread elsewhere) as important, whereas species which are common in Britain but less widespread elsewhere are rated lowly. Again we do not see this as a real disadvantage; our index could easily be transferred to any grid size for which the data exist. By calculating and using such an index one is forced to decide what geographical area one is regarding as important; our index could be adapted to any geographical scale, but only if the information existed to do this. The third main area of criticism which we would accept for this index is that it is based on the Atlas, whose data are biased to an unknown extent, and out of date to another unknown extent. Our answer to this is that the Atlas remains the very best source of data for the whole country, but we await with interest the production of the new breeding atlas. It would be interesting to repeat a similar analysis to that which we have presented for moorland birds for other groups such as farmland species and woodland birds.

We suspect that a common criticism of an index such as the one we have calculated is that it attempts to measure the value of something which is essentially unmeasurable; that it is the approach of a philistine to try to quantify the value of a Greenshank's presence on a remote hill moor. In answer to this we would say several things. First, the fact that the same hill is richer and more valuable with the Greenshank than without it is, we hope, undeniable. Given that you are willing to make this distinction, the obvious question is how much more valuable? One answer, the one which we have given, is in terms of how likely the hillside is to have a Greenshank on it in a national context. The more unlikely it is that a Greenshank will be present, the more valuable is the addition of a Greenshank. However, another approach might be to say that it is simply not possible to assess the worth of the presence of a beautiful bird in the environment; that such a thing is of inestimable worth. We would agree with this view, but only to the following extent. A painting or a racehorse may be worth 'more' to its owner than the monetary value put on it by an insurer, but it is still very useful to have such a measure of value. There is no accepted way for an art historian to compare the 'value' of a Picasso with a Monet, yet the market value is still a useful yardstick of something to the outside world, even if the owner or admirer thinks that it is a wholly inadequate way of measuring its real worth. In the same way, we would agree that indices of conservation value do not wholly capture the worth of wild places and exciting birds, but they are useful in assessing priorities and signalling importance. Such an objective measure is particularly useful when assessing the loss of the object. Just as the insurance value of a painting is a useful measure of how highly rated is a particular work of a particular artist, so the conservation index can do the same thing for different habitats and sites.

The lack of an agreed system of conservation value means that conservationists have been able to claim that almost any area has a unique value. In many ways they do, but this is hardly a sensible way to present a case, and is partly responsible for the opinion among some foresters (and others) that conservationists are woolly-minded idealists. We suspect that the only way to persuade many conservationists to agree about the conservation value of an area is to say that it is under threat; then its value will always be perceived as high. There is a great need for a national inventory of conservation value which is based on objective measures, established outside of the heat of arguments over the future of the sites, which could form a background for site assessment. Our method could form the basis for assessment of conservation value in moorland habitats. Sites which fall in areas of generally high conservation importance should be worthy of more effort to protect them than areas of low conservation importance.

We regard the results of this analysis to be interesting and of some importance. We have shown clearly that the British uplands do not hold a homogeneous set of upland birds. There are great differences between different areas in their species richness and in their conservation importance. In general, species richness increases in a northerly direction within the British Isles, although this trend does not continue from northern Scotland to either Orkney or Shetland. The English and Welsh uplands lack many of the species which characterize the Scottish hills. The very north of Scotland emerges as clearly the most important area in terms of its species richness and the rarity of the species it holds. This encapsulates the future problems which are likely to arise between forestry and conservation interests since the areas which are best for birds are the very areas which are under most threat from future afforestation. As we have already described, no new upland afforestation will occur in England, and little is likely to occur in Wales. In Scotland the centre of gravity of the forestry industry has been moving northwards during the past decade; away from the already heavily afforested southern uplands and the central belt towards the northwest Highlands, where land prices are low. In many ways it may be fortunate for the conservation cause that things have moved in this way, since if the order of afforestation had been reversed, much of the very best land would already have been lost. However, it also means that some of the biggest conflicts between forestry and conservation still lie ahead.

There are many other species of bird which can live on moorlands, apart from those which are largely dependent on the uplands. These have to be briefly considered, too. We have already mentioned that Meadow Pipits and Skylarks are the two most common moorland birds in Britain. Other species include Redshanks, Snipe, Lapwings, Curlews, Kestrels, Teal, Mallards, Pintails, Slavonian Grebes, Little Grebes, Wheatears, Dippers, Wrens, Cuckoos and Oystercatchers. For many of these species, agricultural changes are probably causing declines outside the uplands. In particular, those species, such as waders, which depend on poorly-drained areas, are declining in numbers, particularly in southern Britain. Few of the species which are present on moorland, but are not restricted to it, are of the same national rarity as the moorland species we have already dealt with. Presumably it is their wider habitat tolerances which have allowed their numbers to remain high. Thus although their existence on particular sites is interesting, these other species should not be at the forefront of the arguments between foresters and conservationists.

However, it is valuable to gain some idea of which species fall into that category of bird which is most likely to become increasingly dependent on the uplands as changes in agriculture in the lowlands continue. To do this we have made use of the excellent county bird atlases which are starting to emerge, stimulated by the production of the BTO atlas. As time goes on they will provide useful information for gauging the changes which are occurring in the countryside. From the point of view of this book, they also serve to identify some species which our nationwide analysis did not identify as upland species, but which may nevertheless be largely dependent on the uplands. The Devon and Gwent tetrad atlases are two of the most interesting for us, because these counties cover areas of upland and lowland. For Devon, species such as Snipe, Meadow Pipits, Wheatears, Whinchats and Stonechats are clearly shown to be more frequently present in tetrads on Dartmoor and Exmoor than in the low-lying parts of the county. It is very striking in some cases that although a species may be present in most 10-km squares, the majority of occupied tetrads are upland ones. The Snipe provides a striking example. If the county atlas is blocked up into 10-km squares so that the range is directly comparable with the BTO atlas, then there appears to be no great tendency for the species to be found in upland areas. However, the county atlas, examined at the tetrad level, shows that 60% of occupied tetrads are on Dartmoor and Exmoor. Those 10-km squares in the lowlands which do hold Snipe, do so in fewer of their constituent tetrads than do upland squares. Cursory examination of this excellent county atlas does not, however, suggest that there are a great many other species which fall into this category. The Gwent atlas, although covering a smaller area, supports this contention. Again, species such as Meadow Pipits, Wheatears and Whinchats tend to be found in a higher proportion of upland than of lowland squares, and yet other species such as Curlews, Lapwings, Redshanks and Dippers are found in a higher proportion of lowland than of upland squares. These species are all ones whose futures depend to some extent on what happens to the uplands, but possibly this is more because their former lowland breeding ranges have been eroded than because they are true upland species. It should be stressed that these atlases, like the BTO atlas, do not provide information on bird densities, so that it might still be possible for the more fertile lowland areas to hold greater densities of birds than the upland areas, although this seems unlikely to be true of most species.

At the time of writing there are bird atlases for Bedfordshire, Hertfordshire, Devon, Norfolk, Kent, Gwent, Greater Manchester and the Sheffield area. This, obviously non-random, selection of localities can still provide useful indicators of the changes which are occurring in the wider countryside. Our expectation would be that the distribution of species such as Curlews, Redshanks, Snipe and Lapwings would have shrunk as a result of land drainage. However, no such clear picture emerges from a comparison of the county atlas maps of these species with those of the BTO atlas. In almost all cases these wader species have, in terms of distribution, held their own. This may be due to several influences.

So far we have dealt with the national importance of the birds of the British uplands. Now we wish to put them into an international context. This trend is becoming more and more common in conservation arguments and probably arises from two main factors. First, Britain's membership of the European Community (EC) puts the UK government under international obligation to protect certain listed species which have been recognized by the EC as requiring special measures to protect them. The most publicized recent case, illustrating the power of the EC in such matters, has been that of Duich Moss on Islay, where a distillery wished to drain a roosting site of Greenland White-fronted Geese. Despite the claims that the drainage of the area was essential for the industry and that jobs would be lost if the drainage did not go ahead, the development was halted.

The second reason for the increased use of internationally based arguments in nature conservation is the growing evidence and realization that the fate of most migratory bird species depends on events in more than just one state. A migratory species such as the Sanderling may nest in Greenland, winter in South Africa and pass through a score of countries on migration in between, twice a year. A wider perspective is a much healthier attitude for us to develop, as bird populations do not recognize national boundaries.

If we accept that an international perspective is valuable, how do we assess the international importance of the birds of the British uplands? We will take two methods. The first is simply to see whether any of the species which we have defined as moorland species are also on Annex 1 of the European Birds Directive, which lists those species which should be given special protection by members of the EC. However, the species which are present on Annex 1 appear to be a slightly arbitrary list, and the EC does not form a recognized biogeographical unit. In particular, few of the member states of the EC have comparable areas of upland. The large upland areas of France, Spain, Italy and Greece mostly hold very different birds from those of the UK. Britain's upland bird communities are more directly comparable with those in Scandinavia and Iceland than with those of southern Europe. Britain holds practically all the EC's population of Red- and Black-throated Divers, Great and Arctic Skuas, Twites, Greenshanks and Dunlins. The British Isles contain most of the EC's Merlins, Wigeons, Greylag Geese, Common Scoters, Red-breasted Mergansers and Goosanders. However, all this would change if either Norway, Sweden, Finland or Iceland were to become members. Such an arbitrary, and potentially changeable, method of describing the species of highest conservation priority cannot be viewed with much respect. However, the European legislation to which the UK Government is a signatory does commit the UK to protect those species on Annex 1 of the European Birds Directive. These include Golden Eagles, Short-eared Owls and Merlins. However, although such status provides useful support to conservation efforts, it is doubtful whether this very arbitrary list should form a very important basis for deciding conservation priorities in Britain.

Our second method is to assess the percentage of the total western palaearctic population which Britain holds, since this would appear to give a more rational framework within which to judge Britain's importance. For our figures we have largely depended on the *Birds of the Western Palaearctic*, Volumes 1-5 which, fortunately, cover most of the upland species with which we are concerned. However, for few of the species under consideration is the knowledge of their numbers across the western Palaearctic sufficiently detailed for this to be a very practicable step. It is clear that for many species which have a high profile in the national debate over forestry, such as Greenshanks and Dunlins, the British population forms only a very small part of the total European population. The Greenshank shows how critical are the accounting methods used to derive the importance of any population. The British population of about 900 pairs is (with the exception of the occasional pair in Ireland) the entire EC population. However this figure pales into insignificance compared with the estimates of 17 000 pairs (Norway), 30 000 pairs (Finland) and 50 000 pairs (Sweden). European Russia probably holds even larger numbers and non-European Russia more still. In simply European terms the British populations of Greenshanks cannot be seen as numerically very important, standing as they do at around 1% of the total European population. This then conjures up a problem of interpretation; how should the national and international importance of populations be assessed in conservation terms? To date this issue has not been carefully considered. If the whole of the British population of Greenshanks were in danger of being wiped out by afforestation should we regard this as a total disaster or take a more European view and console ourselves that this is a mere pinprick in the species' European status? So far, conservationists have been keen to use European or EC-based criteria wherever these give best support to their case but a thorough evaluation of the criteria to be used in conservation evaluation has not been undertaken and published. To use mixtures of national and international criteria to set conservation priorities will be a further way of confusing what conservationists are really trying to do.

The moorland species for which Britain seems most important in western Palaearctic terms are probably Golden Eagles, Peregrines, Twite, Red Grouse and Golden Plovers. The last two are included because the British races are distinct. Another species which our analysis has missed is the Curlew. About 30% of the European population are found in Britain, many of them on moorland (Cadbury, 1987).

In this chapter we have defined those species which we regard as Britain's moorland birds and shown that there are predictable trends in their occurrence within Britain. Northern Scotland emerges as the most species-rich area of the country. The conservation importance of the birds of the uplands is discussed and some of the problems in assigning conservation value are described. In the next chapter we discuss how these upland bird populations are affected by new afforestation.

CHAPTER 6

The effects of afforestation on upland birds

In the previous chapter we described Britain's moorland bird communities; these are the species which will be most affected by continuing upland afforestation, since they are the species which are most restricted to the habitat which afforestation will mostly replace. The ecological change due to afforestation is so great that as the forest grows, all the moorland species which inhabited the original land will eventually be displaced. Different species can persist in the young forests for different lengths of time, and some species will do very well for a few years; however, eventually the moorland bird community will have been replaced by a woodland bird community. This might seem to say all there is to say about the effects of afforestation on moorland bird communities, but this is far from the case. Here we deal with the various ways by which the planting of an area can have either disproportionately great or disproportionately small effects on moorland birds. In other words, we deal with the correct methods which are needed to assess the effects of afforestation. We then attempt to define the species which are most at risk from afforestation in Britain and therefore deserve special attention in the arguments over the future of forestry in the uplands. Having described the effects of afforestation on moorland birds and identified those species which will be most affected by the proposed spread of plantations through the uplands, we discuss the overall effects of afforestation on the birdlife of Britain. Given that forests do contain some interesting birds but also displace interesting moorland species, is it possible to say whether there has been a net gain or loss?

Foresters sometimes deny that afforestation has harmful effects on moorland bird communities, by pointing out that some moorland species such as

Hen Harriers often nest in afforested ground. It is true that Hen Harriers are often found in new plantations but this is one of the very few moorland species able to persist in young plantations. There have been no major studies of the length of time which different moorland species persist in young plantations but our experience, and the general view among ornithologists, is that few species persist for more than a few years. Although this subject, like most concerning moorland birds and afforestation, deserves further research, it seems obvious that, at best, plantations will support only a very small proportion of the original upland birds of the area. For example, if a species can persist for five years in afforested land, yet the full rotation is of 50 years, then the best that can be hoped for on the first rotation is that the original population can be supported for 10% of the rotation; a short delay in eventual extinction. The same sort of argument can be applied to the second and subsequent rotations. It may be that some moorland species will be able to adapt to live in restocked areas at the beginning of subsequent rotations (although it would be very surprising if many species did this, and as yet it remains a vague hope rather than a firm expectation) but by the argument described above only very small proportions of the original bird community will be supported in this way.

A commoner argument than the one which states that forests do not displace moorland birds is the one which states that the birds are displaced but this does not do them any harm. After all, the argument goes, there is plenty of room for lots more birds on the adjacent moorlands. This view is incorrect, and its error can be highlighted by the analogy with stocking rates of cattle. A farmer would, justifiably, be unconvinced if told that his cattle would perform just as well if restricted to a smaller acreage. Evidence exists from many bird species to show that similar effects operate on them in natural conditions; as densities rise, then the risk from food shortage, disease, predation or other harmful factors rises too. Thomson *et al.* (1986) have shown for Greenshanks that at higher population densities laying dates are later and clutch sizes smaller. These data suggest that birds displaced by forestry are eventually lost from the population. It is sometimes difficult to believe when one stands on a windswept moor in June with hardly a bird in sight, but the British moorlands are as full of birds, in a different way but to a similar extent, as any jostling seabird colony or southern oak wood.

If upland birds were spread evenly throughout the British uplands, then the loss of 1% of the land to trees might lead to a loss of about 1% of moorland birds, but the real situation is nowhere as simple as that, because of the enormous differences in the quality of the British uplands for birds. For example, there is nowhere in England or Wales where the moorlands hold divers, Greylag Geese, Common Scoters etc. With such big differences in conservation importance, it is possible for small amounts of forestry to have enormous impacts on important bird populations if the forestry occurs in the best places for birds, and also for large amounts of forestry to have relatively small impacts on moorland birds if the forestry is directed towards the areas which are relatively uninteresting for birds. We would argue that this is the main way in which conservation interests and forestry interests could be accommodated; that forestry should be directed towards the land of lower conservation value. This could in theory operate through either voluntary or statutory means.

A similar argument can be applied to smaller areas than the whole country. One study which has tried to address such a question is the work done by the NCC in the Flow Country (Stroud *et al.*, 1987). In this study surveys of birds carried out at many sites scattered throughout Caithness and Sutherland found that the numbers of three wader species differed greatly from site to site. Dunlins, for example, show a very strong preference for areas with systems of small pools, and these pool systems are patchily distributed. Presumably the reason that Dunlins favour these areas is that they are rich in invertebrates, although this has not been studied in great detail. Their dependence on this easily identifiable type of habitat means that Dunlin numbers should be predictable if one knew the habitat in different areas. The work done by the NCC showed that to some extent the numbers were predictable from map information, which enabled them to predict the numbers over the whole of Caithness and Sutherland, and to assess the numbers which had been lost due to afforestation. Similar results were shown for Greenshanks and Golden Plovers.

It is likely that similar results on patchiness of bird distribution would be found in all other upland areas, although the information exists for relatively few of them. Conservationists have learned a great deal from their own unpreparedness when the Flow Country surfaced as a major conservation issue. The survey data simply were not available. Partly as a result, large upland areas have been surveyed in recent years by both the NCC and RSPB, but it is still true that detailed information exists for few areas outside the Flow Country. The reasons for this lack of knowledge are at least two-fold. First, the British uplands cover an enormous area, approximately equal to the area of lowland agricultural land. Thus the task is dauntingly great. Second, conservationists have been torn between the need to identify specific areas of high quality in order to designate them as SSSIs or nature reserves, and the need for more extensive work which will identify the relative importance for birds of different habitats within a large area. At first it was probably thought that the same fieldwork could do both jobs but the realization has been growing that this is not the case. To provide information on the very best sites requires high-quality information from just those sites, but to provide general correlates of bird-habitat relationships, a truly representative and large sample is needed. Thus the work in the uplands has to some extent been an unsatisfactory compromise between two different approaches. This is evident in the work described by Stroud *et al.* (1987) which took place in the Flow Country. The NCC survey sites, from which extrapolations were made to the whole of the Flow Country, were a very unrepresentative selection of the available land-forms; more than half of the NCC survey sites were situated on landform A, the very best land for waders, which covered less than a third of the area suitable for breeding waders.

Despite the shortcomings of the survey work to date, some generalizations are likely to emerge from recent work in the near future. Most upland workers accept that well-managed heather moorland holds important densities and species of upland birds and that grass moorland is relatively poor for birds (Bibby, 1986; Hudson, 1986; Cadbury, 1987). Few data exist to demonstrate this as yet. However, if true, this provides a very easily applied rule of thumb with which to direct forestry onto the areas of the upland which are least important for birds. Restrictions of planting on areas of heather would be one option.

So far in this chapter we have only considered the effects of the loss of moorland to afforestation for the area which is actually planted, but it has been suggested that this may not be the end of the story. Could plantations have effects outside their boundaries? The answer to this question is that they certainly could do but that this has not yet been convincingly demonstrated in Britain. The types of way in which forests could influence the bird communities outside their boundaries (henceforth termed edge effects) are various. The removal of grazing animals may occur over larger areas than those planted and result in a growth of vegetation. Burning of heather is likely to be greatly curtailed near to forests, which results in an accumulation of long, rank heather which may provide nesting sites for some species (e.g. Short-eared Owls) but is less nutritious for Red Grouse and supports fewer insects. Perhaps the most obvious possible edge effect is that the forest may provide shelter and nesting sites for predators, either mammalian species such as Foxes and Stoats, or avian ones such as Carrion Crows, which will cause problems for the adjacent moorland birds. It has also been suggested that there may be a psychological block in the minds of some moorland species which will prevent them from nesting close to plantations (and this might have survival value for them in helping them to avoid predators). The main importance of edge effects is that their existence would mean that afforestation was having a greater effect than could be measured merely by the area of moorland which had been lost to trees.

The first attempt to test whether edge effects exist in British upland forests was conducted at six sites in Sutherland and three in Durham (Stroud and Reed, 1986). The study collected information on bird numbers and vegetation communities at different distances from the forest edge and then investigated whether a given vegetation community would contain more or fewer wading birds if it was situated next to the edge of a forest than if it was distant from the forest edge. Stroud and Reed claimed that their data showed that this was the case; the numbers of waders in a given vegetation type were lower close to plantations than far away. This important claim, although based on a very small sample of sites, received widespread publicity. Unfortunately the analysis on which this finding is based contains not only an inherent bias which would tend to produce their barely statistically significant result, but also important miscalculations (Avery, 1989). Under these circumstances the claim that edge effects have been shown to occur has to be set aside until a rigorous and correct demonstration of such an effect is published.

As a result of the NCC's work, the FC commissioned research on edge effects from the RSPB. This work has surveyed 62 sites in Caithness, Sutherland, Inverness, Moray and Nairn, covering 248 km of forest edge and a total area of nearly 40 000 ha. The findings of this work were complicated. Looking at the birds of the areas adjacent to plantations, it was found that in some study areas the numbers of some birds changed with distance from the forest edge. In some cases wading birds such as Dunlins and Golden Plovers were rarer close to the forest than they were far away from it, but Curlews were more numerous close to the forest edge than far from it. However, these findings were not true of all areas. The bird data should not be looked at in isolation, though. Bird numbers are usually related to differences in habitat features, and although moorlands may look uniform to the untutored human eye, there is considerable variation in vegetation height and composition which might influence the numbers of birds. It was therefore necessary to allow for the effects of vegetation on bird numbers and then look again for any effects of plantation proximity. When this was done there was practically no effect of distance on the bird numbers; in other words, for a given vegetation type, the same numbers of birds will be found at all distances from a forest edge. One possibility still remained; that the growth of the forest might have influenced the vegetation on the adjacent unplanted land. This might be the case because of the trees providing shelter, increased transpiration drying out the nearby land, or perhaps from changes in moorland management such as a reduction in moor-burn near inflammable forests. It is difficult to assess this possibility without long-term studies from particular sites. However, map features provide a way of gaining information on habitat features which cannot have been influenced by the presence of forestry plantations. Analysis of map features for the Sutherland study area indicated that they changed with both forest age and with distance from the forest. Although not necessarily conclusive, this suggests that the trends in vegetation with forest age and with distance from the forest edge have probably arisen due to differences in planting strategies of foresters over the past 30–40 years rather than during the growth of the forest (Avery, 1989).

Experiments were also done to assess the possibility that predation might be higher closer to a plantation than further away from it (Avery *et al.*, 1989). This is feasible, since the rise in vole numbers associated with the early years of afforestation might attract Foxes into an area, gamekeeping pressures are likely to be reduced, and as the trees grow they may provide nesting habitat for species such as Crows. A large-scale experiment was performed in Sutherland on sites next to forestry plantations. Chicken eggs were set out at 200-m spacing along four transect lines parallel to the forest edge, and then these sites were visited eight days later to check whether or not the eggs had disappeared. It was found that the eggs placed near to the forest edge did disappear more often than those further away from the forest; in fact, the fall-off in predation rate when averaged across all sites was approximately linear. However, the predation rates were closely related to vegetation differences, which also varied with distance from the forest edge. When the effect of the vegetation differences was allowed for statistically, it appeared that predation in this artificial situation was unrelated to proximity to the forest edge. This suggested that predation on real nests might also be unrelated to forest proximity, although to study real predation rates on the nests of real birds is a real problem.

The subject of the possibility of edge effects has diverted attention from more important and undeniable direct effects of afforestation. There is considerable irony in the fact that it was RSPB research, funded by the FC, that first identified the shortcomings of the NCC's work and then went on to contradict the NCC's conclusions. It has also been clear that some conservationists have been keen that edge effects should exist, despite the fact that if they do not exist it means that the birds are less harmed by afforestation of an area. Two issues are worth considering further; do edge effects exist anywhere, and what would it mean if they did? Both the RSPB and the NCC have done most of their research on edge effects in Caithness and Sutherland. This is for a number of reasons, but partly because of that area's exceptional moorland bird populations. However, it is possible that the area is exceptional in other ways which would make it less than ideal as a model area to study edge effects. The mechanisms by which edge effects might manifest themselves are various, but at least two, change in management after afforestation and predation, might be better studied elsewhere. The Caithness and Sutherland moors are not intensively managed grouse moors like those found further south in Scotland or in the Pennines. They are not so regularly burned to provide the patchwork of heather stages which benefits grouse. Thus it could be argued that afforestation will affect the management of the adjacent land much less strongly in the Flow Country than in other parts of Britain. Perhaps the cessation of moor burning which usually accompanies afforestation in areas such as northern England would have effects on the adjacent moorland birds through changes in vegetation. Similarly, it may be, we do not know, that changes in predator numbers due to afforestation might also be greater in areas further south in Britain, where densities of Foxes and Crows may be higher than in the north and where it is possible that the cessation of keepering may have greater effects than in the Flow Country. Thus there still exists scope for further work on edge effects.

But what if edge effects were found to exist? This would be a useful additional political weapon to use in arguments against afforestation because it would mean that the area affected by any forestry scheme was greater than the actual area planted. However, the existence of edge effects would have other consequences. One consequence of the existence of edge effects would be to make site protection more difficult and costly because each upland SSSI or nature reserve would need an extra buffer zone around its edge. But the most far-reaching repercussions would be in the area of forest design. To minimize edge effects for a given area of plantation, it would be necessary to plant large, uniform plantations with short boundaries. Foresters would probably argue that there was little point in planting small blocks with gaps between them or in leaving any unplanted areas within the forests. Thus the existence of edge effects would be a reason to favour the large, blanket afforestation which conservationists have criticized for so long.

For a species which has small territories or home ranges, and in the absence of edge effects, it is reasonable to assume that if x% of a relatively uniform area is planted with trees then x% of the population will be affected by afforestation. However, the situation is different for species with large home ranges. Two good examples which have been studied in some detail are the Raven and the Golden Eagle, but first we will describe the theory behind the effect. Quite simply, for species with large hunting ranges, their home ranges may be made unsuitable if their food supply is reduced below a certain threshold level which need not, of course, be as low as no food at all. An analogy would be that you cannot buy a home of your own unless you have enough money, but that does not have to mean that you are literally penniless. So it could happen, for example, that all the territories of a particular species could be made unsuitable if only 20% of an area was afforested if this meant that every territory had 20% afforestation and 90% of a territory was needed for the pair to survive.

This sort of effect is most likely to be noticed with species with large ranges, although it might well happen in all species. In species with very large territories many territories fall into the partly afforested group. Five species which have been studied in detail in this respect are Ravens, Golden Eagles, Merlins, Red Kites and Buzzards.

There have been two studies of Ravens which have had very different findings. The first study was carried out in Galloway, an area where afforestation has been particularly extensive (Marquiss *et al.*, 1978). The study used

Despite this Merlin's success the species has been declining for many years. Attached to heather moors, afforestation is a real threat – but tree nesting is becoming increasingly common and tree nesting pairs can do better than vulnerable ground nesters.

historical data which stretched from 1946 to 1975, a run of data which is available for very few species in Britain. From 1946 until the 1960s the population remained stable but then began to decline until in 1974–1975 the population had reached 55% of its former level. The decline in the Raven population was associated with afforestation of the area. For 22 territories where the date of desertion was accurately known, Ravens tended to desert their territories when they were afforested. This in itself is compelling, but not totally watertight, evidence that afforestation is causing the desertion of territories, since it could be that some other harmful factor (such as poisoning of sheep carcasses) was also increasing during the period so that the link with afforestation is a spurious one. This suggestion is not too far-fetched, because the data of Marquiss *et al.* (1978) show that during the period for which they have data the chance that a Raven territory will be deserted in a year of afforestation increases markedly; and the only cases of territory desertion which occurred in years of non-afforestation also occurred late in this period. However, Marquiss *et al.* (1978) also produce evidence which shows that as the percentage of a Raven territory which is afforested increases, the productivity of the pair of Ravens decreases. Marquiss *et al.* (1978) showed that Ravens eat a large amount of sheep carrion and so the decline in productivity is likely to be due to the removal of sheep prior to afforestation. Mearns (1983) showed that the inland Ravens of Galloway continued to decline in numbers during the late 1970s and early 1980s. He suggested that this was due partly to continuing afforestation but also to changes in sheep husbandry, with a greater pro-portion of ewes being lambed and overwintered on lower ground and in buildings. Mearns showed that the coastal population of Ravens was also declining. This suggests that factors other than afforestation may be contribu-ting to the decline, although it is possible that the decline in Ravens of inland territories due to afforestation has caused the decline in coastal numbers by reducing the number of recruits to the coastal breeding population. All in all the studies of Ravens in Galloway form the most complete picture of the effects of afforestation on a moorland bird. They illustrate the difficulties of proving beyond reasonable doubt that afforestation is having a harmful effect on a species which has a large home range. The weight of evidence clearly favours the suggestion that afforestation has been the main factor responsible for the decline in Raven numbers in this region, although other factors may also have contributed to the decline. The most convincing way to dispel this doubt would be to show that the numbers of Ravens had not declined over the same period in an otherwise similar control area which had not been afforested. Marquiss *et al.* (1978) state that in the English Lake District, an area of low levels of afforestation, Raven numbers remained fairly constant.

Ravens have also been studied in Wales (Newton *et al.*, 1982). The results of this study were in marked, and interesting, contrast to those obtained in southern Scotland and northern England. Overall there was no decline in the Raven population during the period of the study, and there was no evidence to show that breeding success of individual Raven pairs declined as the percent-age of their territory which was afforested increased, even though the amount of territory afforested ranged from 0% to 60% closed forest within 3 km of the nest. In Wales, Raven territories which were around 50% afforested, within either 1 km or 3 km of the nest, produced equal numbers of young as Ravens in territories without afforestation. In Galloway, similar amounts of forestry had marked and measurable effects. For example, in Wales, Raven territories with 40–60% closed canopy forest had an occupancy of 72%, whereas in Galloway and the Cheviots, territories with this much forestry in them were occupied approximately 40% of the time. Newton *et al.* (1982) suggest three reasons why their findings might differ from those of Marquiss *et al.* (1978). First, in Wales the tree cover was less than in southern Scotland and Northumbria. This is certainly a reason why, if the strong relationship between afforestation and desertion of nest sites which had been found in the first study had been replicated, the effects of afforestation on a regional scale would have been less. However, those effects were not replicated; the important finding of the Welsh study was that for a given amount of closed canopy forest, the effects of forestry on Ravens in Wales were much less harmful than the apparent effects in the northern study. The second suggestion of Newton *et al.* (1982) was that the Welsh forests were generally younger than those in the first study. Ravens can feed in young conifer plantations on invertebrates and rodents, so this might delay the effects of forestry on them, but the fact that occupancy is so much higher in Wales than in Galloway and the Cheviots for a given amount of closed canopy forest, where Ravens cannot hunt, suggests that there must be something more than this factor operating. Newton *et al.* (1982) also point out that in Wales there were more sheep outside and inside the forests than was the case in Galloway and the Cheviots, so that food was still abundant despite afforestation. They show that the variations in Raven breeding success between areas and from year to year were related to differences in the amount of carrion available. It seems highly likely that the very high densities of sheep are the main reason why Welsh Ravens can better tolerate afforestation of their territories than southern Scottish or northern English Ravens.

The results of these two detailed studies of Ravens are very interesting. They bring home the point that detailed studies from many different areas are necessary to fully assess the effects of land use changes. Unfortunately there are few species on which such good studies have been carried out. The differences between the results of the two studies underlines the fact that the effects of afforestation are likely to be complex. It is not possible to say that afforestation always leads to reductions in Raven numbers or breeding success, even though it is most strongly implicated in the decline of Ravens in Galloway and Northumbria. Which of the two studies is most likely to be typical of the effects of afforestation on upland Ravens? This is a question which would be best answered by more studies rather than by speculation, but it is perhaps possible to guess that the Welsh study might apply to most of Wales and the large Raven populations in the Lake District and southwest England, whereas the findings of Marquiss *et al.* (1978) might be more applicable to Scotland. If this were the case then it would mean that future afforestation, which will mostly occur in Scotland, will be in the areas where Ravens will be least able to cope with its effects.

The second species with large home ranges that has been studied to see how afforestation has affected its numbers is the Golden Eagle. This species is in some ways ecologically similar to the Raven in nesting on cliffs and feeding to a large extent on sheep carrion in the winter. Marquiss *et al.* (1985) describe the fate of a very small population of Golden Eagles in Galloway. Three pairs recolonized Galloway in the early 1940s. A fourth pair recolonized the area in 1965, during a period of expansion of afforestation. Two pairs have ceased breeding successfully in recent years and these are the two pairs with the highest amount of forestry in their territories; 61% and 43%. For both these pairs the areas in their territories which have been left unplanted are those areas at highest elevation, which Marquiss *et al.* (1985) suggest are the areas of lowest food availability for Golden Eagles. This is supported by the fact that the fourth pair, which actually colonized its territory when 19% of it was afforested (and now 32% is afforested), which continues to breed successfully, retains much of the lower ground in its territory in an unplanted state. This suggests that Golden Eagles can tolerate large parts of their territories being afforested only if these areas are those which are relatively poor for hunting by Golden Eagles. It is not clear at the moment whether it would be easy to predict which areas of a territory could be afforested without harm to a pair of Golden Eagles nor whether this would in fact be possible.

This very limited work on a tiny population of Golden Eagles was followed by a more wide-ranging study of Golden Eagles by the NCC in several study areas in Scotland. This showed that Golden Eagles had declined in all five study areas since 1956–1960 but that the area with the highest percentage decline was mid-Argyll, where forestry had been widespread. Watson *et al.* (1986) provide no information on the timing of the desertion of individual territories in relation to the timing of afforestation in contrast to the work of Marquiss *et al.* (1978) on Ravens in Galloway, so it is much more difficult to be sure that it is afforestation which has caused the decline in numbers. Mid-Argyll is compared with two other areas in Scotland, northwest Sutherland and Lochaber, where agricultural data show that the decline in sheep numbers has been similar to that in mid-Argyll but that afforestation has been negligible. In these two areas Golden Eagle numbers have dropped by 5% and 12%, which suggests that a decline in Golden Eagle numbers by around 9% would have been expected in mid-Argyll regardless of any effects of afforestation. A similar decline in Golden Eagle numbers to that observed in mid-Argyll was observed in the eastern Highlands. This was attributed to changes in the management of deer in this area and not to afforestation. These data provide weak evidence for an effect of afforestation on Golden Eagle numbers. They do show that Golden Eagles have declined in areas which have not been afforested, where afforestation cannot be causing the decline, and also in areas which have been afforested, where it remains unproven that the decline is caused by afforestation. It would be very surprising if afforestation did not lead to a decline in Golden Eagle productivity and occupancy of territories, but at present it is impossible to show that this happens. It is also impossible to assess whether afforestation is the greatest threat to Golden Eagles in Scotland or only one of the adverse factors.

The Merlin is another species with a large home range which has been studied in some detail. Merlins are declining in numbers in several parts of Britain. For example, in Northumberland Newton *et al.* (1978, 1986a) found that Merlins were declining in numbers. Some sites had lost their Merlins as a result of afforestation. However, the studies showed that the productivity of the remaining Merlins was too low to maintain the population. One suggested reason for this was that predation by Foxes might have increased with the increased area of land which was afforested. This would be a true edge effect.

In Wales, Bibby (1986) studied Merlins and found no very clear relationship between Merlin site occupancy and the area of forestry around the nest. Heather was the most important predictor of whether sites were or were not occupied; sites with most heather were the ones with the highest level of occupancy. Grassland areas were less frequently used. The relationship between occupancy and the amount of conifer plantation in the nest area was weak. Compared with unoccupied sites, occupied nests tended to have less forestry within 1 km of the nest site but more between 1 km and 4 km. Breeding success was unaffected by the amount of conifer plantation near the nest. Merlins have also declined in numbers in areas where no afforestation has occurred; the Pennines (Yalden *et al.*, 1979), Orkney (Meek, 1988) and Shetland (P. Ellis, personal communication). This suggests that afforestation cannot be the only factor, and may not be the major factor, affecting their numbers in Britain. However, the losses of suitable habitat due directly to afforestation are unknown and may be large. There are no published studies of Merlins from the Scottish mainland which could show how afforestation in areas such as Argyll, or the Grampian or Highland Regions have affected this species.

Species such as Ravens, Golden Eagles and Merlins are distributed over large areas of the British Isles, some of which may become heavily afforested and some of which are likely to remain lightly afforested. Compared with the Red Kite, which is restricted to a small breeding range in mid-Wales, these species' fates are likely to be governed by a whole range of factors which might interact in different ways in different parts of their ranges. The Red Kite has such a restricted distribution that a land use change in just one small area of the country might be enough to lead to the British extinction of this species.

Newton *et al.* (1981a) investigated the effects of land use changes in mid-Wales on Red Kites. They concluded that there were good and bad effects of afforestation on this species but that, overall, afforestation had probably been a neutral factor. Red Kite numbers had increased slowly during a period of growth of forestry in Wales. Afforestation, mostly of open sheepwalk, probably provided Red Kites with a more diverse prey community, and in particular increased the numbers of voles and birds available to them. Loss of sheepwalk would have reduced the amount of sheep carrion available, but Newton *et al.* (1981a) point out that this reduction would not necessarily be proportional to the area of sheepwalk afforested. Red Kites showed no clear relationship between their breeding success and the degree to which their territories contained forestry; they appeared to perform better in the early stages of nesting in the more afforested localities, yet this advantage was neutralized by them performing less well in the later stages of breeding, so that overall there was no effect. A small number of pairs were affected by disturbance by forestry operations in the years of afforestation. It appeared that some pairs had been displaced from their territories by high levels of forestry, yet these losses were compensated by gains in other areas. Increasing levels of forestry might impinge more on Red Kites, but it is probably true that at present there are sufficient apparently suitable areas for displaced birds to move into that any major effect seems unlikely under present conditions.

For completeness we will briefly discuss the effects of afforestation on Buzzards in Wales. These too were studied by Newton *et al.* (1984). Buzzards, whether on farmland or upland, seemed completely unaffected by the amount of forestry in their territories. The mean number of young produced each year was similar in areas of no forestry to that in areas of forestry or woodland of up to 90%.

These examples from species with large home ranges bring to the foreground the problem of planting patterns. Blanket afforestation is often used as a term of denigration by conservationists but we would stress (as did Petty and Avery (1990) and Avery (1989)) that in many credible situations blanket afforestation should be the preferred pattern on conservation grounds. Wherever the effects of forestry operate in such a way as to be supraproportional to the percentage of land planted, then it will pay in conservation terms to concentrate a given amount of forestry into the most compact block possible. This is true for several reasons. It will minimize the amount of forest/moorland edge for a given area of forestry. In those cases where a threshold level of afforestation is enough to make an area unsuitable for a species, then by increasing the area of forestry in regions which are already heavily afforested, where it can do no more harm, other, hitherto unaffected, areas are spared. It seems highly likely that many species will react to forestry in a similar way to that shown by Ravens in Galloway; a decrease in breeding success and nest site occupancy as the area afforested increases and a total desertion of territory before the area is fully afforested. Different species will react to forestry in different ways; however, this general pattern of effects is likely to recur. These findings point to a possible ploy by conservationists in influencing forestry development; in areas which are already highly afforested, it might be scientifically justified to press for further afforestation, which fills in the gaps, if this can take the pressure off other less severely affected areas.

MOORLAND REMNANTS

Where foresters leave unplanted areas within a forestry scheme, these are like small moorland islands in a sea of trees. Are such areas worth preserving? Each case must be treated on its individual merits, but since it is often the case that the areas left unplanted are unplantable, perhaps because they are too wet for ploughing, they will often hold greater wildlife interest than the average of the original moorland. Some consideration has been given to whether such moorland remnants will hold their wildlife value after planting. This question is similar to that of edge effects but can be considered to be slightly different because different mechanisms could lead to reduction in value of habitat islands compared with open moorland adjacent to plantations. There is only one study which has looked directly at this question (Rankin and Taylor, unpublished), and it has been widely quoted by NCC staff (Reed, 1983; Stroud and Reed, 1986; Stroud *et al.*, 1987). Rankin and Taylor worked in northern England and southern Scotland and studied 29 sites ranging in size from 8 ha to 850 ha. The study found that large sites held more moorland species than small sites. Small sites always held common species, such as Meadow Pipits and Skylarks, but lacked the rarer species such as Merlins, which were found on large sites. From this evidence Rankin and Taylor concluded that species such as Merlins will not nest on small enclosed pieces of moorland. This may be the case, but such a conclusion is unwarranted from the data that are presented. Rankin and Taylor's conclusions have been criticized by Bibby (1988) and Petty and Avery (1990). Obviously, the chances of finding a rare species on a small area of moorland are very low. Thus to be sure that rare species actively avoid small sites it is necessary to sample a very large number of small sites. In Rankin and Taylor's case the smallest site was approximately 1% of the size of the largest site, so one would need a minimum of 100 such sites to even start comparing similar areas. Put another way, there is a very small chance that a species like the Merlin would be found on any particular 8 ha patch of moorland on a moor where Merlins nested; so their absence from a few such small islands is of no significance. Certainly there is no basis in using Rankin and Taylor's work to suggest that moorlands have to be above a certain threshold size in order to support a representative moorland bird fauna.

However, this misuse of species–area relationships to predict minimum habitat requirements does not mean that small moorland remnants surrounded by conifer plantations will hold their initial bird communities. That question is best investigated by long-term studies of individual sites, although unfortunately the information is needed now in order to assess the effects of afforestation, not in 20 years time when the data might be available. The only alternative to long-term studies is to have very large sample sizes of sites; and the greater the size range of sites, the bigger is the required sample size. This creates a very big practical problem. In most upland forests there are some unplanted enclosed areas, but they are rarely very numerous. Thus it will probably be impossible to find anywhere in Britain on which such a study might be soundly based.

There is evidence from botanical studies that small, enclosed moorland islands can change in vegetation with time (Lindsay *et al.*, 1988) and although these changes have not been shown to affect bird numbers they may well do so.

WATER QUALITY

The way in which forestry might have most potent effects on areas outside the forest itself is by affecting water quality. Upland forests are generally located at the heads of river systems, so anything happening may be transmitted downstream for tens of miles, affecting not only the rivers but also the lakes and reservoirs which they feed. This is an area where research has not concentrated wholly on birds, and one where the effects of afforestation can be judged using much wider environmental criteria. The bird conservation interest of upland waters is usually fairly low. Dippers, Grey Wagtails and Common Sandpipers are three of the characteristic species of upland streams throughout Britain, but upland rivers are also used by Grey Herons, Mallards and Teal, and in some parts of Britain by Goosanders, Red-breasted Mergansers, Ospreys, Kingfishers, Oystercatchers, Redshanks, Greenshanks and Lapwings. Upland lakes usually have a low biological production but in some areas are used by species such as Black-throated and Red-throated Divers, Greylag Geese, Wigeons, Common Scoters, Common Gulls and Black-headed Gulls.

The effects of afforestation can conveniently be discussed under three headings; physical changes, nutrient enrichment and acidification.

Ploughing and drainage for forestry have immediate effects on hydrology, especially on wet soils. Speed of runoff is greatly increased initially, and this can give rise to higher peaks of water flow in times of rain and lower troughs in between rains (Robinson, 1980). The flash floods could lead to flooding of nests of birds such as divers, which must nest very close to the water's edge. Flooding is an important cause of nest failure in Black-throated Divers in Scotland. Interestingly, the dry year of 1988 was one of the best years for Black-throated Divers for several years, because flooding was much less common, but there is no evidence to suggest that divers have suffered more in afforested catchments than elsewhere. As the trees age, the amount of runoff from forests declines because of interception of the water by the trees and increased transpiration. This can lead to reductions in total runoff of 20–30% compared with those of open moorland (Hornung and Newson, 1986). Changes in the pattern or amount of water flow can affect invertebrate species which are adapted to particular current velocities. At the extreme, invertebrates and fish eggs or fry can simply be washed away from streams in spate. If water flow decreases to the extent that streams dry up or become sluggish, fish and invertebrates will be killed.

Ploughing increases the silt load of streams to a variable extent depending on slope, soil type and ploughing technique. Excessively steep drains can result in serious gully erosion, removing hundreds of tons of material, and leaving 1 m deep drains at 3 m or 4 m depth. Robinson and Blythe (1982) estimated that a peatland catchment lost 120 tonnes/km^2 due to erosion in a five-year period, compared with 15 tonnes/km^2 in the five-year period prior to draining. Contrary to initial expectations, siltation continues through the forest cycle. Even mature forests' catchments can have high levels of erosion. This may be due to the suppression of plant growth within and beside the watercourses by shading by conifers. Bare ground continues to erode. Management within the forest such as road construction, digging gravel from river beds, extracting timber down watercourses (because they had hard gravel beds) and driving vehicles into streams to wash them have all exacerbated the

Dippers are the most visible link in the complex acidification process which links power stations, fish and trees to the birds of upland streams.

problems of sedimentation, but have also largely now been stopped or their effects reduced.

The effects of sedimentation are varied. At high levels, plant growth is either shaded out or swamped by sediment, gills of fish become inflamed, oxygen supply to fish eggs and invertebrates is reduced, filter-feeding invertebrates become choked, and visually hunting predators such as Kingfishers, Herons and Otters may find it difficult to find prey.

Planting conifers right up to the edges of streams creates deep shade once the trees are grown. Sitka Spruce are particularly shady, and the light penetration under a closed conifer canopy can be as low as 10% of that in open environments (Smith, 1980). Shading reduces the growth of streamside and aquatic vegetation. Invertebrate species diversity may be reduced (Harriman and Morrison, 1982) and therefore both trout numbers and numbers of birds might be expected to be lower in shaded streams. The amount of invertebrate fall into the stream from the surrounding land may be lower for coniferous forests than for either broadleaves or moorland, and the rate at which conifer needles are broken down is slower than for broadleaves. Microclimate is also affected by shading, so that temperatures are buffered by the enclosing tree canopy. Welsh stream-water in an afforested catchment in Wales averaged 2°C cooler in summer but 1°C warmer in winter than for moorland catchments (Roberts and James, 1972), though whether the overall effects on birds of these

Black-throated Diver, a suggested victim of the effects of acidification, but in north Scotland where they breed, soils are less susceptible and air pollution less than in Wales or southern Scotland where acidification has been studied. Fluctuating water levels and disturbance are probably bigger threats than acidification but recent RSPB research suggests their low productivity may not threaten the species' future in Britain.

differences would be beneficial or harmful is hard to say. It is now normal forestry policy to plant broadleaved trees along watercourses to minimize these harmful effects. In the young Fountain Forestry plantations of Caithness and Sutherland, thousands of young Birch have been planted in recent years along lochsides and streams. In Dalby Beck in North Yorkshire, four pairs of Dippers held territories on a one-mile stretch of water which was 90% shaded by Alder along its length.

Forest management may result in increases in the amounts of a variety of chemicals in fresh water. Phosphate applied as fertilizer can be lost to stream water (Harriman, 1978; Malcolm and Cuttle, 1983) and this can stimulate algal growth in standing waters (Richards, 1984). However, no specific effects on birds have been identified. The release of naturally occurring compounds as a result of forest operations may be more important than applied chemicals. Of more importance to humans than birds are the high levels of nitrogen from clearfell sites, which may exceed standards for human drinking water (Likens *et al.*, 1970). The changes in invertebrate composition can be complex as a result of eutrophication; some species gain and others suffer, as with many other examples of environmental pollution.

The effects of afforestation on the acidification of surface waters is complex but it is now clear that runoff from catchments afforested with conifers is more acidic than open moorland catchments. These effects are most extreme in areas of base-poor soils, such as the mudstones of the Llyn Brianne area near the head of the Tywi River in Wales. There, streams draining forest blocks are more acidic and have higher aluminium levels than those draining open moorland (Stoner *et al.*, 1984). Ploughing for planting results in the exposure and oxidation of soil minerals. It might be possible to minimize such effects by using minimal cultivation techniques such as scarifying in place of ploughing.

However, the more important effects occur in older forests. The major source of acidification is industrial pollution in the atmosphere. Most rain in Britain is acidic because of this, with the lowest levels of acidity in rain in northwest Scotland and the highest in eastern England. With the prevailing westerly winds the UK exports most of its acid rain to Europe. Conifer plantations exacerbate the effects of acid rain in several ways. The large surface areas of the conifer foliage result in 'scavenging', which increases the concentrations of sulphates and chlorides in runoff from plantations when compared with open moorland. Transpiration by the trees reduces the amount of water available to dilute the acidity, and water running over conifer trunks may pick up extra acidity from the surface of the bark. Thus the acidity of water which has passed through a forest canopy is slightly greater than that of the initial rain. This effect is greater for larch than for Sitka Spruce, but in deciduous woods some neutralization of the water takes place.

The effects of acidification have been studied in some detail and are strong enough to penetrate up the food chain to birds. One of the best studies is that of Dippers in Wales (Tyler, 1987). The RSPB surveyed Dippers on tributaries of the Welsh River Wye in 1982. Interesting differences in Dipper numbers were found between different tributaries which were partly explicable in terms of physical differences between the rivers. However, once physical differences were taken into account it was found that the more acidic rivers held fewer Dippers than the less acidic ones (Ormerod *et al.*, 1985b). The more acidic rivers also supported fewer caddis flies and mayflies, which were shown to be important food items for Dippers during the breeding season (Ormerod, 1985). This association of stream acidity with low Dipper numbers was also shown to hold true within particular rivers over time. The River Irfon in Powys increased in acidity after 1963, and by 1982 its Dipper population dropped from nine pairs in the upper reaches of the river to just two pairs. This change coincided with extensive conifer afforestation of the catchment (Ormerod *et al.*, 1985b, Ormerod and Edwards, 1985). The River Edw acted as a control site where acidity had increased only slightly and the drop in Dipper numbers had also been very minor. Large-scale surveys of rivers throughout mid- and north Wales in 1984 confirmed that Dippers tended to be absent on the more acidic streams, which held few caddis fly larvae and mayfly nymphs (Ormerod *et al.*, 1986a). Dippers were rarely present on acidic streams draining catchments which were more than 50% afforested. To add another piece to the jigsaw, studies of breeding success showed that Dippers on acidic streams nested later, with smaller clutches and broods, and lower chick growth rates, than Dippers on less acidic streams. The Dippers on the acidic waters never attempted to nest twice during the season, unlike the birds of less

A Timberjack skidder uses the hard bed of a stream to extract Douglas Fir logs; an operation that would not take place today. The potential harm to the watercourse from siltation has only been generally recognized by foresters in the last five years.

acidic streams, which often did. The effects of acidification on Dippers are large enough to measure, which makes one wonder what more subtle effects may be occurring which have not yet been noticed.

Similarly detailed studies on Grey Wagtails in Wales have not found such obvious effects of acidification on them, despite the fact that both species are invertebrate-feeders and can be seen feeding side by side in many Welsh upland streams. The reasons for this difference may lie in the fact that Grey Wagtails are much less dependent for their food on invertebrates which live in the stream bed itself. Grey Wagtails appear to feed more on insects which are derived from the neighbouring vegetation. This may explain why Grey Wagtails appeared to prefer streams which passed through either open moorland or broadleaved woodland rather than through coniferous forests. There was no evidence that Grey Wagtails avoided the more acidic streams nor that their breeding success was lower on acidic streams (Tyler, 1987).

The effects of high acidity on fish are well documented (Drakeford, 1979, 1982; Stoner *et al.*, 1984; Harriman and Morrison, 1982; Stoner and Gee, 1985; Newson, 1985) and some rivers in south Wales and Galloway have become too acidic to support fish life. These effects are likely to affect fish-eating birds such as Ospreys, Goosanders, Red-breasted Mergansers and Grey Herons in some areas, though such harmful effects have not so far been documented.

It appears unlikely that there is any short-term easy solution to the problems of acidification. In the long term, reductions in air pollution provide the answer. But remedial steps are worth exploring, of which the most direct and simple is liming of river catchments to counteract the acidity. Obviously this has to be done carefully since a rapid switch to high alkalinity could do as much harm as the high acidity to sensitive invertebrate populations. However, at present the problem seems to be finding a way of supplying lime which is responsive to water flow. Turnover of water can be rapid in upland streams and so lime can quickly be lost. What is needed is a medium which releases lime in proportion to runoff, so that the harmful effects of 'acid incidents' when the river is in spate can be counteracted. Experiments are under way with liming of strips along watercourses, liming the open moorland of the catchment and liming the whole forest from the air

CHEMICAL USE

Chemicals of various sorts are used in forestry and can find their way into watercourses, from where they can spread outside the forest. The aerial application of phosphate as fertilizer can often result in drift of fertilizer onto adjacent ground. However, it is the persistent herbicides and insecticides which are more alarming.

Few herbicides are used widely in forests, and many are applied very locally. 2,4,5-T was extensively used in British forests up to the late 1970s, and although not shown to be directly harmful to animals, does contain dioxin. Insecticides are most likely to be harmful to birds, because they are formulated to kill animals and may be ingested by insectivorous birds, and thence by raptors. Use of insecticides in forestry to date has been limited. For economic reasons, foresters in Britain have adopted the approach of only using insecticides in cases where otherwise trees would die, which meant that up until 1970 such use was restricted to small outbreaks of Pine Looper Moth attack. The British uplands thus avoided large-scale application of DDT, as happened in American forests.

Ironically the FC tried to use biological control against one of the early outbreaks of Pine Beauty Moth at Rimsdale in Sutherland in 1977. *Bacillus thuringiensis* was used but proved a failure. The next year the organophosphate insecticide Fenitrothion was used successfully. Between 1977 and 1987 Fenitrothion was used in seven years, and on up to 4860 ha/year. This is the only recent large-scale use of aerial insecticide in forestry in Britain, and has involved detailed monitoring of the effects on birds. No harmful effects on breeding success of small passerines could be detected after spraying with Fenitrothion (Crick and Spray, 1987). However, sublethal poisoning has been shown in Canada (Hamilton *et al.*, 1981). In contrast the effects of spraying on non-target invertebrates was severe, with noticeable kills of aquatic invertebrates (Morrison and Wells, 1981). Since Pine Beauty Moth attacks appear to be worst on the deep peats of Caithness and Sutherland and in large Lodgepole Pine plantations, the need for further use of this insecticide might be obviated by judicious site and species selection.

One poison is used routinely in forestry. This is the anticoagulant warfarin, used to kill Grey Squirrels. Although often regarded as a southern problem, Grey Squirrels are present in many upland forests, and as well as being the single most damaging agent to broadleaves, also damage conifers, particularly Scots Pine. Even when used in the legally compulsory special hoppers at the specified low concentrations for out-of-doors use, warfarin kills non-target species such as small rodents. It is not, however, a threat to birds because they are immune. Warfarin may not legally be used in Scotland and much of Wales because of the danger of killing the native Red Squirrel.

The use of dangerous chemicals in British forests does not appear to have resulted in major damage to wildlife. Compared with the levels used in agriculture, the amounts of insecticides used by foresters are very small. However, there is little room for complacency. Accidents do happen, as is demonstrated by the case of young trees which had been treated with Gamma-HCH (to guard against bark beetle) and were placed in a stream to keep them moist; this resulted in the death of fish for a considerable distance downstream. Accidental spillages of diesel into watercourses have also caused local problems. However, it is possible that other, more subtle, effects of chemical usage would have been overlooked. Whereas piles of dead fish are easy to notice, effects such as a long-term reduction in breeding success are much more difficult to identify without the sort of detailed and sensitive monitoring which is absent from the uplands. Another cause for concern over the use of pesticides in forests is that this is one of the few ways of introducing them into the uplands, where pesticide usage is generally low.

GAINS AND LOSSES

Some species seem to both gain and lose from afforestation and so it is appropriate to try to assess in which direction the net effect operates. Examples of such species include the Hen Harrier, Short-eared Owl and Black Grouse. All three species were ones which our analysis selected as meeting our criteria for being labelled as moorland birds. However, all three are species which also live in young forestry plantations. In many parts of the country, young plantations are the very best places to see these species.

The Hen Harrier benefits from the lack of persecution and increased food supply in the new forests, yet cannot persist in them after the canopy closes. What is the overall effect of afforestation on Hen Harriers? Although young plantations are only available at a particular site for a limited period of time Hen Harriers produce more young and may have greater adult survival in young plantations than they do on open moorland, where they are persecuted by some gamekeepers and have more difficulty in finding food. Forest-living Hen Harriers therefore contribute more representatives to the next generation than do moorland-living Hen Harriers, but for a shorter time.

A consequence of afforestation is fragmentation of what were hitherto large areas of open moorland. It is possible that this might have adverse effects on some moorland species by reducing the amount of dispersal from one

moorland unit to another. If moorland areas become very small and isolated, then localized disasters (e.g. fires) might cause the extinction of populations of some species of birds on such sites. In cases where the site has been cut off from other areas of similar habitat, recolonization is impossible, whereas in large, continuous stretches of habitat, dispersal from nearby areas will quickly lead to the recolonization of any site. We do not know of any cases where such events have actually occurred but would imagine that they are possible in very highly forested areas such as Galloway and Kielder, where moorland areas have been enclosed by forestry. A factor which will probably militate against such effects is that many of the birds of moorland are not resident on moorlands throughout the year. The other side of the coin is that fragmentation of the open upland habitat means that the distances between areas of woodland decrease. This is likely to aid the spread of woodland species. One possible example is the Crested Tit, a species which has never spread into apparently suitable areas of woodland on Deeside. The increased areas of conifer plantations in the area, which Crested Tits have colonized, may allow this species to spread to new areas which it would otherwise not have been able to reach. The range increase of the Hen Harrier may also provide evidence for such an effect. The Hen Harriers breeding in Wales are mostly moorland birds, but the southward spread of the species from Scotland seems to have been greatly aided by the large areas of young conifer plantations which allowed them to spread first to southern Scotland and then to northern England, from where the Welsh birds have presumably come. It seems likely that moorland-breeding Welsh Hen Harriers may have the young forests of the north to thank for their existence. This role in assisting species to extend their ranges into suitable areas which they otherwise would not have reached is an important one. In the case of the Hen Harrier, it will be interesting to see whether or not the recently established populations in southern Scotland, northern England and Wales will persist when the forests which have probably helped their establishment have grown too high for them to use. Will there be a permanent legacy of the expansion of upland forestry in the past 20 years or not?

We have argued here that the effects of forestry are complex. This means that in any attempt to limit the undeniably harmful effects of afforestation on moorland birds, it will be necessary to take into account the complexities of the whole process. It is too simple to say that forestry is bad for moorland birds and therefore we wish to limit the amount of forestry which goes on, because it is not just the amount of forestry which determines the size of the problem. The location and the pattern of planting are also likely to be important. In very simple terms it would be possible to rearrange the existing forestry in Britain so that it had a very much smaller or a very much greater effect on Britain's moorland birds. Thus, although the questions of how much forestry should go on in Britain will always be important, we would like to see more thought given to where and how forestry should take place.

WHICH SPECIES ARE WORST AFFECTED BY FORESTRY?

We defined (Chapter 5) the species which are most restricted to moorlands in an attempt to isolate those which are potentially most affected by afforestation. We have also discussed the types of effect which afforestation is likely to have on the different species and the few studies which have looked at the effects of afforestation on birds directly (this chapter). Are we then able to say which British species are most at risk from afforestation? The answer must be that we are not, because the information about the effects of forestry on moorland birds is small. For few of the moorland species which we have considered have detailed studies been made in any part of their range, let alone detailed studies in comparable areas which either have or lack afforestation. In particular there is little information on the indirect effects such as might be mediated through water quality or increased predation. However, the level of public discussion about the effects of afforestation on birds has seldom been restrained by this lack of hard facts, and it is important to try to identify those British species most at risk from afforestation with the limited information available. Some intelligent guesswork is required, and these are our guesses.

How should we try to identify those species which are most at risk from afforestation? There are two main criteria. First, forestry must be known or expected to harm these species. We have already seen that few of the smaller moorland birds can survive for long within afforested areas, but the effects on waterbirds or on species with large home ranges are much less certain. A useful indicator of whether further afforestation will be harmful to a species is how it has responded to the afforestation which has already occurred. It seems reasonable to assume that species which have undergone considerable expansions in numbers and range within afforested areas in the past 30 years cannot be very seriously affected by afforestation. Second, the species must live in places which are likely to be afforested. Species can be disqualified on these grounds if they inhabit either parts of the country or habitats which have low risks of being planted with trees. We will use this second criterion as our first sieve of the moorland species before considering which of the remaining species are likely to suffer most from the afforestation which might occur in their ranges.

The first species which we can eliminate from our original list of 31 are those whose ranges and habitats lie outside the areas which are likely to be afforested. These comprise those species with truly montane habitats and also those for whom the majority of their British populations are restricted to Orkney and Shetland. In the first category come Dotterels, Ptarmigans and Snow Buntings, whose habitats are safe under present forestry practices, and there is little sign that the changes in technology and economics which have driven the growth of forestry in Britain since 1919 are about to alter to such an extent that this will become untrue. For these three species it is unlikely that forestry will have indirect effects that are so far-reaching that they will creep sufficiently up the hills to affect them. These then are safe (a rash prediction?). Species which are largely restricted to Orkney and Shetland and the most inhospitable Outer Isles, where forestry has not spread and is unlikely to because of the harsh conditions, comprise Red-throated Divers, Whimbrels, Great Skuas and Arctic Skuas. There is little to suggest that the British populations of these seven species have been, or under continuing conditions will be, more than trivially harmed by afforestation.

The 24 species with which we are left can all be regarded as being vulnerable to the effects of afforestation to the extent that they live in areas where trees might be planted. Of these we would regard Eiders and Mergansers as having only a low vulnerability, because of their largely coastal distributions and the fact that their numbers have increased during the past 30 years coincident with a great expansion in the area of afforested land which has largely occurred within their initial geographical range. The Twite is rather an unknown quantity, but can be removed from the list of particularly threatened species on the grounds that much of its range is coastal, and large proportions of the total population occur in areas where afforestation is unlikely to occur; Shetland, Orkney and the Pennines. We have not heard any claims that the Twite is in danger from afforestation. For Black-headed Gulls and Common Gulls unknown but fairly large proportions of their breeding populations are coastal and in sites which, although they may be under threat, are not, in general, threatened by afforestation. The Peregrine is not a species which, as yet, has been shown, or even suggested, to be particularly harmed by afforestation, and its population has greatly increased during the last two decades coincident with large-scale afforestation.

Hen Harriers, Black Grouse and Short-eared Owls are more problematical. They may benefit from afforestation in the future, as they have probably done in the past, by spreading into new areas when they are planted. These benefits will be highest if the newly planted areas were predominantly grassland rather than heather moorland before afforestation. It is difficult to predict how the population sizes of these species will compare in 40 years time with now, when all of the young plantations which are now favourable for these species will have grown to maturity and not hold them any more. It is possible that their expansions in ranges which was aided by afforestation will have led to a sufficient spread in distribution that the new areas colonised will compensate for the old areas which are lost under trees; that will be the real test and remains to be seen, but at present these species are gaining. For Hen Harriers the data do not exist to prove it, but we would guess that the activities of gamekeepers are still a greater harmful influence on the species than is the afforestation of the uplands.

The Buzzard is unlikely to be greatly affected by afforestation if the data from Wales are representative of the rest of the country. This seems to be a species which is likely to benefit from afforestation once the trees are tall enough for nesting in areas which previously held few nesting sites. Afforestation may, in some areas, favour Buzzards by reducing the amount of persecution by gamekeepers.

The Merlin is declining in all parts of its range, in afforested and unafforested areas alike. Afforestation is one of the factors which may be contributing to this decline. However, there would be considerable scope for an increase in Merlin numbers, despite the projected increase in afforestation, if only the other unknown factors causing the decline were known and dealt with.

The Wood Sandpiper only has a tenuous hold in Britain and its numbers are certainly not limited by the availability of suitable habitat at the moment. If this species were to become established in Britain in large numbers, then its potential area of colonization would be curtailed by afforestation in northern Scotland, but it remains to be seen whether this species really will develop an increasing, spreading population or whether it will remain an interesting but very temporary addition to the British breeding list.

The Goosander has not, as far as we can tell from the published records, been harmed by afforestation. We have come across no suggestions that it has been affected either for good or ill. The Goosander population has increased considerably in numbers, even though its stronghold includes southern Scotland and northern England, where afforestation rarely affects the best habitats for Goosanders, maturely wooded river valleys, and there have been no noticable effects on this species acting through water quality and food supply.

So far we have excluded 20 of our 31 species. The remaining 11 are those which we would expect to be the most affected by the continued growth of afforestation in Britain; these are Golden Eagles, Black-throated Divers, Wigeons, Common Scoters, Greenshanks, Golden Plovers, Red Grouse, Ring Ousels, Greylag Geese, Dunlins and Ravens. Of these, there are four waterbirds for which no link between afforestation and declines in numbers has been proven; Black-throated Divers, Greylag Geese, Wigeons and Common Scoters. Of the others, it is clear that loss of upland habitat under trees will reduce the numbers of Dunlins, Greenshanks, Golden Plovers, Red Grouse, Ring Ousels, and Ravens. For Golden Eagles the evidence linking forestry to decline in numbers is less clear-cut, but is still enough to establish serious grounds for concern. Four of the species, Greenshanks, Golden Eagles, Black-throated Divers and Greylag Geese (if feral Greylag populations are excluded) have very similar British ranges, being largely confined to the northwest Highlands, where large-scale afforestation has been spreading in recent years. These species can be considered as being at particular risk from afforestation, because new afforestation is quite likely to be concentrated in the areas which are their strongholds. They are species for which a relatively small part of the forestry enterprise can have major impacts. Afforestation south of the Great

Glen would have relatively little (but not nil) impact. For Common Scoter, the situation is even more extreme. This has a very restricted range, with most of the British population occurring in the Caithness and Sutherland Flow Country. This group of five species could, in an ideal world, be easily protected from the effects of afforestation by preventing any new afforestation in the northwest Highlands. Their ranges overlap sufficiently, and are restricted enough to make this possible. The other species, Ring Ousels, Golden Plovers, Dunlins, Wigeons and Ravens, are widespread in Britain, and the effects of afforestation on them are likely to be more gradual, less noticeable and much more difficult to counter. Their fate may be to have their populations whittled away bit by bit as each new area to be planted removes another few pairs.

If our guesses about the effects of forestry on British moorland birds are correct, then we have been able to narrow down the species most likely to be affected to a shortlist of 11. This list could provide a basis for guiding conservation action towards the areas of highest priority. For example, the importance of the northwest Highlands for many of the potentially affected species is highlighted. This could be one factor in helping to shape a reserve acquisition or site protection plan for protecting moorland birds. In this area many of the potentially affected species occur together.

Signs that British conservationists are moving towards a consensus on the species of highest priority are now emerging. The publication of the *Red Data for Birds in Britain* (Nature Conservancy Council and Royal Society for the Protection of Birds, in press; Bibby, Housden, Porter and Thomas, 1989) should mark the beginning of a new stage of bird conservation; based for the first time, on an objective assessment of conservation priority. The *Red Data for Birds in Britain* lists 119 species of bird which are of high conservation importance in Britain. Their inclusion in the book was on the basis of several clearly defined attributes. Species were included on grounds of international importance if their populations (breeding or wintering) were more than 20% (an arbitrary figure) of the northwest European population. Rare breeding species were included if their populations were less than 300 pairs in Britain. Species whose ranges (breeding or wintering) were so localized that more than half of the population was found in less than 10 sites were also included, as were breeding species whose populations had at least halved since 1960. Nine species were included as special cases, if it was widely thought that if only better information were available then they would probably fit one or more of the other categories.

Excluding 17 species which breed irregularly or in very small numbers in Britain, 80 British breeding species meet at least one of these criteria, and only one species, the Roseate Tern, meets all of them. Eighteen of our moorland species are included in this list (Red-throated Divers, Black-throated Divers, Common Scoters, Hen Harriers, Golden Eagles, Merlins, Peregrines, Red Grouse, Black Grouse, Dotterels, Golden Plovers, Whimbrels, Greenshanks, Wood Sandpipers, Great Skuas, Short-eared Owls, Twites and Snow Buntings), thus supporting our contention (Chapter 5) that the British uplands are of national importance for birds. Twelve species which are associated with commercial conifer plantations are also included in the list; Hen Harriers,

Goshawks, Black Grouse, Capercaillie, Barn Owls, Short-eared Owls, Night-jars, Woodlarks, Redwings, Firecrests, Crested Tits and Scottish Crossbills. Considering the much greater extent of moorland than plantation in Britain, this represents a surprisingly even split between the two habitats.

It is interesting to note that all nine of the species which gain entry solely in the Special Category are ones dealt with in this book: Hen Harriers, Merlins, Black Grouse, Dotterels, Golden Plovers, Whimbrels, Greenshanks, Barn Owls and Nightjars. These are species for which the evidence is not strong enough to include them in any of the other categories. This certainly confirms the view that the birds using coniferous plantations and moorlands are under-studied. If all of these were excluded from *Red Data for Birds in Britain*, in the event of further knowledge coming to light, for example, the difference between the contributions of moorland and conifer plantations to the list of birds would be even more similar.

This list should not be used too naively to compare the conservation value of different habitats. After all, in areas where Golden Plovers will be harmed by afforestation it is unlikely that Firecrests will benefit greatly. However, two useful lessons can be drawn from the analysis. First, it supports our contention that interesting and naturally rare species gain, as well as lose, from afforestation, so that the conservation arguments about further afforestation are bound to be complicated ones. Second, there are important moorland species whose populations in Britain are under threat from afforestation. Of the 11 out of 31 of our moorland species which we consider to be most threatened by afforestation, six of these are included in *Red Data for Birds in Britain*; Black-throated Divers, Common Scoters, Golden Eagles, Red Grouse, Golden Plovers and Greenshanks.

However, examining the number of species of international importance in each of the two categories is also instructive. Six of the species defined by us as moorland species have British populations of international significance; Red-throated Diver, Golden Eagle, Peregrine, Red Grouse, Great Skua and Twite. Only one species found in the forests, Scottish Crossbill, is of international importance, and this status lies vulnerable to taxonomic changes. What this analysis shows is that there are more species of international importance found in moorland habitats than in conifer forests. This adds further weight to the arguments of conservationists that the spread of forestry is doing real harm to important bird populations which are not compensated by the gains which accrue from the new forests.

There is no doubt that afforestation has done considerable damage to important bird populations in some parts of Britain. In the next chapter we take four areas which have experienced afforestation and look at the gains.

© PS

CHAPTER 7

Case studies

In this chapter we look at four areas where there has been extensive afforestation. They have not been selected at random: all have had planting above the national average, meaning that changes to birds are likely to be emphasized. More information is available for them than on average, and three out of the four cases are well known to us personally. Two of the four areas are older than the average British commercial forest and might be expected to provide a better idea of how the vast areas of middle-aged forest will develop and how the new ideas for the management of second rotation forests are likely to work. These case studies are not meant to describe the complete ornithology of each area, but rather aim to pick out relevant information on the way bird distribution has changed with afforestation.

The selected areas are discussed from south to north, which is in declining average age of forest. Thetford Forest in East Anglia's Breckland is rather atypical of the new forests; barely above sea level, it cannot be described as an 'upland' forest but does qualify as a 'new' forest on the poverty of its soils and the scale of its planting. It was amongst the first of the big FC forests to be created. Planted on heathland of exceptional ornithological and botanical interest, it has more rare breeding birds than any other planted forest, and a unique record of systematic study.

The forests of the North Yorks Moors are more typical. There are many areas where early planting took place on relatively good valley slope land prior to 1939, followed by extensive post-1945 plantings on poorer soils. Species change, in this case from early Douglas Fir to later Lodgepole Pine, and so does the forest's character, from relatively varied and intimate early plantings to far more extensive later schemes.

Galloway is perhaps the best example of the enormous post-1945 plantings. It is also the most heavily afforested region in Britain, and as good a representative as any of large, upland spruce forests. It epitomizes the style of forestry of the technological era, also being greatly affected by the two major problems of windthrow and deer. Although harvesting has begun, in contrast to the first two cases a minority of areas have so far been affected and there are large areas of unbroken closed canopy forest.

Caithness and Sutherland is the only region of Britain where new afforestation continues to dominate forestry, with very little harvesting yet taking place. It is also the focus of conservation concern over the effects of afforestation, particularly the low-lying peatlands of the 'Flow Country'. In this extraordinarily wild area, tree-planting technology and bird conservationists are in direct collision.

THETFORD FOREST

Thetford Forest is in southern England and rises to less than 60 m above sea level, but in many other ways it is a typical new forest. Scots and Corsican Pine grow on impoverished soils, which restricted the development of agriculture on the heaths of the Norfolk-Suffolk border known as the 'Breckland'. This area of sand deposited over chalk and boulder clay remained a wilderness even after the vast reed beds of the Fens had been drained and converted to productive farmland. The human population of the Breckland was always sparse and the people poor (Clarke, 1925). Providing work was an important reason for the establishment of one of the FC's first forests here. Planting began in 1922, and by 1939 Thetford was the largest of the new forests. Planting of bare ground totalling just under 20 000 ha was complete by 1960, and most of the forest had closed canopy.

The young pine were planted three or four feet apart in shallow plough furrows. The biggest problems were Rabbits and late radiant frosts. Returning to Kings Forest after its use by the army during the war, Head Forester J.J. Smith organized the catching of 77 079 Rabbits in five years from 6000 acres of thicket-stage pine (Smith, 1955). Bare, rather than grass-covered, soil was found to reduce the frost problem, but frost limits the establishment of conifers other than pine. Although early growth was variable (Lack, 1933, 1939), by mid-rotation differences were less obvious: the establishment of this forest was a notable early success for the FC. There were no natural obstacles on the flat heathland and it was possible to create a forest of exceptional regularity. By 1960 the straight roads of the forest grid led for miles down quiet avenues of tall pine. The only open ground left consisted of extra-wide fire-break rides and the last few areas of new planting: there were no bogs or rocky outcrops to break the forest here. Already producing timber in the 1960s, in 1983 Thetford was the first FC forest to pass 200 000 m^3 timber production and 400 ha restocking per annum.

A bird fauna of conifer high forest has developed. In addition to widely distributed and characteristic Coal Tits, Goldcrests, Wrens and Chaffinches a number of rarer species are well established. The Pine belts of the Breckland were one of the few places that Crossbills had bred continuously in Britain: in the 1930s places like Mayday Farm on the Bury St Edmunds-Brandon road were famous with egg collectors. In the vast expanse of mature plantations Common Crossbills became one of the commoner birds of conifers. In Thetford forest they do not seem to be as cyclical in numbers as they are in the Border spruce forests. Pines tend to cone more regularly and in smaller quantities (Staines *et al.*, 1987) than the spruce cone crops that support high population peaks. The Siskin, the other conifer species that has expanded its range most in the new forests, also breeds in Thetford but is not as numerous as in many upland spruce forests.

Raptors are still not numerous: persecution by Pheasant keepers had eliminated most species by the time the forest was planted, and the impact of DDT was particularly severe in East Anglia. The 1970 Norfolk bird report records no proven successful breeding of Sparrowhawks in the county, and although now frequent in Thetford they are still noticeably less common than in more northern upland forests. In 1975 a pair of Goshawks, the female sporting a falconer's bell on its leg, bred in the central block of Thetford Forest. The nest was kept secret and the nesting area protected from disturbance by FC staff, and young were fledged in every year from 1975 to 1981, except 1976 when the nest holding an exceptional six young fell out of the tree and all were killed (B. Pleasance, personal communication). Most years produced three fledged young. Inevitably, the secret got out and today these must be the best-known Goshawks in Britain. Few parts of the Thetford Forest are more than two miles from a public road, and the large Goshawk nest is far easier to find in the open canopy of a pine forest than in dense, spruce foliage of more northerly breeding forests. A battle of wits ensued between nest robbers stealing eggs or chicks for sale for falconry and the FC, RSPB and police, following the first nest robbery in 1982, watched by an audience of thousands of bird-watchers. Most nests were robbed and failed, but in 1986 a nest robber was caught in the act and prosecuted and a heavily protected nest succeeded in 1987.

Less obvious than the nest robbing, but equally disturbing, is that the large number of young produced during the 1970s has not led to an increasing breeding population. It is likely that the tendency of young Goshawks to hunt at Pheasant pens has resulted in high losses following dispersal from the breeding territory. Despite the dramatic change in attitudes towards birds of prey, it only takes a few human predators to wipe out large, rare, wide-ranging birds like Goshawks, and it is doubtful whether the species can become firmly established at Thetford at the moment. The story of the Thetford Goshawks underlines the real importance of the seclusion of large, upland forests to raptors. This species has benefited from the initial indifference of foresters, which has increasingly turned towards active protection.

Thetford is unusual amongst the older forests in having Long-eared Owls throughout the forest. Tawny Owls are reasonably well distributed though not numerous, and tend to be birds of the forest edge. Elsewhere it is Long-eared Owls that are found only around the forest edge. Interspecific competition (Petty, 1979) has been suggested as the reason for this but in Thetford both

Long-eared Owl threat display; this young bird fell out of its nest and was found hanging upside down 2 m from the ground in the middle of the main block of Thetford forest. By the time a ladder was fetched it had disappeared – presumably back up the tree. Quite how Long-eared Owls manage to live throughout Thetford but not other conifer forests is a mystery that needs solving.

species of owl occur in the same compartment (R. Hoblyn, personal communication). There is no shortage of suitable nest platforms for Long-eared Owls, which can use squirrel dreys, old Carrion Crow nests and Wood Pigeon platforms. Tawny Owls have been found nesting on the ground at Thetford, which is an exception to the rule of owl distribution in the new forests, and the reason for the rarity of Long-eared Owls in the middle of most mature forests remains an interesting mystery which may be investigated one day.

A number of other rare or scarce species breed in the mature forest. The most unusual is the startling Golden Pheasant, a feral species with the largest of its half dozen or so British populations in Thetford forest. The male is a rhapsody in scarlet and gold and should be easy to see, but Golden Pheasants are actually very elusive, skulking in the dense Snowberry thickets planted by keepers for common Pheasants. They are most numerous to the south of Santon Downham in the Highlodge and Brandon Park blocks, where up to 100 cocks have been seen drinking together (R. Hoblyn, personal communication).

A few Firecrests breed in mixed woodland of mature Douglas Fir and broadleaves near Santon Downham. Although Douglas Fir can grow well on the sandy soils, young trees are very vulnerable to frost and almost impossible to establish. Thetford has, however, an unusually large area of broadleaves. Although old trees of any species are scarce, the conifer plantings of the 1920s and 1930s were ringed by belts of Beech and Oak. Hawfinches, Lesser-spotted Woodpeckers and Nightingales are all scarce birds dependent, in Thetford, on the existence of broadleaves.

That such a diverse assemblage of birds, including so many scarce or rare species, has built up so soon after the establishment of high forest conditions is unexpected but has been mirrored in other new forests. The Breckland forests hold many interesting birds and are now quite well watched by bird-watchers. Species such as Crossbills, Siskins and Goshawks cannot so easily be found in other parts of East Anglia but all are increasing in numbers nationally. The fate and future of the original inhabitants of the Breckland heaths is exactly the opposite: far more specialized in their requirements, Stone Curlews, Night-jars, Woodlarks and Red-backed Shrikes have declined throughout the 20th century. Red-backed Shrikes are on the verge of extinction as a British breeding bird, Stone Curlews and Woodlarks are endangered, and the continued decline of Nightjars is serious enough for eventual extinction to be a real threat. The effect of afforestation on these species is of central importance in assessing its overall impact on the conservation value of the Breckland for birds.

Man had relatively little impact on the Breckland up to 1920. Although most of the area had been farmed in the 13th, 14th and 15th centuries (Clarke, 1925), it reverted to wilderness after the Black Death. Arable agriculture

Typical view through Thetford forest today: 20 years ago this ride would have been lined with wall to wall pole-stage pine. Now, retained older trees are increasingly important as more and more is felled – not just for today's Goshawks but also as tomorrow's clearfells for Woodlarks and Nightjars. The tower is a high seat for shooting the increasing numbers of Roe Deer.

ebbed and flowed with the economic fortunes of the times in and around the poorly defined area of 80 000–100 000 ha of Breckland. Scots Pine wind-breaks planted from the late 18th century onwards to check the blowing sand dunes that occasionally blocked rivers and overwhelmed villages were the first major permanent change to the area. They were the Common Crossbill's toehold in the area through the 1920s and 1930s. The peculiar lithology of Breckland separates it from other English heaths: sand overlays chalk so closely that the soil can change from acid to alkaline over a hundred yards where the chalk outcrops are near the surface. The area's flora and prehistory are as important as its birds, making it of exceptional conservation interest (Ratcliffe, 1977). Grass heath is the commonest vegetation type and there are also large areas of bracken and, locally, heather and sand-dune sedge are dominant. Shifting cultivation and the burrowing of millions of Rabbits disturbed the soil and maintained a sparse vegetation cover with a lot of bare sand.

By the time reliable records of birds begin, the rising popularity of shooting for sport had already taken its toll. By the end of the 19th century Breckland was split between large estates, and its main use was for Pheasant shooting. Britain's last native Great Bustard was killed near Swaffham in 1838. Wild, sparsely populated country, with vast numbers of Rabbits should have been a paradise for Buzzards, harriers and Red Kites, but wild birds of prey are scarcely mentioned by Dutt (1906) and Clarke (1925). Nethersole-Thompson and Nethersole-Thompson (1986) mentions occasional Montagu's Harriers on the river fens. We assume that gamekeepers eliminated whatever large birds of prey were once there. The pine belts checked the moving sand but the heaths in between remained pretty much unchanged. Birds like Stone Curlews that were neither predator nor prey to the shooters continued to thrive (Clarke, 1925).

The afforestation of the central Breckland from 1922 onwards was, as far as it is possible to tell, by far the most significant habitat change to the area since the last glaciation. We know rather more about this forest both before and after planting, thanks to Clarke's vivid descriptions and Lack's work on the newly planted forest (Lack, 1933, 1939). Clarke considered the original 8100 ha centred on Santon Downham to be some of the poorest land in Breckland, which may well have meant it was some of the best for birds. He found 30 pairs of Stone Curlews on Santon Downham Heath. It was all planted, and hearing a Stone Curlew call on a restock in May 1987 was a rare event. Elimination of Rabbits and tree planting resulted in the immediate loss of birds of the barest heathland, Stone Curlews, Ringed Plovers, Lapwings, and Wheatears. Other open-land species, Skylarks, Meadow Pipits and Whinchats survived in the plantations till they were eight or nine years old (Lack, 1933, 1939), but by 15 years, scrub species like Whitethroats and Willow Warblers were replaced by birds of the forest, Chaffinches, Goldcrests and Coal Tits (Lack and Lack, 1951). Between 1960 and 1970 the proportion of younger crops was very low. Only 300 ha were planted or restocked in 1962–1967, less than 2% of the forest area and less than a single year's restocking in the mid-1980s. Slightly over 20% of the original Breckland had become unsuitable for birds of open land.

From 1922 to 1945 afforestation remained the only significant agent of habitat change. However, changes in agricultural techniques and the near-extinction of Rabbits by myxomatosis set in motion a chain of events that has led to the forest being the main refuge for several of the most important birds of the Breckland heaths. High fertilizer inputs mean that the easily worked, light land can produce high yields every year, and about double the area of the forest is now intensively farmed arable land. Although the ground is bare in spring the crop grows rapidly, and late rolling and other cultural treatments destroy the nests of Stone Curlews, Lapwings and Ringed Plovers on cereal fields. They may survive on the rarer root crop fields. There are no holes on farmed fields for Wheatears or Stock Doves to nest. Several large airfields established during the Second World War are still in use and some heathland has been built on. Less than 20% of the original area remains. The largest chunk is the 6070 ha of the Army's Stanford practical training area.

However, the heaths have changed too (Webb, 1986). Grazing by domestic stock has declined with the move towards specialist arable farming over most of East Anglia, but it was myxomatosis, which reached the area in 1954, that did most damage. The scratching and digging of Rabbits maintained the sort of bare heathland that is now exceptionally rare; the Norfolk Naturalist's Trust reserve at Weeting Heath, where Rabbits are fenced in, is the best example. Many other heaths have become overgrown or the vegetation structure has changed. On the Stanford training area, grazing by sheep keeps the grass low, but it is a tight sward with little bare ground. Heathland birds have declined more than can be explained by the simple and obvious loss of habitat caused by agriculture or forestry. Woodlarks in particular could have been lost to the area by now had they not moved on to the open ground generated by clearfelling of the forest; all Breckland Woodlarks now occur in the forest.

A few compartments were felled early following windthrow but most survived at least 50 years, deep-rooted in the sand in the least windy part of England. Many reached the optimum economic felling age of between 55 and 65 years. The main reason for premature felling was fungal disease, and Thetford has been more affected than any other of the new forests. Disease has been important to birds because large areas of open ground have been created earlier than expected. Two species of fungus are responsible. *Heterobasidium annosum* destroys the roots and spreads into the stem to rot the timber. It survives in roots left in the ground after felling from which it reinfects the young restocking. The problem is most serious on alkaline and old agricultural sites, and, following severe losses in the first restocking, stumps are uprooted and piled in rows to prevent reinfection of the most vulnerable sites. In Britain, the operation is unique to the East Anglian pine forests. It means that two distinctly different types of restock habitats occur: conventional sites with open areas extending to tens of hectares and destumped sites with 3-m-high rows of tangled stumps breaking the open ground into 70–100 m broad strips and providing perches and nesting sites for birds. The Resin Top Fungus, *Peridermium*, which attacks the tops of Scots Pine, arrived later as a serious problem and exacerbated already low stocking caused by *Heterobasidium*. By 1983 the rate at which many Scots Pine stands were gaining volume barely exceeded losses by death from fungal disease, and felling accelerated to harvest the timber before more was lost. Although most clearfells are 10–30 ha, in some places such as Wangford Warren, older restocking approaching canopy closure has aggregated with younger restocks and fresh felling into uninterrupted open areas of up to 200 ha.

Another factor peculiar to East Anglia which affects the character of the restocks as a bird habitat is the use of a mechanical furrow planter. For this to operate, larger branches must be cleared or chopped up, so Thetford restocks do not have the dense cover of branchwood normal on upland spruce felling sites. Felling and restocking disturb whatever ground vegetation there was under the mature pine trees, but vegetation becomes re-established very quickly: completely bare soil on a de-stumped area in March can have 100% grass cover by the end of May. Ploughing does not give the two to three years freedom from competition usual on upland sites. The variability of vegetation type on restocks reflects the variability of the soil: at Wangford Warren a single 20-ha compartment five years after restocking varied from bare ground, through low-growing sand sedge, dense growth of the grass *Calamagrostis epigejos*, to a small area of 2-m-high Birch regeneration.

Although quickly achieving a complete cover of the ground, grass growth is frequently low and not very dense. There are also more general vegetation trends: growth tends to be sparsest, mainly short grass and bracken, in the south of the forest. Clearfells in the northern part of the forest may be covered by dense Bramble, established under the first rotation pine crop, and grass growth is also ranker in the north. Protection of the trees from competing vegetation is particularly important here because of frost and because restocking plants are normally tiny seedling Corsican Pine rooted in a small paper pot full of peat. Atrazine is a soil-acting herbicide for killing grass and is applied to most sites in a band over the ploughed furrow with the young trees (Sale *et al.*, 1986). Because it suppresses the germination of colonizing plants, it leaves a band of completely bare soil. Glyphosate is used to kill Bracken and Bramble and is usually applied to the whole area rather than just bands. It does not tend to suppress vegetation as completely as Atrazine. The comparison in upland forests of restocks that are and are not grazed by deer shows the importance for birds of species and structure of recolonizing vegetation and even fine variations may make a big difference to former heathland birds' use of the Breckland restocks.

The restocks in the Breckland are uniquely important. They are of conservation importance because of the unexpectedly large numbers of heathland birds that colonize them and they are of scientific importance because they are the only clearfelled areas in the new forests for which there is a quantified assessment of the birds that were there at the time of the original planting. We owe this to the pioneering work of David and Elizabeth Lack who used a transect method to study the effect of tree succession on birds between 1931 and 1951 (Lack, 1933, 1939; Lack and Lack, 1951). Fifty years later the trees are being felled and the youngest restocking is growing towards thicket stage. A point count study (Leslie, Dolton and Avery, in preparation) between 1984 and 1987 for exactly the same crop stages means that we can compare the birds of similar sites over the span of a whole rotation. The methods used have changed and it is not possible to pin down the precise areas Lack covered from the records he left. Felling had not started by 1984 in some of the blocks he worked, such as Parsonage Heath, and breaking up even-aged blocks means that a transect approach would have been less suitable today than when there were large, homogeneous areas to walk. However, with plots in, for example, the Wangford Warren and West Tofts blocks, it is almost certain that a few of the places Lack actually worked on were surveyed again, perhaps even at the same growth stage.

Leslie *et al.* encountered 41 passerine species and 13 non-passerines on Breckland restocks. Most of them were common scrub species, with Chaffinches, Tree Pipits, Willow Warblers and Whitethroats being particularly common. Skylarks are ubiquitous on young restocks, as on both heathland and new planting, but Meadow Pipits, also numerous breeding birds on new planting, do not usually breed on restocks; although they are common in winter there has been only one recent breeding record, in 1984 (R. Hoblyn, personal communication). Tree Pipits are possibly the most numerous birds on restocks, rapidly colonizing fresh felling sites and becoming more numerous as the trees age.

Birds typical of broad-leaved woodland were rare on the restocks; Marsh Tits were absent and Chiffchaffs, Redstarts and Garden Warblers were rare. Some uncommon species which were seen were doubtless just passing through (Hen Harriers, Hawfinchs) but other interesting species held territories or displayed on restocks (Whinchat, Stonechat, Wheatear).

Several species of the original heathland occur on restocks, but some do not, and there are some new birds, too. Of the three species characteristic of the bare heaths, Stone Curlews, Lapwings and Wheatears, only the last breeds on restocks and then rarely. Stone Curlews, by far the rarest nationally, have received most attention. Stone Curlews went on nesting on planted compartments for some time after they were planted, and for longer on wide fire break rides (Lack, 1933), Nethersole-Thompson, and Nethersole-Thompson (1986) quote a case from 1925 when three nests were found in a conifer plantation where the trees had grown to a height of 15–30 cm. Such behaviour from a bird of open country was unexpected, and has led to a belief that they are nesting on restocks, but there has not been a case of Stone Curlews breeding on a restock in the 1980s. They persisted in the detached southern block of Kings Forest, which is one of the youngest parts of Thetford, into the 1960s. The close proximity of Berners Heath may have contributed to their continuing for so long there. Lapwings occur on restocks more than occasionally (R. Hoblyn, personal communication), but were not found on any of the study areas of Leslie *et al.* Wheatear, which have become less common in Breckland generally because of the lack of Rabbit burrows for nesting, breed only very rarely on restocks (R. Hoblyn, personal communication) and were seen only occasionally by Leslie *et al.*

Red-legged Partridges are often found on the youngest restocks, even deep in the middle of the forest. Another slightly unexpected change is the frequency of Curlews on restocks, which were not recorded at all by Lack. In

1987 the restock population was about 25–35 pairs, which makes it a significant part of the tiny inland lowland population, and the most important inland population in East Anglia (Smith, 1983).

The Woodlark is one of the rarer birds of the clearfells. Not well known to most people, Woodlarks seem like Skylarks but in fact are very different; they are more thickset, with a very short tail which is obvious in flight. Their song is superb, far more rounded and melodious than that of Skylarks, and possibly the most beautiful of any British bird. Disturbing a Woodlark on a fine morning in early spring can launch it into the sky, where it hangs as a tiny musical dot for as long as half an hour. Robertson (1954) paints a superb pen picture of this bird.

Woodlarks do not ever seem to have been common birds in Breckland: Clarke (1925) noted only that they were more common than elsewhere in East Anglia, and Lack (1933, 1939) recorded only odd birds; two on 15 sites studied in 1931. On this evidence Woodlarks would actually seem to be more numerous on restocks today than on the open heathland and new planting of the 1920s and 1930s. The range and population of Woodlarks in Britain has fluctuated over the past century (Sharrock, 1976) but there has been a continuous decline since a peak in the 1950s, and by 1983 Sitters (1986) estimated a total population of only 210–230 pairs. This makes it one of Britain's rarest passerines, rarer at the moment than Dartford Warblers or Bearded Tits for example. There were 45 pairs in the Breckland in 1983 and all but two pairs were on forest restocks. Most pairs of Woodlarks in Breckland have probably been located by thorough survey since 1975 (Figure 7.1, after R. A. Hoblyn in Sitters (1986) and personal communication for 1985–1988). Between 1971 and 1974, surveying was less comprehensive and heathland birds in particular might have been missed.

Up to 1976 most of the Woodlarks associated with the forest were on the marginal land of the wide, cultivated fire-break along the Brandon to Thetford railway track. The heathland population was very small, the highest count being 11 pairs in 1976. However, from about 1974 onwards, Woodlark numbers started to increase on restocks, and in recent years have peaked at just under 50 pairs (Figure 7.1). The small number on the marginal land disappeared as the fire-break was planted with pine once the danger from steam trains ceased, but the heathland population had also died out by 1984. In contrast to other British Woodlarks, the Breckland population has not dropped noticeably following bad winters. Where the Breckland birds winter is not known but no Woodlarks are seen on or near the forest from November to January inclusive (R. Hoblyn, personal communication).

Intensive research on Woodlarks at Thetford in 1986 and 1987 (C. Bowden, in preparation) has been aimed at identifying the species' key requirements. This has shown which elements of vegetation structure are important; both the obvious factor of tree growth but also the subtler variations in ground vegetation on the young restocks, especially grass cover. Woodlarks use clearfells from almost the moment they are felled till the trees are a maximum of seven years old. They favour the earlier stages and are commonest on restocks less than three years old. Recently there has been an increasing bias

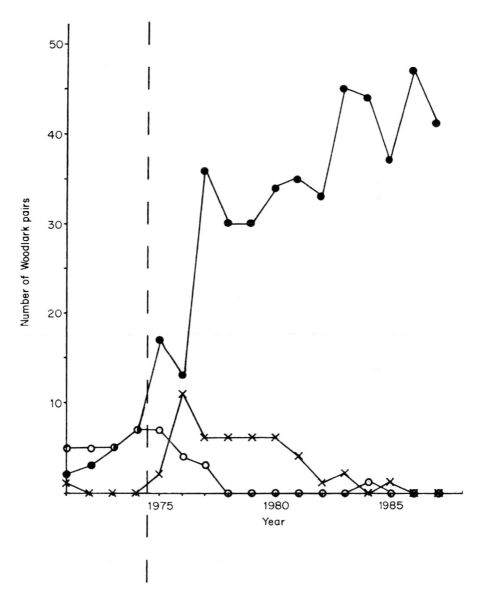

Figure 7.1   Woodlark population in Breckland.
Key: ●, *young plantations*; ×, *natural heath/Breck*; ○, *marginal land.*

towards the younger age classes. Far more of these are now available because of the rapid acceleration in the felling programme from 1979 onwards, and the biggest increase in Woodlarks in the forest took place in 1977. The increase in Woodlarks was not synchronous with a particular increase in felling rate and has not kept pace with increasing amounts of apparently suitable habitat. However, it looks as if Woodlarks, whilst able to use older clearfells, are showing a preference for the newest sites. They appear to avoid areas with longer, thicker grass which by year 4 may cover over 70% of many restocks.

Herbicide treatment limits growth of grass, and in 1988 a massive experiment was undertaken to see whether additional herbicide treatments would keep conditions suitable for Woodlarks for longer. The trial covered 20 compartments, over 300 ha, all of which had had two years treatment with the herbicide Atrazine. Two years treatment is normal, but half the experimental compartments were given an additional application in 1988 and the other half left untreated as controls. Atrazine not only controls the grass but, because it inhibits germination, also leaves the completely bare soil that is important to Woodlarks. Sites with Woodlarks as opposed to those without tend to have kept a proportion of bare soil longer, as well as having a much smaller proportion of long grasses. The Atrazine experiment is one of the largest ever in habitat manipulation for a bird in Britain. The results of this work (C. Bowden, in preparation) extend well beyond Woodlarks: they may tell us more about why semi-natural heathlands seem to have become unsuitable for so many of their characteristic birds.

One of them is the Nightjar, which augmented the Stone Curlew's unearthly shrieks with the monotonous rise and fall of its churring to make the Breckland of the 1920s a poor place for sleep. Little Lodge Farm must have been a particularly noisy spot, with 30 pairs of Stone Curlews across the Little Ouse on Santon Downham heath and Nightjars so maddening that Clarke tells of a farmer shooting several to shut them up. Today the Stone Curlews are gone but Nightjars have returned to the new clearfells and can be heard in Santon Downham village itself. Nightjars have always been common in the area and still are today. Like Woodlarks, Nightjars are now uncommon on heathland. The 1981 national survey (Gribble, 1983) located only 27 territorial males, 13% of the Breckland population, on heather and bracken heaths. None were found on grass heaths. One-hundred-and-seventy-two territorial males, 86% of the population, were found on forest restocks. A total of 201 Nightjar territories were found in the Breckland, about 10% of the British population in 1981.

It is hard to judge whether there are more or less today than in 1925, but three surveys of the forest in 1974, 1978 and 1981 proved the Thetford Nightjar population to be the only one in Britain that had certainly increased during the 1970s (B. Pleasance, unpublished). Only the 1981 survey included non-forest habitats, so it is possible that, like Woodlarks, Nightjars shifted from the natural heathland to the newly available clearfell sites. The rise in numbers is so great that it is more likely that there has been a genuine rise in population which has closely followed the increasing area of young forest as large-scale clearfelling got underway. In contrast to Woodlarks, the proportionate rise in

Nightjar numbers is not far off the increase in apparently suitable habitat. Table 7.1 shows the numbers of Nightjars in each survey year against the area of suitable young forest habitat at the time.

TABLE 7.1

Survey year	Suitable habitat (ha)	Male Nightjars
1974	1611	90
1978	2437	123
1981	3519	168

(from B. Pleasance, unpublished)

Between 1974 and 1981 Nightjars increased by 87% and suitable habitat by 118%. Nightjars colonize felled areas immediately; nests have been found within 100 m of working chainsawyers. The oldest occupied plantation was 15 years old but Nightjars usually stop using a site at about 10 years old and at a tree height of 3 m. They apparently occupy most available habitats, and the future structure of the forest is likely to be of particular importance to them: although Thetford will never return to the solid, high forest of the 1960s, the felling rate and thus available restock is already decreasing. This is a potential problem for both Nightjar and Woodlark populations which is being considered as a part of a current research project. Density on the central part of Thetford, measured as distance between churring males rather than as true territory, is slightly less at 18 ha of restock per male than on the semi-natural heaths of Minsmere, coastal Suffolk (10.3 ha), Dersingham, North Norfolk (9.3 ha), and Leziate near Kings Lynn (16.6 ha). Nightjars are known to feed well away from the nesting area so it is not certain that territorial density is a significant indicator of habitat quality.

Nightjars are present for a longer season than in the North Yorks Moors forests discussed in the second case study (Leslie, 1985) and there were three double broods out of 34 nests found at Thetford (B. Pleasance, unpublished). The birds' rapid adaptation to forest sites is interesting but the really important discovery from these three censuses is that Nightjars do seem able to increase their numbers in Britain despite a long and well-charted decline nationally (Norris, 1960; Sharrock, 1976; Gribble, 1983). As with Woodlarks, detailed habitat structure does seem to be important and a research project to follow on from the Woodlark study discussed above began in 1988 to try and discover the key factors. Clarke said that 'in appearance, in song and in habits, the Nightjar somehow seems alien to our English bird-life, but it is nevertheless one of the most attractive of our summer visitors'. He was right, and it is not surprising that there are a growing number of Nightjar addicts amongst British birders. This is a bird that must not be allowed to slip away without a fight.

It looks like the fight could soon be over for the Red-backed Shrike. Another

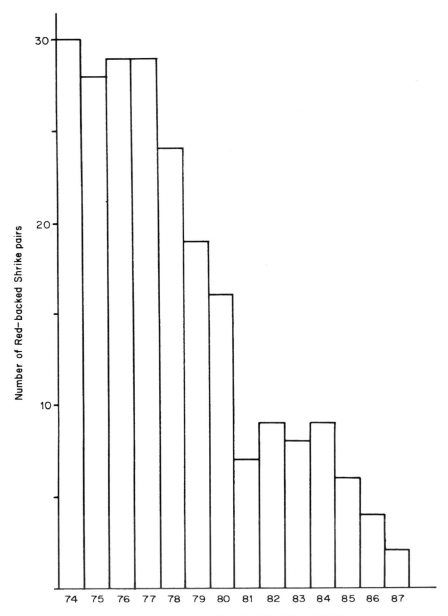

Figure 7.2   The decline of Breckland Red-backed Shrikes.

attractive and interesting summer migrant, it is almost certain to be extinct as a British breeding bird within the next few years (Bibby, 1973). Unlike Nightjars and Woodlarks it is not dependent on forest sites, although it has nested successfully on restocks in recent years, nor does habitat loss or change in the breeding area seem to be the cause of its plight. Breckland has always been one of its strongholds and was an obvious centre at the beginning of the 1970s when Sharrock's (1976) estimate of 50 pairs nationally was probably slightly low. However, by the beginning of the 1980s a Breckland population that had 10 years earlier been in the 20s was down to near 10 pairs (Figure 7.2), and the scattered but widespread pairs from the North Norfolk coast to the New Forest had disappeared completely.

Although a few sites in Breckland have been lost through scrub being cleared or housing development, most traditional nesting places remain intact and apparently unchanged and the availability of restocks has increased. Although human disturbance, from egg collectors and by accident (the only successful nest in Britain in 1987 and 1988 was in a corner of the FC's most popular public car park in Thetford forest), was a factor in the past, no nests are known to have been lost to it since about 1980. No-one knows the reason for the decline of the Red-backed Shrike: it could for once really be changing climate. The forest has almost certainly been a neutral factor in this decline but has by chance become the very last refuge. Against all the odds a pair of Red-backed Shrikes returned to St Helens in 1988 and successfully reared three young. These birds were one of the British bird spectacles of the year, with probably 10,000 people seeing them. It is reasonable to speculate that this could be the last British nest for some time.

Against this perspective of some heathland species adapting to the forest environment it is sobering to remember that the forest was possibly the first really important and permanent change to the unique habitat of the Breckland since the last ice age: this and the Flow Country are perhaps the places now afforested that would have had the lightest cover of trees before man's intervention. The forest has proved quite inhospitable to the group of birds that occupied the very barest heath, the most interesting of which, the Stone Curlew, is now endangered in Britain. It has allowed some interesting and rare species like Goshawks and Golden Pheasants to become established in East Anglia and for Siskins and Crossbills to become quite common. However, the gain of these birds could not have offset the total loss to the area of Stone Curlews, Woodlarks and Nightjars had this been the result of afforestation.

The paradox in assessing the impact of the afforestation on birds has come through later changes. After 1945 agriculture swallowed up a larger part of the semi-natural heathland than forestry. With the exception of Stone Curlews, which maintain a fraught and tenuous existence on arable fields, forestry has on the whole been the friendlier environment for both birds and flora. Ironically, it is possibly a better habitat today than most of the remaining semi-natural heathland because of the disturbance to the ground caused by harvesting operations which favours not only birds like Woodlarks and Nightjars but also rare Breckland flora. It was the loss of the Rabbits that transformed the heathland; many areas are moving rapidly towards woodland, whilst others now have the tight grass sward of sheep grazing, especially the largest remaining area, the Stanford training area.

We cannot resolve this paradox: there is no doubt in our minds that were Breckland today in the state that the early FC foresters found it in 1922, no-one would dream of changing it: it would be treated with the reverence afforded the New Forest and vigorously defended from change. On the other hand, in the real world of the 20th century the forest has by chance ended up being the home and refuge of several declining British species. Since 1945 the forest has increasingly become a positive rather than negative force for birds. During the 1980s the chance element in this has started to decrease: ideas for altering the structure of the forest are being applied and the results of a far-sighted review of the future of the forest have been published (Simpson and Henderson-Howat, 1985). Although this at the time missed the importance to birds of maintaining sufficient young restocks, planning for landscape did introduce a greater degree of age–class diversity to future felling which may now be further modified to take account of new knowledge on the needs of Woodlarks and Nightjars.

However, for its most important birds the forest remains a surrogate for semi-natural heathland. If it has achieved one thing, it has given birds like Woodlarks and Nightjars an alternative to heathland that seems to have become unsuitable since Rabbits declined. Stone Curlews have not been so lucky. A large area of heathland, nearly 1/5 of Breckland, still exists in various states of overgrowth. The restoration of this heath to something nearer its previous state strikes us as being one of the top bird conservation priorities for lowland England in the 1990s. Fortunately, much is already in national ownership, and light cultivation or other positive management (the fostering of Rabbits perhaps) to break the tight sward of sheep grazing would not conflict with the national priority of training the army on the Stanford PTA. With all the heaths returned to their former glory Breckland could again be an absolute paradise for birds, with several high forest species a worthwhile addition to populations of Stone Curlews, Nightjars and Woodlarks well secured for the future.

THE NORTH YORKS MOORS

As with Thetford, planting began in the North Yorks Moors shortly after the formation of the FC. There has been little new planting since the mid-1970s and felling is well under way. Large-scale planting in the North Yorks Moors was spread over a far longer period than it was at Thetford, working from the brown earth soils of the dale slopes, through heathland podsols of the lower moors to gleys of the higher moors. The North Yorks Moors forests are true upland forests and are similar in their combination of older, better sites and younger, poorer soils to many of the earlier FC areas. The majority of the new forests are within the North Yorks Moors National Park, which was used as a convenient study area boundary in Leslie (1985).

Seventeen per cent of the National Park's 1432 km^2 is forest, largely new

The better soils of the earliest North Yorks Moors planting shows what can be achieved: a variety of conifer and broadleaved species, and with age and windfirmness allowing the development of varied age class. Dalby dales, Forestry Commission.

forests, although there is some ancient woodland and older private estate forestry. The area is located in northeast England, between the river Tees to the north, the sea to the east and the low-lying Vale of York to the south and west. The North Yorks Moors have a central plateau rising to 424 m and dissected by deep valleys. The southern part of the area is a complex of scarp and dip slopes. The dominant soil type is a podzol with a pronounced ironpan (a sandy soil which can become waterlogged because the ironpan impedes drainage) covering much of the central plateau. Gleys (clayey, wet soil) occur on flat ground and in the dales. Brown earths are found in the dales too, and there are some rendzinas where the calcareous bedrock outcrops. Heather covers much of the moorland, but the area of Bracken is growing. Average rainfall is 762 mm (30 inch) on low ground, rising to 1015 mm (40 inch) over higher ground. Despite the extensive gley soils, the moorland is of a characteristic dry, east coast type, and the very small amount of standing water and wetland must restrict distribution of some moorland waders and wildfowl.

Afforestation is concentrated in the southeast corner of the National Park, with additional areas on the steep scarp of the moorland edge extending right the way to Cleveland in the north. As the forest developed from the best brown earths, both tree species and growth rates changed. The first plantings mixed Douglas Fir and Sitka Spruce, with Corsican Pine and Scots Pine on the rigg top rendzinas, later supplemented with Beech. The ironpan soils of the heathland plateaus were planted with Scots Pine, which failed or grew very slowly; most Scots Pine is only yield class 6–8 compared to 16 for Douglas Fir on the best sites. Some Scots Pine areas were ploughed in and replanted, as new techniques for establishing trees were developed (Zehetmayr, 1960). Sitka Spruce went into check on these heathland sites and Lodgepole Pine became the staple for 1960s planting on the poor soils of the high moors.

The poor tree growth on the exposed, infertile soils of the high moors combined with the growing influence of the National Park Authority to end new afforestation, leaving a substantial heather moorland core to the National Park. Whilst the higher-elevation plantations have little natural diversity, the better soils of Dalby and East Cropton retained a legacy of broadleaved remnants dating back to their historic status as the royal hunting forest of Pickering. To the west, in the smaller blocks around Helmsley, commercially worthless Oak and Birch woods were underplanted with Western Hemlock during the late 1960s and 1970s. The lower reaches of most watercourses are Alder lined, and even where surrounding forest soils are acidic, limestone measures are so widely distributed that most have a high pH.

The older forest areas on better soils are now well into the second rotation, with extensive felling and restocking dating back to the early 1970s. It is possible to chart the arrival of some bird species that have moved into the maturing forest but assessing changes in the bird populations of open ground habitats is difficult. Heather moorland is now known to hold high densities of the species of dry moorland, including Red Grouse, for which large areas are still managed, Golden Plovers, Curlews, Lapwings and Merlins (Campbell, 1983). Some Redshanks and ducks breed on the very limited wetland, but Dunlins are absent.

So too are the larger birds of prey; human persecution over many years has prevented successful breeding by Hen Harriers, Short-eared Owls and Common Buzzards. Levels of persecution are by their very nature difficult to pin down. However, there were seven known poisoning incidents on FC owned or leased land between 1983 and 1987. This is despite FC efforts to prevent persecution, including the investigation of all incidents with the help of the Ministry of Agriculture. In only one case was a dead bird found; poisoned bait, two Rabbits, a goose egg and two hen's eggs were found, and in two cases domestic cats and dogs were poisoned. It is hard to conceive the pressure on birds in the area generally if this is happening on land where efforts are being made to prevent it. There are many recent records of Hen Harriers around the forest edge in early spring, and also rumours of destroyed nests. In contrast to other parts of Britain the species does not appear to have exploited the young forest for breeding. Crag nesters, Peregrines and Ravens, were never numerous because of lack of suitable nesting sites (Nelson, 1907).

The national conservation value of the heather moorland of the North Yorks Moors has only recently been recognized, partly because other traditionally richer areas have declined, but mainly because of the lack of information prior to recent surveys (Campbell, 1983). Merlins and moorland waders, especially Golden Plovers, are most important. Bibby (1986) showed that Merlins breed around the edges of moors, hunting out over them and that in Wales the presence of heather was a key element in their breeding distribution. There is a great deal of edge in the North Yorks Moors because many dales run far into the central plateau. High Golden Plover densities of 1.9–3.8 territories/km^2 (Campbell, 1983) compare with the 2.37 territories/km^2 of the best of the four wader zones identified by Stroud *et al.* (1987) in Caithness and Sutherland. Densities of Curlews and Lapwings are also high; RSPB moorland surveys in 1983 also covered sites in Lanarkshire and Perthshire. Curlew densities in Perthshire, of 3.7–7.4 territories/km^2 were higher than the 3.0 territories/km^2 of the North Yorks Moors. However, Golden Plover densities were much the highest of the three areas, and overall the North Yorks Moors had the most diverse wader assemblage with, for example 2.4 Golden Plover, 2.8 Curlew and 3.6 Lapwing territories per km^2 on Spaunton Moor, as well as Snipe and Redshanks on the wetter ground.

There is insufficient information to prove any change in populations of moorland birds resulting from forestry: historical records are not quantitative, and it is not possible to identify any species that could have actually disappeared from the area following afforestation. It cannot be assumed that changes to birds have been proportional to the area of land planted. Not all the moors planted were heather. The earliest planting at Dalby was on land previously managed as Rabbit warren, with a predominantly grass vegetation. Plantings after 1945 on Broxa, Newton House, and Wykeham and Langdale High Moors were certainly on heather sites, but earlier plantings of Cropton and Wykeham forests were on a mixture of grass, heather and bracken. Campbell (1983) found low densities of the key species of heather moorland on the grass and bracken Levisham Moor.

Too small to attract persistent persecution, the Merlin is the most important species of the heather moorland. Surveys in 1983 and 1984 (Mather, 1986) found almost as many pairs in the North Yorks Moors as in the whole of the declining Welsh population (Bibby, 1986). There are few Merlins in the southeast quarter of the area. This is where there is most forest, and the lower moorland edges are heavily planted. There is a circumstantial link between the distribution of forest and loss of Merlins, but as it is not known whether there were Merlins before planting it is no more than that.

In contrast, demonstrating change in some forest species is possible because they were not present at all in the past and are now common. Most obvious are the two passerine conifer specialists, Siskins and Crossbills.

The Siskin is of most interest because its song flight is hard to miss and its population does not fluctuate between years as dramatically as that of the Crossbill. The Siskin is a common bird in winter, but was exceptionally rare in summer before 1960; breeding was under-recorded by general ornithologists, and this is an interesting illustration of a common problem in the study of forest birds. Records collected by forest-based ornithologists were unfortunately not published (mainly G. Simpson 1961–1967, and R. Leslie 1978–1979, all in FC records, NYM Forest District, Pickering), and show a much wider evidence of breeding than Sharrock (1976) or Mather (1986). Mather's records also reflect the better coverage of the small conifer forests of the Yorkshire Pennines compared to the extensive areas of the North Yorks

Moors. Mather (1986) records two summer Siskins in the North Yorks Moors forests in the 1950s, but no more till the late 1960s. Simpson found Siskins only as winter visitors up to 1964, but in that year a frequently singing male stayed through the summer in Newtondale. In 1966 there were three locations between May and July, one again in the Newtondale block of Cropton forest, another on the western side of Cropton, near Spiers, and another further east in northern Dalby. Fieldwork for Sharrock (1976) located only one confirmed and one possible record for the whole area, both to the north of the main forest area.

During 1978 and 1979 casual observation located singing Siskins in eight places separated from one another by more than 1 km between 1 May and 31 July (R. Leslie, unpublished). No systematic survey of Siskins was undertaken, and records tend to reflect forestry activity, for example the forest office at Spiers, Keldy forest cabins in central Cropton, and, most of all, Dalby village, where there were several individuals and song in May 1978. Siskins were found in pine, mixed larch and pine, and spruce and Douglas Fir. Siskins are now a well-established forest breeding bird rather than an 'increasing tendency of a few pairs to breed' suggested by Mather (1986). A spring male Siskin is a real gem in the forest, its coal-black cap contrasting with the bright lemon of head and breast. Its twittering, butterfly-like song flight enlivens more and more of the new forests. Bearing in mind the lesson of the Nightjar survey discussed below, it would be interesting to see what distribution a thorough survey, which should take place as part of the 1988–1990 breeding atlas, will show.

It is harder to tell what is happening to a Crossbill population, because the species is harder to locate, fluctuates wildly in numbers between years and can nest during more months than most birds. Up to the late 1950s Crossbills were as rare and irregular as in most of England, occasional breeding following irruptive invasions (Mather, 1986). Since 1962 (G. Simpson, unpublished), Crossbills appear to have been present in the North Yorks Moors forests in every year for which there was some recorded observer coverage, except 1965, and throughout the year in many years. There are many records of proven breeding, ranging from a family party in Newtondale in April 1964 to very newly fledged young being fed by adults at Keldy on 27 March 1978 (R. Leslie, unpublished) and several nests found in Wykeham forest in 1981 and 1982 (M. Francis, personal communication). As a general impression the numbers of Crossbills seem to have increased over the years: the largest flocks seen by Simpson were 25 in May 1967 and 23 in October 1962. During January and February 1980 there were about 100 in the Dalby valley, and sightings of flocks of 20–30 birds were greater in both number and frequency during 1978–1982 than 1962–1968.

However, the spectacular population explosion of 1977–1978 in the Borders following the massive 1977 spruce cone crop was not matched in the North York Moors, although Crossbills were noticeably commoner in both 1978 and 1980 than in other years. Tree species may account both for the lack of great fluctuation and the regular occurrence of the species: mature Sitka Spruce, which cones erratically and heavily, is at present rare in the area, and Norway

Spruce did not cone nearly as heavily here as in the Borders in 1978. On the other hand, mature Scots Pine, which cones more regularly, is common and may account for evidence of fairly regular breeding by small numbers of Crossbills. There are two interesting questions concerning Crossbills in the North Yorks Moors forests. First, is the species established in the sense that it breeds in most years? The balance of evidence would tend to suggest that this is the case in this area. Second, is the population self-sustaining? This is no more possible to answer than for any other part of Britain at present; an east coast location and the relatively frequent occurrence of Parrot and Two-Barred Crossbills (Mather, 1986) suggest that topping up by immigrants from Europe may still be important.

The most thoroughly studied colonist of the forest is, however, a bird of the earliest successional stages, the Nightjar. The results of systematic distribution surveys during 1980 and 1981 actually influenced the development of forest planning concepts by emphasizing the importance of the young restocking habitat to a scarce breeding bird. By 1945 Nightjars had declined considerably in their traditional moorland edge habitat (Yorkshire Bird Reports in Mather (1986) and Leslie (1985)). By the early 1970s the Nightjar's national decline was well documented (Stafford, 1962; Sharrock, 1976; Gribble, 1983) and a maximum of seven breeding locations around Scarborough and one or two on the Cleveland escarpment were all that were known in the North Yorks Moors. However, records of forest breeding stretched back to 1939 (J. Simpson, in FC Records). The surveys found what was really a lost forest population, with 40 territorial males located in 1980 and 46 in 1981. Ninety per cent were on forest sites in 1980 and 80.4% in 1981 (Figure 7.3). The balance occupied more traditional heathland areas of heather and bracken with scattered Scots Pine. Coverage for the surveys was comprehensive; of a total of 86 sites visited in 1980, only 33 were occupied. Forest Nightjars were distributed almost equally between first rotation and restock sites. Because new planting had largely ceased by the late 1970s most afforestation sites were well into thicket stage by 1980 and Nightjars used failed or unplanted areas; although some sites appeared to have as little as 1 ha open land, even the widest roads and rides were not used for breeding without some gap in the main crop area.

The North Yorks Moors have a greater range of altitude and soil type than most British Nightjar breeding areas (Gribble, 1983). With sites ranging in altitude from 75 to 311 m above sea level, there was no significant difference between the altitudinal distribution of Nightjars and all sites inspected. There were only three pairs in the sheltered Dalby Dales despite large areas of young restocking, but exposed, forest/moorland edge sites in poorly stocked Lodgepole Pine and checked Sitka Spruce in Silton, Cropton and northern Dalby forests held six Nightjars in 1981. Similarly, with free-draining podzols and brown earths and wet gleys widely distributed, Nightjars showed no clear preference, with the greatest number occurring on the gleys. Nightjars probably occur most frequently on dry, sandy soils because they are most likely to carry a sparse, low vegetation over several years. The windswept higher moors seem an unpromising prospect for a bird feeding on flying insects

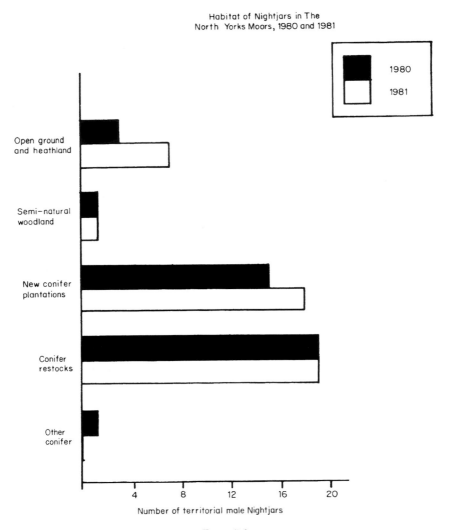

Figure 7.3

and the even distribution over such an altitude range suggests that temperature is not a major climatic factor at a local level.

Habitat, and in particular vegetation structure, appeared to be the major factor influencing Nightjar distribution, although the study was not sufficiently rigorous to pin down the key factors. Most Nightjars were found on sites with trees between 1 and 2 m but they also occurred on freshly felled sites prior to restocking and amongst older, taller trees where stocking was low or there were substantial failed patches. The smallest occupied open areas were failures or open ground within first rotation crops of about 0.5 ha or more. The two areas with four or more birds, Blackpark in Cropton and the research plots and Pine Looper clearfell in Wykeham, were both recently restocked clearfells of over 20 ha.

Exactly why Nightjars no longer occur in their traditional moorland edge habitat remains a mystery; it is possible that because the moorland edge bracken beds favoured for nesting are no longer cut for animal bedding, their structure may have changed. Bracken is now spreading from the dale slopes out onto the heather moorland and is an increasing natural threat to the character of the National Park. The shift towards woodland habitats is a national phenomenon, with over half the British population now using them (Gribble, 1983). The continuous provision of open ground within the forest and young stands of trees is clearly a key factor in the future of the species in the North Yorks Moors. Even sparsely stocked first rotation areas will in most cases eventually close canopy and the availability of new clearfells will be increasingly important. Unlike the case with Nightjars in Thetford, fluctuations in the total area available may not be critical, as these Nightjars do not appear to fully occupy apparently suitable habitat. Illustrating the potential for ecological continuity of forest habitats, the site at Skelton Banks in Cropton Forest where J. Simpson found Nightjar nests in the young planting in 1938 reached rotation age and was felled in 1982. Following ploughing and replanting, Nightjars were present in 1983 (P. and J. Ottaway, personal communication).

Although peripheral to the changes caused by afforestation the story of the Pied Flycatcher in the North Yorks Moors forests is particularly interesting as an illustration of how one habitat factor, in this case nest boxes, appears to have attracted and then, when the scheme decayed, lost the species. The bird box schemes in the North Yorks Moors were another part of the national project from which the boxes at Nagshead in the forest of Dean survive today (Campbell, 1967). The first assessment sheet (in FC records, NYM Forest District, Pickering) is for 1945 and the last for 1955, all for the scheme along the Dalby Beck. The Beck was and still is bordered along most of its length by a narrow band of Alder. In 1950, 20 of the 156 boxes were occupied by Pied Flycatchers. By 1980 all the boxes had fallen into disrepair and no Pied Flycatchers were seen during fieldwork for a BTO waterways bird survey in 1981 and 1982 (R. Leslie, G., M. and J. Haw). Another scheme in Newtondale ran during the 1960s, with six Pied Flycatchers nesting in 1962 (J. Simpson, FC records), but again this scheme was defunct by 1980. However, some new boxes in Duncombe Park near Helmsley (Fellowes, in Mather (1986)) had 20 pairs in 1975 and 1976; it seems very likely that the simple provision of boxes could get Pied Flycatchers breeding again along the considerable length of Alder-lined watercourse in these forests, and hopefully this is an experiment that might be undertaken soon.

Perhaps the greatest imponderable in assessing the impact of forestry on this area is the history and future of birds of prey: the absence of several species because of human persecution certainly reduces the overall value of the moorland habitat from its biological potential. Had more of the easily observed larger species like the Hen Harrier been present, there might have been more information on the changes resulting from afforestation. However, levels of persecution have remained high and the relative compactness and heavy keepering of the moors may also explain why more birds of prey did not establish themselves in young plantations (although there were a string of Montagu's Harrier breeding records at Dalby and at Wheeldale (North Cropton) from the late 1930s to the mid 1960s (Mather, 1986)).

The potential of the older forests for woodland birds of prey is now good: the smaller forest species are now very well established. Sparrowhawks are particularly numerous and it is difficult to spend more than two or three hours in the forest without seeing one. Tawny Owls are well established. Chainsaw-yers found several nests at the bases of large conifers during the late 1970s and several were seen on restocks during the Nightjar survey. However, although Common Buzzards are seen annually and the forests now provide ideal nesting sites, they have not bred and seem unlikely to do so whilst illegal poisoning continues.

The scanty information we have on the effects of afforestation in the North Yorks Moors shows that the forest has attracted some birds previously rare or absent as breeding species, and there is no firm evidence for the elimination or significant reduction of any open-land species. Taken at face value, recent surveys actually show gains in both forest and moorland. Although the latter is attributable to better ornithological coverage, it does show that while afforestation may have reduced some birds without our knowledge, an extremely valuable moorland bird assemblage remains intact. On the other hand, gains following afforestation are clearer in the cases of Siskins and Crossbills because they were previously absent as regular breeding birds and are now well established; historical records are reasonably reliable for easily identified species, making it far easier to demonstrate gains for the forest than the proportionate decrease of open land birds.

The Nightjar story shows one benefit of diversified second rotation forest, but also hints at possible adverse changes within existing semi-natural habitats: subtle unintended changes resulting from a change of land manage-ment method may set the bracken advancing or drive the heather back and threaten birds like Merlins or Nightjars as much as planned development such as afforestation or agricultural improvement. Rather easier to pinpoint but equally difficult to deal with is the continued persecution of raptors to protect game: a small minority of keepers acting illegally can prevent wide-ranging, large birds like Hen Harriers or Buzzards becoming established and, with forest birds well established, it is attracting these birds back to the area that could most enhance its conservation value. The paradox of persecution is that another aspect of management for game, traditional heather burning, is the key to the excellent populations of moorland waders, especially Golden Plovers.

Afforestation in the North Yorks Moors cannot be shown to have done great harm to birds of open land and has not eliminated any species. It has attracted some species that were previously rare or absent and allowed some woodland species to become numerous. Overall, the present level of afforestation has enriched the avifauna of the region, and as extensive new planting has now been finished for a decade this general situation will not change: the interest in the forest is likely to continue to improve and hopefully the notification of the best areas of moorland as SSSIs will make the resources for their positive management available. Further unintended deterioration of heather moorland is now the greatest threat to the area's birds, and relaxation of persecution followed by recolonization by open-land birds of prey its greatest opportunity.

Had afforestation taken place only post-1945, and without the latterly restraining influence of the National Park, the present situation in the North Yorks Moors might have been very different: much of the land of pre-war plantings would have been retained in agriculture and improved. On the other hand, afforestation could have expanded far further into the heather core of the National Park and into areas today known to hold high densities of moorland birds. By the mid-1970s, land that would yield little more than GYC 6 looked economically unattractive, particularly in the face of fierce opposition from the National Park Authority, but in the late 1960s, when the northern end of Cropton and the Newton House block were planted, achieving programme targets in reality outweighed economic return and the National Park was less powerful than later. Far greater areas of the moors could well have been planted; although it did not happen in the North Yorks Moors, agricultural restrictions on land availability and far lower environmental knowledge and concern eventually caused this situation in Scotland, culminating in the Flow Country.

GALLOWAY

The place where post-war afforestation has progressed furthest and fastest is Galloway; the promontory of southwest Scotland sticking out into the Irish sea. Gentle hills of sedimentary rocks rise to the less fertile central granitic core of the 845 m Merrick/Rhinns of Kell/Cairnsmore of Fleet range. Galloway's two local authority districts, Wigtown and Stewartry, make up the western half of the Dumfries and Galloway Region.

The region has the highest proportion of forest cover (21%) in Britain, and Stewartry is the only district in the uplands with over 30% forest cover. Between 1921 and 1940 conifer planting in the region was just over 5000 ha (FC regional survey statistics, after Locke (1987)). After 1945, planting rocketed to the rough equivalent of a Kielder Forest every decade: 27 686 ha in the 1950s rose to 38 022 ha in the 1960s and 43 204 ha in the 1970s. The region still has amongst the highest levels of current new planting in Scotland.

Much of the planting was concentrated in Galloway. Very large forests were planted to the west of the Merrick, Glentrool Forest, and to the east of Cairnsmore of Fleet, Bennan Forest. Earlier FC areas have been added to more recently by extensive private planting. A very high proportion of the rolling, gentle Galloway hills was physically plantable. Within the space of a single rotation, far more than half of the afforestable land of central Galloway that lies above the better farmland and below the altitudinal limit for economic tree growth has been afforested (Peterken, 1981). The reasons for the rapid progress of forestry were the same as for almost every area of major planting from Thetford to the Flow Country: the balance between what agriculture and forestry could achieve in the particular physical and economic conditions of the place and the times tipped towards forestry. In Galloway's case the soils were nothing special, mainly gleys and peaty gleys. The elevation and exposure of the land acquired for planting was typical of the sort of land generally available for forestry after 1945. It was the extreme oceanic climate of the area and the relative scarcity of heather that made forestry such a good prospect because this was ideal Sitka Spruce country. Even at the height of the Lodgepole Pine era in the 1960s, only 5609 ha were planted in Dumfries and Galloway compared to 24 056 ha of Sitka; in the 1970s Sitka Spruce planting was 36 310 ha out of a total conifer area of 43 204 ha.

The era in which these forests were established combined with the regularity of the land led to featureless forests; most of the photographs illustrating 'blanket forestry' in the NCC's *Nature Conservation and Afforestation in Britain* (Nature Conservancy Council, 1986) are from Dumfries and Galloway. It was possible in places like Bennan to plough from valley bottoms across streams, up to lochsides and right over the top of the lower hills. Sitka Spruce increasingly dominated the conifer species planted, and the geometrical grid of rides adds to the artificiality of the young forest and constricts views from older stands to strictly rationed lines down the straight rides. Galloway has been used as the type example of the large-scale 'tree farming' style of afforestation, but in fact it is the extreme, and as such is especially important as an indicator of the possible impact of similarly high levels of planting elsewhere.

By the start of large-scale afforestation Galloway was dominated by grass moorland; heather declined from the middle of the 19th century onwards as sheep replaced traditional cattle (Gladstone, 1910), and the quality of the moors for Red Grouse still seems to have been deteriorating in the first half of the 20th century before forestry came in and transformed much of what was left. There were some good areas of heather but these hills were not renowned for their moorland waders: Golden Plovers are quite common but do not match the densities of the heather moors in northeast Yorkshire and the Pennines. Dunlins occur but are scarce. What attracted birders in droves in the 1950s, including many egg collectors, were the raptors and carrion eaters, especially Peregrines, Ravens and Britain's southernmost Golden Eagles which re-established themselves in 1945 after a long absence. Buzzards and Merlins were reasonably common and afforestation may have helped one formerly common species, the Hen Harrier. By 1945 all these birds were far commoner than at the end of the 19th century but scarcer than at its start: the carnage of intensive keepering was as bad here as anywhere in the country. In Ayrshire, just to the north, 350 Hen Harriers were killed in five years on Lord Ailsa's land: it is hardly surprising that they, White-tailed Eagles and Golden Eagles became extinct in southwest Scotland. Even Buzzards became very rare and Peregrines were also vigorously persecuted, although, as Ratcliffe (1980) has shown, this never had as much impact on their population as pesticides in the 1950s and 1960s.

Ravens, Golden Eagles and Buzzards did well in Galloway because of the abundance of carrion. Marquiss *et al.* (1978) showed that the breeding density of Ravens was higher on poorer-quality land and at greater altitude. This is the direct opposite to the findings of Newton *et al.* (1986b) for Sparrowhawks but it is understandable: Ravens are dependent on sheep carrion and it is on the poorest ground that most sheep die. The land on which most sheep die was also the most economically fragile, the hardest to farm and the most likely to be sold when the opportunity offered by forestry came along. In Galloway that was the infertile acidic granite core of the central range of hills and it held the highest density of Ravens in southern Scotland. Because of their dependency on sheep carrion, Ravens are also the species most likely to be affected by the removal of sheep for afforestation. They have declined dramatically throughout southern Scotland and Northumberland over the past 30 years (Marquiss *et al.*, 1978; Mearns, 1983) and provide the clearest case of a wide-ranging species declining as a result of afforestation. In contrast to upland waders, whole territories of which can be changed by a single, medium-sized afforestation scheme, Ravens use the uplands on a grand scale, and it is the broad level of afforestation at a regional scale that is going to affect them.

By 1975 throughout southern Scotland and Northumberland only 55% of the regular Raven nesting territories of the 1950s were occupied and by 1981 the decline had continued to 70%: only 1/3 of the original population survived and over some quite large areas there were no breeding Ravens left. The core of the Galloway hills, where a significant area of higher-elevation land remains unplanted, still holds one of the few remaining concentrations in south Scotland, but the lower ground to the east, which had Ravens in virtually every 10-km square before 1970, now has hardly any (Mearns, 1983). Marquiss *et al.* (1978) showed a clear relationship between the afforestation and Ravens ceasing to breed. Even where they went on breeding they produced later and smaller broods when more than 25% of former sheepwalk within 5 km of the nest site was afforested. Nearly half the abandoned nesting territories had over 50% of land within 3 km afforested, emphasizing the practical implications for the uplands of the 21% overall forest cover of Dumfries and Galloway.

A minor cause of Raven decline has been the loss of nesting sites to Golden Eagles. They too are heavily dependent on sheep carrion in Galloway but elsewhere produce more young where live prey make up a higher proportion of the diet. The re-establishment of the Golden Eagle as a regular breeding species in 1945 built up to a peak of four pairs by 1966. However, from about

1970 breeding success deteriorated. Two out of the four pairs failed to breed successfully from 1974 to 1982. Marquiss *et al.* (1985) linked this decline to the loss of open hunting ground through afforestation; territories that have consistently failed over the last decade have a far higher proportion of planted ground around the centre of the territory than the pairs that have bred successfully. Open ground within 2.5 km of the centre of the territory was, in 1983 38.0% and 54.3% for the territories that failed and 78.3% and 64.8% for the two that have consistently succeeded. However, the difference in the proportion of land below 305 m elevation planted by 1979 was far greater: for the failed territories it was over 80%, for the successful under 35%.

There seemed to be a fairly clear link between forest and Golden Eagle failure, in particular with respect to the importance of the middle ground for live rather than carrion prey. However, the resumption of nesting in 1983 in one of the apparently deserted territories was not a one-off, and at present there again appear to be three viable territories. There had not been any significant change in forest structure: clearfelling has not got underway on a large scale in the higher hills used by the species. There will have been some further afforestation, but with over 80% of land below 305 m already planted, this must of necessity have been limited. The Golden Eagle story remains inconclusive rather than proven either way: the Galloway population is in any case a rather small sample size for examining a complex problem like this.

The Buzzard is the third species that eats a lot of carrion, and its story is different again. Detailed population figures are not available but Buzzards seem to have maintained their numbers though changing their habitat: before afforestation their distribution was apparently limited mainly by persecution, which kept them to the higher ground, much of which has been planted and does not produce much sheep carrion now. However, the safety of the forests for nesting has allowed Buzzards to feed on the better farmland previously excluded by human pressure. As well as sheep carrion they also live on small, wild mammal prey, and are starting to use restocks as they become available. In the lower ground of Glentrool Forest they nest in conifer retentions of less than 1 ha.

What is really most surprising is not that some carrion eaters have declined with the removal of sheep but that they have declined so little: Newton *et al.* (1982) estimate available sheep carrion in their Welsh study area at 190 kg/km^2 per annum. Most sheep die in late winter and early spring, the hungriest time of year for both predators and prey when numbers of voles and other small live prey are likely to be low. Discussion of the Galloway studies has largely ignored the possible effect of open land reappearing at clearfelling. This does seem to be important for Buzzards feeding on the vole crop of the clearfell. However, even at levels of population where density leads to an increase in mortality, deer are unlikely to provide anything like the quantity of carrion of traditional shepherding and we do not know how accessible such carrion would be to birds anyway. Golden Eagles do live in forested places on the continent of Europe but forests in mountainous areas are far more diverse than the first rotation in Galloway, with more permanent open ground. They are scarce in the regular unbroken forests of countries with gentler terrain like Sweden and Finland.

The Peregrine is still a very important species in Galloway. Peregrines increased from 28 to 67 pairs between 1974 and 1982 (Mearns and Newton, in preparation). Of all the birds of prey, their fortunes have been the least connected with forestry. After 1945, as the forests built up, persistent pesticides first caused a massive drop in population followed by the equally dramatic recovery recorded by Mearns and Newton after use of the worst offenders was curtailed. Human persecution has remained a significant threat to Peregrines but it is difficult to untangle this or other lesser effects like forestry from the broad trend. The other factor that would be expected to minimize any effect of forestry is that Galloway Peregrines live almost exclusively on feral or domestic Pigeons passing through rather than living in the area; local land use is of little importance to these transient birds or their Peregrine predators.

Afforestation could affect Peregrine nesting sites, which Mearns and Newton show to be an important factor in breeding success in Galloway. Peregrines select the biggest cliffs possible but as these are not in good supply inland and as the population has increased, some pairs have ended up nesting on the sort of small heathery crag that could easily be obscured by trees planted too close. There is no information on whether this has actually been a problem but in contrast to much of Scotland, where large, nesting cliffs are common, in the rolling countryside of Galloway every rocky outcrop has the potential to hold a raptor nest. Predation, both human and from wild animals, is greater at these sites, some of which it is possible to walk up to without any climb. It is also impossible to tell whether afforestation has had any effect on the levels of this predation.

Two other important raptors have definitely been affected by afforestation in Galloway. Planting has covered large areas of Merlin hunting habitat, and whilst densities were not as high as in the best areas for the species, like the North Yorks Moors, available records do show that a significant number of pairs must have been displaced, especially from very large blocks like Bennan. Although our knowledge of this species is not good it is possible that as much as half the population found between 1968 and 1972 (Sharrock, 1976) has been lost. Watson (1979) studied two nests in forests, one area in late thicket stage and another where Merlins nested at the edge of an area of late thicket stage. Merlins brought the majority of their prey from the direction of open moorland below 250 m elevation, and it was dominated by birds of open land, especially Meadow Pipits and Skylarks. Woodland birds made up only 5.5% of total prey. Little prey was brought from the direction of the higher moorland above 250 m elevation. Although this agrees with Bibby's (1986) finding that Merlins tend to be birds of the moorland fringe, ranging out towards the middle of moors some of the remaining Galloway Merlins do nest up to 350 m elevation (R. Roxburgh, personal communication). In Wales, although over 50% of Merlin food by weight was Meadow Pipits, a number of forest or restock species were taken, including Goldcrests and Siskins (Bibby, 1987).

*Bennan forest is typical of a 1950s' forest - rolling hills covered with pole-stage Sitka Spruce.
Broken only by rides, it is neither a beautiful place to walk nor any use to moorland birds - but
forest birds are surprisingly good and include Crossbills, Siskins, Sparrowhawks and Black
Grouse.*

At least 15 pairs continue to nest regularly in the Galloway hills, and because Merlin are declining generally in Britain, any further loss is of conservation importance. Merlins are hard to get a grip on: Bibby showed in Wales how little relationship there was between territories known to have been used in the recent past and the number of breeding pairs in any one year. Much the same applies in Galloway with pairs moving around for no apparent reason, perhaps one year nesting in heather on the open moor and the next in a tree on the forest edge (R. Roxburgh, personal communication).

Bibby's (1986) Welsh findings also show that Merlins can decline for reasons other than forestry, and the Galloway hills have been subject to a long-term degradation due to overgrazing, but it seems likely in this case that the loss of most territories in recent years has been due to the direct effect of growing forests reducing open land for hunting. A subsidiary effect on the species has been the boom in Peregrines in the last few years: Merlins do not nest on cliffs, but they do use the heathery crags for which Peregrines are now competing. Both Merlins and Kestrels have lost as a result.

In contrast, afforestation allowed Hen Harriers to more firmly re-establish themselves in Galloway. The large tracts of unkeepered ground provided by forestry allowed them safe havens for breeding: the young forest provides them with food but this is not the crucial factor in their survival (Watson, 1977). Their story has, however, been one of boom and bust over the past 30 years. In at least five regions of Britain, numbers increased with expanding afforestation. In Sutherland and Caithness their population may still be increasing. In Wales they are on the verge of extinction less than 30 years after first becoming established in the country. In Galloway, Hen Harriers are now less numerous than when they were at their peak but there is still a substantial population. It is likely that the high proportion of forest now past canopy closure limits hunting, but nesting birds are found in surprisingly small gaps in stands well into thicket stage (R. Roxburgh, personal communication). The security from human predation provided by the forest remains an important factor in breeding success. Unlike the Merlin, the Hen Harrier has once been found breeding on a restock (Petty and Anderson, 1986), and it is one of the rarer species that might benefit from large restocks.

Hen Harriers are not the only birds to have benefited from forestry in Galloway: the characteristic avifauna of the new conifer forests has built up here as elsewhere. Not knowing Galloway as well as the other three study areas, we went and took a brief look: we spent three hours in the Bennan forest in early May, much of it on the Raiders Road forest drive, and sampled a typical area of dense 1950s spruce plantation. The birds included Crossbills, Siskins, Goldcrests, Chaffinches and tits; all were numerous. There were Kestrels, Wrens and Willow Warblers on the limited area of restock. In addition, during this short visit there were two Blackcock and two Sparrowhawks along forest roads, and a Buzzard on a small area of cleared windthrow. On a large loch there were still Goldeneyes, but despite the late date they showed no interest in the boxes provided for them along the shore.

Common Crossbills and Siskins were, as in most other parts of Britain, occasional scarce breeders in the region (Gladstone, 1910). Crossbills only bred after irruptions and many years went by without any records at all whilst Siskins were more regular but rarer nesters in 19th-century conifer plantations. As our visit illustrates one can now expect to see Crossbills on an average day's birding in the forest. Siskins are now so common in spring that they could be the most numerous bird of these forests, outstripping even Chaffinches and Willow Warblers. They are not confined to older trees: they occur in any age from late thicket stage onwards.

Sparrowhawks are common. The work of Newton *et al.* (1986b) on this population has provided some of the most valuable basic information on bird ecology in the new forests. It includes the spin-off study of their prey, small passerines, in the forests of southwest Scotland (Moss *et al.*, 1979ab). Sparrowhawks need trees for nesting, and reasonable-sized woods provide a wider range of opportunities for nest site and reduce the risk of disturbance or predation, especially by humans. However, although ideal for nesting, the Galloway forests of the 1970s lacked open ground and edges, and Sparrowhawks left the forest for most of their hunting.

Black Grouse are still reasonably numerous but there is too little information to assess their recent population history. Loss of open ground may well have contributed to their more general decline, and this is one species worth further study, with the possibility of applying similar habitat improvement techniques to those developed in Wales.

Birds of streams and rivers have suffered a two-pronged attack in this area, from afforestation and acidification. Galloway watercourses do not tend to be well defined: their banks rise gently without any distinct break in the ground and few steep gorges. Smaller watercourses are now shaded out by trees for much of their length. The granite core of the Galloway forest is one of the places most vulnerable to acidification in Britain, with virtually no capacity to buffer acidic precipitation. Water quality varies between stream systems and the pH is particularly low in the Fleet catchment, which runs through the western side of the Bennan forest. Juliet Vickery (personal communication) found a significant correlation between Dipper density and stream pH. The density and species of invertebrate fauna also vary with acidity, and Dippers hold far longer territories on streams with low pH. Common Sandpipers and Grey Wagtails do not seem to be affected by acidity but like Dippers, they do not occur on streams shaded out by conifers. Several Galloway lochs have lost their trout populations or are at a marginal pH for trout survival.

Some of the earliest British liming experiments were tried here and ran into problems with the low water residence time of these upland lochs: quick-acting lime put into the loch will raise pH but is only effective until the water of the loch changes, usually no more than a few months. Slower-acting limestone chips cannot neutralize acidity fast enough to have any effect on acid incidents at peak flow. Now the application of lime to surrounding moorland or forest is being tried. The few lochs in these forests tend to be quite large and surprisingly rich in birds. Loch Moan in Glentrool has a large Black-headed Gull colony, and the gulls commute to surrounding farmland to feed. There are also Greylag Geese. Goldeneyes favour relatively nutrient-poor and acidic lochs and are quite common as non-breeders, and it cannot be long before they breed here.

Because of the speed of afforestation, there are massive areas of closed canopy forest in Galloway but very little clearfell. What little we know of these early restocks gives a better indication of the potential diversity of the forest in the longer term than their present situation. Their birds have not yet been studied systematically. Whitethroats seem to be as numerous as on many other upland restocks, and this habitat must now be rather important for the species in Britain. 1988 was an exceptional year for Grasshopper Warblers and we heard one in a damp flush in Galloway. There is a tiny population of Nightjars in Bennan Forest which was established on the open site of an old nursery but has now spread out into adjoining cleared patches of windthrown spruce.

An important bird in the early second rotation Galloway forests is the Barn Owl. This is one of Britain's most attractive birds and has declined greatly this century (Shawyer, 1987). Barn Owls did well throughout the region's new plantings in the 1950s, 1960s and 1970s, being limited mainly by availability of nesting places. They are still widespread in southwest Scotland (Taylor *et al.*, 1988). Some continued to nest in dilapidated cottages on the retained

farmland of the Glentrool Valley. Most of these buildings are now in the later stages of collapse and increasingly precarious for the owls, and boxes have been put up in the forest (G. Shaw and A. Dowell, personal communication). Initially these were quite elaborate A-shaped wooden boxes, but now a far larger number of large plastic drums with a hole cut for access have been put up.

The project was successful: Barn Owls used the boxes and in 1987 eight were occupied. The winter of 1987-1988 was mild and vole numbers climbed. When the spring came 30 boxes were occupied by Barn Owls and 19 pairs succeeded in raising young to fledging stage; at the time of writing, 51 chicks have been ringed and there are still three broods with young. Vole numbers were so high during the summer that the young Barn Owls were sitting in the drums surrounded by uneaten prey. Most of the newly-established pairs are first-year birds, many of which had been ringed the previous year at farmland nesting sites below the forest. Numbers will not stay at this level when the vole cycle goes into decline, but Barn Owls now seem to be an established part of the second rotation scene in Glentrool, and the prospects for encouraging them in other forests in the region by providing the 'missing link' of a nesting site seem excellent.

Other habitats of value to birds are scarce in the first rotation forest. There are reasonable areas of broadleaves on the lower ground of the new forests. Old Oak woods like the one near Loch Trool have Pied Flycatchers and Redstarts. Broadleaves are spreading onto the lower, richer restocks, a few of which have been taken over by Birch. However, most of the forest is at higher elevation and remote from seed sources. The infrequency of gullies along watercourses meant that trees had been completely eliminated from large areas of open hill before the forest was planted. Large areas of conifer forest have no established broadleaves at all. Planned open space is an equally rare commodity in first rotation plantings. From the top of the retained farmland to the altitudinal plantable limit, worthwhile open space is confined almost entirely to a few wide fire-rides.

Forest design can be changed. It will be a long job in Galloway, but starting from such a low baseline means that even limited improvements can significantly benefit birds. There are two main management problems, money and windthrow. Broadleaved planting is limited largely by available money but this probably does not matter much: small groups of broadleaves are now being widely planted, and if every tree shelter actually sprouts a tree, then broadleaves will have been established in blocks which have been without any until recently. As well as concentrating in visible places like Clatteringshaws Loch, they spread up into deer glades and watercourses in remoter parts of the forest. Introducing broadleaves is likely to be a long process, and there is no urgent need to get them everywhere in that their introduction brings in a habitat that simply did not exist before. In contrast, there is more urgency in clearing young crops back from watercourses which may still hold much of their pre-planting fauna but are in imminent danger of being shaded out and lost. Something we do not know is how well watercourses recover when conifers are removed after shading: the best bet is not to try it in the first place. Again, significant areas have already been treated but priority needs to be given to making sure that no existing pre-thicket crops are allowed to grow on to shade out streams.

The problem is greater for older crops: it is not only that costs of clearance per metre of watercourse increase rapidly with tree size but also that opening up pole-stage crops in Galloway will almost certainly lead to windthrow. This applies equally to opening fresh edges away from watercourses, and the problem of creating age class diversity presents particular problems here because so much of the forest is hazard classes 5 and 6. Strict planning with the expectation of carrying it through near enough as envisaged is not realistic: the wind will certainly blow and wreck precise boundaries or felling sequences.

This does not mean it is not worth planning; as at Kielder, edges can be left in the hope that they will stand. Some will, some will not. There are other opportunities: high pulp prices locally mean that the penalty of felling very young crops 10 or even 15 years before rotation age is far lower than normal. In pure nett discounted revenue terms this may not be ideal, but for age class diversification it is a gift: it could introduce restocking into blocks that only closed canopy 15 years ago, vastly accelerating the cycle of habitat availability.

The broadleaved area should expand to become a significant area of native broadleaved trees, intermingled with open space for birds like Black Grouse, Buzzards and the riparian species. Some of the lower yield class, high-elevation conifer crops could move towards tree-line broadleaved scrub, a habitat the potential of which we can only guess at in Britain today, and although Galloway is a long way south, it is one place where there is forestry land ownership right the way to the high tops.

So, present forest design is poor but a start has been made on improvements and the potential for creating a forest far more valuable for birds in the long term is good. The extent of the forests is rather more permanent. As we have discussed, large blocks of forest are not necessarily a bad thing, because at present levels of forest cover, concentrating afforestation in one place leaves other areas of open ground completely unaffected. Valid for areas of 10 000–

20 000 ha, the Galloway situation is rather different in that a large proportion of a whole region has been planted. This spectacular change in habitat has eliminated open ground in the middle altitude range around most of the central Galloway hills. For once it is possible to demonstrate a clear effect on one wide-ranging species of bird, the Raven. We have less firm information for the Merlin but it has clearly suffered too. As in other cases, the forest has increased the diversity of the region's avifauna by allowing forest birds previously absent to come in, but it has now reached a level where there is a demonstrable and serious decline in open-land species to the extent that extinction of some at a regional level is a possibility.

The comparison with mid-Wales shows that quite high levels of afforestation may have little effect on Ravens, Kites and Buzzards. In Galloway, planting has gone that much further, further than would be seen, with our present perspective, as being a reasonable balance between land uses. The change has been exacerbated by its speed, both in its impact on human perception and in that the potential cushion of restocking for open-land birds comes only after most of the area has been covered by closed canopy forest for several decades. In contrast to Breckland and the Flow Country, the question in Galloway is not whether there should have been forestry; in the overall perspective of upland land use it would always have been a preferred area on both economic and ecological grounds. It is rather the question of extent. Available evidence supports the charge that too high a proportion of this region has been afforested. With the vast area of the open uplands there is no need for any one place to be so heavily planted unless we as a nation feel we need a vastly higher forested area than at present. The evidence from Wales suggests that quite a high level of forestry can be reached with little impact on wide-ranging birds like the Raven.

There is no practical prospect of reversing this situation: forests were planted because of the fragility of agriculture and it would be necessary to re-establish sheep farming on a grand scale to bring back birds like the Raven. There is, however, a practical lesson for the development of forestry elsewhere: over much of the uplands it may be possible to have the best of both worlds, a limited decline in the conservation value of the open-land avifauna and the addition of forest species; but there are limits (possibly quite high, as in mid-Wales) to how much can be planted before the former starts to suffer badly.

THE FLOW COUNTRY

Nothing has brought the conflicts between forestry and conservation into the public arena more than the dispute over the Flow Country of Sutherland and Caithness. It seemed that almost every week through 1987 and 1988 one of the Sunday papers ran a feature on this issue. Understandably the arguments about birds have sometimes taken a back seat to the good copy to be made from castigating celebrities for their heavily subsidized investments. However, the fact that Terry Wogan owns some trees in the Flow Country is not, we think, enough by itself to mean that afforestation here is a bad thing. Although

Ploughing around a dubh lochan system in the heart of the Flow Country. Softness of the ground determines where the ploughing stops. The pools are left untouched – but such a small area may be of little use to birds. Local drainage in the peat is very limited so the pools could go on holding water – but the way a raised bog works on the larger scale is not clear and ploughing could pull the plug on the whole system.

it is difficult to persuade the general public that these apparently barren wastes in the far north of Scotland can be good for birds, and that forests can be bad for them, the case for conservation is at its strongest in the Flow Country. But through all the politics and controversy the validity of the arguments in favour of both birds and trees have been questioned. What is the strength of the conservation case and is it strong enough to justify a halt in tree planting?

The name 'Flow Country' is a fairly recent invention, first published by the RSPB in the 1985 report which really launched the public controversy. It graphically describes the unique flat, wet basin of western Caithness and eastern Sutherland. Huge, raised peat bogs, growing into shallow domes over thousands of years in the cool, humid atmosphere are topped by strings of tiny pools, deep cracks in the peat, called 'dubh (the Gaelic for black) lochans'. The depth of the peat at the middle of the bog can be as much as 10 m. Individual raised bogs, even small groups, occur in many parts of Britain. What is extraordinary about the Flow Country is the extent and size of the bogs, which stretch one after another in an unbroken series for over 20 miles. As remote from Edinburgh as it is from London this is a little-known land, monotonously boring at first sight but with a peculiar magic when the sun glitters from a sinuous string of pools. Well off the tourist track today, it is a place with a growing number of devotees, ourselves amongst them.

Water is the key to the Flow Country's rich birdlife and the best areas for birds are the low-lying flat bogs which support many species of moorland bird. The RSPB selected the 800-foot contour as an arbitrary boundary for what is now known as the real Flow Country. This definition, made largely on political grounds, has turned out to be an excellent definition in terms of bird distribution; similar habitats exist in other parts of Caithness and Sutherland but not in the same quantity as in the Flow Country proper. To the west the character of the land changes markedly from low-lying flat ground to some of the most dramatic mountain scenery in Britain, including peaks such as Arkle, Foinavon and Canisp. The hills to the south are less spectacular but similarly break the continuity of the flat blanket bogs of the Flow Country. They are covered in shallower peat, very different from the deep peat of the Flow Country. The distinction is important: the argument which was initially confined to the 184 000 ha of the Flow Country proper (Bainbridge *et al.*, 1987) was extended by the NCC to all peatland in Caithness and Sutherland, something like 2/3 of the 764 094 ha of the two counties (Stroud *et al.*, 1987, Lindsay *et al.*, 1988). Whatever the conservation arguments for the international importance of the peat landforms themselves (Lindsay *et al.*, 1988), the distinction is vital for birds: it is the bogs of the Flow Country proper that are really exceptional for birds.

Forestry in the far north actually has a far longer history than is generally realized: the FC began planting soon after its formation in 1919 and has continued ever since. The deep peats proved an intractable technical challenge until the late 1950s, when some experimental plots were planted. By the 1970s deep peats were being planted, but mainly with pure Lodgepole Pine, not an attractive silvicultural or economic prospect despite the availability of large areas of cheap land. However, the limited area planted, amounting to 300–400 ha most years in and around the Flow Country proper, did not attract much attention. It was the much larger scale of the Fountain Forestry operation, which peaked at nearly 4000 ha in one year, that was an obviously massive, rapid and threatening change to the bogs of the Flow Country.

There were two key factors in Fountain Forestry's sudden expansion into this area; the difficulty of obtaining land elsewhere and the progress that was made in the technology which allowed the afforestation of formerly unplantable land. The lack of land for planting in the late 1970s was largely due to the restrictions put by the DAFS (Department of Agriculture and Fisheries Scotland) on the release of land from agriculture to forestry. The DAFS did not object to planting in the Flow Country, because by the late 1970s large areas of this bleak, infertile land were not being grazed or were only grazed at very low intensity; this is still the case. There was still the problem of getting worthwhile trees to grow on the deep peats. Although the earliest Sitka Spruce planted on mineral soils at Borgie showed the area to be climatically capable of growing trees the bog peats were extremely deficient in nutrients (Zehetmayr, 1954; Wood, 1974).This remained a problem long after wide-tracked ploughing tractors had been developed to solve the other big problem, site drainage. Lodgepole Pine will grow under normal fertilizer regimes of phosphorus and potassium but gives a poor yield, generally no more than GYC 8, and, however cheap the land, this does not give an acceptable rate of return.

However, by the late 1970s FC experiments planted in the 1950s and 1960s were showing that pure Sitka Spruce could be grown on these sites with frequent, and expensive, additions of nitrogen. Again, this is not an economically attractive option. In amongst these experiments were some stands of Sitka Spruce mixed with Lodgepole Pine or larch, and it was found that these mixtures would grow, without added nitrogen, at GYC 12–14 by 20 years old and were in pole stage, whereas stands of Sitka Spruce with otherwise identical treatment were in pre-thicket check. This was the key breakthrough. The prospect of GYC 12, close to the average conifer growth rate for Scotland, on cheap and available land, was good enough for Fountain Forestry to move in and, as a private company independent of annual government cash allocations, forge ahead with planting much faster than the FC could contemplate.

Much has been written about the trees of the Flow Country. Depending on the point of view, it is possible to select references suggesting that few trees will reach a harvestable size at all, and others making out that this is one of the best possible places for tree growth. In fact, the Flow Country is very far from the best place in Britain for commercial tree growth but the prospects are also better than those of a number of other areas planted during the past 20 years. It is true that there is a very real element of risk in the Flow Country planting: no example of the staple Lodgepole/Sitka mixture has yet reached maturity but on the other hand the experiments on which the system is based are much older than many which were successfully put into practice by the FC in the past. Windthrow will occur; it is a question of when, not whether. However, rooting in deep peats is better than that in peaty gleys, added to which the Flow Country is protected from the prevailing wind by the west Sutherland mountains meaning that it is likely to be less at risk than the large areas of forestry down Scotland's exposed west coast; the projected average hazard class of 5 looks sensible, meaning that trees will reach a worthwhile size before they blow over.

The unexpected can always strike, and did so in one of the worst pest problems to hit British forestry when the Pine Beauty Moth attacked Lodgepole Pine in Sutherland and Caithness in the 1970s. The effects were rapid and dramatic. Several hundred hectares have been killed by complete defoliation. In areas such as Strathy Forest and Rimsdale the bare dead trunks still stand as a warning of the damage that can be done to human aspirations by mere insects. Pine Sawfly, which can retard growth rates and sometimes result in death of the trees, has also made its presence known in these new forests. Research by FC staff showed that the problems of Pine Beauty Moth damage were much greater for Lodgepole Pines which were planted on deep peats than those on the more mineral soils (Leather and Barbour, 1987); no doubt because life on the Flow Country peatlands is so difficult for trees that any extra stress, even perhaps of such a nature which normally would occur without damage, causes reductions in tree growth potential. Conservationists have been keen to point out the harsh facts of life facing trees in this area, but the fact remains that money has poured into Caithness and Sutherland in order to plant trees. The reasons that afforestation can be expected to be profitable in areas which are not very good for growing trees are explained in the next chapter, but have been based on cheap land prices and a favourable fiscal system.

We have stressed the fact that the whole of upland forestry in Britain has been about risks. The risks in the Flow Country are certainly higher than average, but how do they compare with the original selection of Sitka Spruce as the principal species of the vast pre-war plantings at Kielder? Much of the emphasis on the risks to trees in the Flow Country has been based on the premise that to protect the birds it is necessary to discredit the viability of forestry and thus stop the investment of government money which will stop planting. Although politically this may well have been true at the time the Flow Country dispute blew up, it should not be the basis for a factual analysis: if the bird conservation value of the Flow Country is high enough, it should be conserved regardless of how well the trees might grow.

Looking at most current maps, the area around Altnabreac Station, on the railway line from Inverness to Wick and Thurso, appears to be one of the most desolate in Britain, with no habitation for literally miles in every direction. However, a passenger alighting from the train at Altnabreac today finds himself in one of the largest conifer plantations in Britain. The vast majority of the trees in this area have been planted in the past few years, since the late 1970s, by Fountain Forestry. This plantation can be looked on in different ways depending on one's standpoint. It is difficult to remain unimpressed by the sheer size of the achievement. Vast areas of trees have been planted in some of the most inhospitable land in Britain. A network of rough forestry roads has been built, forest blocks have been fenced and huge ploughs have scored deep furrows through the thick peat. Millions of Sitka Spruce and Lodgepole Pine have been planted. It is a salutary lesson in the speed with which ecological change can occur when economic pressures are strong and a company is technically efficient. If large-scale forestry can have such a dramatic impact here, and in so little time, is anywhere in the uplands safe?

A conservationist arriving at Altnabreac station, however, would survey the scene differently. Here is a disaster for the conservation cause. Unarguably this area is one of the most important for upland birds in Britain and it is now under great threat. There is also the fact that this was until recently one of the largest areas of relatively untouched moorland in Britain. The sense of loss of a unique wilderness is strong, and underlies the thinking of many of the conservationists involved in this issue. Both the NCC and RSPB have invested considerable time, money and effort into arguing against the spread of afforestation in the Flow Country. Each has produced evidence that this area of northern Scotland is nationally and internationally important for moorland birds (Royal Society for the Protection of Birds, 1985; Bainbridge *et al.*, 1987; Stroud *et al.*, 1987; Lindsay *et al.*, 1988). How strong is this evidence?

An initial estimation of the importance of the Flow Country for birds can be based on the richness of upland species which it holds. The exercise based on the breeding Atlas (Chapter 5) forms a useful background to this. One of the randomly chosen plots in our analysis fell inside the Flow Country and happened to have the second highest species richness and easily the highest conservation importance of all the plots in our sample. Was this just chance or is the Flow Country really that much better than the rest of the British uplands? To test this we have worked out the species richness and conservation importance of all 96 Atlas squares which fall mainly into Caithness and Sutherland (Figure 7.4). This analysis showed that 67% of 10-km squares in Caithness and Sutherland have higher than the national median number of moorland species (12), even though many of the Caithness squares would not have been classified as moorland squares in our previous analysis. One 10-km square in Sutherland held a remarkable total of 25 moorland species; lacking only Eiders, Dotterels, Whimbrels, Wood Sandpipers, Great Skuas and Snow Buntings from our list of 31 moorland species (all of these, except Snow Buntings, were recorded in Caithness and Sutherland during the Atlas fieldwork, and one of us saw a Snow Bunting on a Sutherland peak in May 1986). Three other squares contained 23 species. In terms of the index which we developed in Chapter 5, 80% of Caithness and Sutherland 10-km squares are above the national average for conservation importance. This analysis is enough to confirm what conservationists have been claiming vociferously since the early 1980s, that Caithness and Sutherland are truly exceptional in national terms for their moorland bird communities.

When only that part of Caithness and Sutherland which is regarded as the Flow Country by the RSPB is considered, the outstanding nature of the area is even more evident. The average number of moorland species found in the 10-km squares which fall mostly within the RSPB Flow Country is 18, compared with 15 in the whole of Caithness and Sutherland and 12 in our national sample. Using our index of conservation importance, the Flow Country squares have a median value of .23, which is considerably greater than that for the whole of Caithness and Sutherland (.13) or for Britain nationally (.08). All of the Flow Country 10-km squares have a conservation importance for moorland birds (as defined in Chapter 5) which is above the national average, and this relatively small area (22 10-km squares) contains 9 out of the top 10 in conservation importance for Caithness and Sutherland. This analysis confirms not only the national importance of Caithness and Sutherland for moorland birds but also the extraordinary importance of the much smaller area known as the Flow Country.

The above analysis, because it is based on Atlas data, might be regarded as being out of date, but recent evidence supports its findings. Surveys carried out by the RSPB in 1986 and 1988 supported the view that this area held the richest assortment of upland breeding species in Britain. Although the sample plots cover a wider area than that of the core Flow Country, on nearly all plots Greenshanks, Dunlins and Golden Plovers were found. However, there is much else besides; Red-throated and Black-throated Divers, Wigeons, Common Scoters, Greylag Geese, Arctic Skuas, Hen Harriers, and Merlins are widely found (Avery, in preparation; Stroud *et al.*, 1987). Several northern species which rarely breed in Britain have either bred or been seen in suitable

Species richness of moorland birds in Caithness and Sutherland

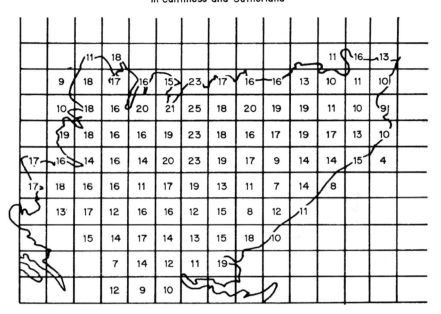

Conservation index of moorland birds in Caithness and Sutherland

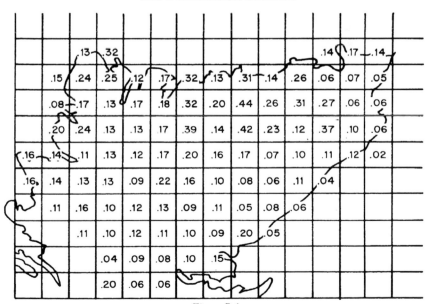

Figure 7.4

breeding habitat in this area. These include Scaups (personal observation), Whooper Swans (personal observation), Temminck's Stints (personal observation, Stroud *et al.*, 1987), Dotterels (personal observation), Wood Sandpipers (Stroud *et al.*, 1987, Nethersole-Thompson, 1986) and Red-necked Phalaropes (Stroud *et al.*, 1987); even a Pectoral Sandpiper has displayed over a remote cluster of dubhlochans (Nethersole-Thompson, 1986). A day spent bird-watching in the Flow Country in May or June is exhilarating; one feels that almost anything could be found, and a greater variety of moorland birds will be seen than anywhere else in Britain. The birds can be fickle, though; in June 1969 Derek Ratcliffe walked for 15 miles through the Flow Country bogs without seeing a Greenshank (Nethersole-Thompson, 1971).

Another way of assessing the national importance of an area is to calculate the percentage of the national populations of a bird species which occurs in the area. This is a common way of ascribing national or international significance to small sites which have fixed and non-arbitrary boundaries, such as estuaries or reservoirs, but there is a large element of arbitrariness in doing this for large and artificially defined areas such as administrative units. The NCC has calculated the percentage of several species which nest in Caithness and Sutherland and shown that for several species the percentages of the total British populations are high. However, this gives little information on the relative importance of the area independent of its size. Because Caithness and Sutherland cover a big area, it is not surprising that they contain many birds. Without providing information on the percentage of the total area of the country or of the appropriate habitat, such figures for large areas are relatively meaningless; after all, 100% of Britain's Golden Plovers occur in Britain, but does that mean that all of Britain is important for Golden Plovers? Another disadvantage with this approach is that there is insufficient information to estimate accurately the numbers of several of these moorland species. Thus it is interesting to see that the estimated Dunlin population of Britain has varied widely over the past few years; such fluctuations raise doubts over the accuracy of the current (temporary?) figures. There has also been confusion over the numbers of birds present in the Flow Country itself. RSPB figures (based on NCC estimates [Royal Society for the Protection of Birds, 1985]) conflict with the NCC's most recent figures. For example, RSPB reports suggest that the Greenshank population of the Flow Country prior to afforestation was 720 pairs (Royal Society for the Protection of Birds, 1985, Bainbridge *et al.*, 1987) whereas NCC figures suggest that the Greenshank population of the whole of Caithness and Sutherland was only 760 pairs prior to afforestation. The fact that foresters have not usually challenged the validity of these estimates may perhaps indicate that they have chosen to ignore the conservation case rather than assess either its soundness or even consistency. The important point remains that estimates of the bird populations show that Caithness and Sutherland hold large proportions of the populations of several rare bird species.

All the available evidence shows that the Flow Country is of unique importance on the British mainland for moorland birds, and that therefore forestry in this area is possibly more damaging to the conservation of birds of

A Greenshank, symbol of the Flow Country battle, settles on its eggs. Sparsely scattered across the vast flat wilderness of the Flow Country, Greenshanks are good indicators of the very best wet areas. Also a bird of the forest bog, they have retreated from the native pinewood bogs of the Spey valley and are now known mainly from open moorland in Britain.

the open uplands than if it happened anywhere else in Britain. It is therefore tempting to blame the foresters for this destruction and assume that they knew exactly what they were doing. This would be unfair. From the published information and the general level of public awareness, we would say that this is a perspective which is only possible with the benefit of hindsight. Neither of us was aware of the significance of the Flows before 1984. It is worth pointing out that the great value of the Flow Country for birds has only been realized very recently. The great importance of the area was not fully realized by either the NCC or the RSPB before forestry moved in and started to change the area. The NCC selected a number of sites as being nationally important for their peatlands or their upland grasslands (Ratcliffe, 1977b). Four grade 1 and three grade 2 peatland sites were identified in Caithness and Sutherland. Of these, southern Parphe, Knockfin Heights and A'Mhoine lie outside of the Flow Country. Forsinard and Badanloch Flows, Strathy River Bogs, Dubh Lochs of Shielton and Blar nam Faoileag lie inside. Five grade 1 and seven grade 2 upland grassland sites were identified. All of these sites lie outside the Flow Country. The sites identified as important upland grasslands or peatlands in Caithness and Sutherland covered an area of just under 70 000 ha, of which only 16 150 ha was peatland. Considering British blanket bogs alone, for which Caithness and Sutherland are regarded as internationally important, only 8300 ha were identified by the *Nature Conservation Review* as being of sufficient quality to rate as grade 1 or 2 sites out of a national total of 53 730 ha. There are no peatland nature reserves in the area even now, which has allowed foresters to claim that conservationists have been acting more to stop forestry than to protect bird populations. Thus it could be argued that whilst the *Nature Conservation Review* did identify parts of Caithness and Sutherland as being of high conservation value, the emphasis was put largely on upland grassland sites, and the importance of the peatlands, in either a regional or a national context, was underplayed. This is hardly surprising, because identifying areas of nature conservation value is an extremely difficult job, but the bitterness of the arguments over afforestation in this area stems partly from conservationists' realization that they had not spotted the value of the area in advance of Fountain Forestry's arrival on the scene.

Foresters can be given the benefit of the doubt that they may not have predicted either the damage to wildlife which their activities would do, or the hornets' nest which they might disturb. We will never know whether prior knowledge of either of these things would have slowed down the rate, or reduced the extent, of afforestation in the Flow Country. It could at least be hoped that if all afforestation in the area had been carried out by the FC, rather than by a public company, the public outcry would have had an effect rather earlier than it actually did. Throughout the press campaign of the late 1980s, when the Flow Country was the major conservation issue in Britain, Fountain Forestry was able to continue ploughing and planting areas which had already received Forestry Grant Scheme approval in a way that the FC simply would not have got away with while spending public money. As it was, the FC sold some of its land in the area during the course of the public outcry but this had little effect on planting levels, since the buyer was Fountain Forestry.

It has been argued that Caithness and Sutherland represent an area of international importance for their bird communities. Is this claim justified? We would state that this is stretching the evidence a little, and is the sort of argument which shows the conservation cause in a bad light. The Flow Country blanket bogs really are far from being habitats of the same international standing as South American tropical rainforests or the Serengeti plains (with which they have been compared by the NCC (Pienkowski *et al.*, 1987)) nor would we rank them as the most internationally important area for birds in Britain; Britain is internationally important for moorland birds because historical deforestation left it with more open upland than most other European countries. However, its location at the western extremity of Europe makes it more significant for wintering birds, especially wildfowl and waders, and for its breeding seabirds. Sites like the Cairngorms, the Wash, Islay, St Kilda and the Ouse Washes have at least as good or probably better claims to international importance for birds than the Flow Country.

Part of the case for the international importance of the Caithness and Sutherland peatlands comes from the fact that the UK government has international obligations under various international conventions. One of these, the Ramsar Convention on Wetlands of International Importance, sets quantitative criteria for identifying important wetlands. The NCC argues that the Caithness and Sutherland peatlands meet these criteria. This is true in the same sense that Sutherland and Ross and Cromarty probably also meet the criteria, and northern Scotland certainly does.

If we consider such large areas, almost all parts of Britain should be considered as Ramsar sites; after all, the districts of Caithness and Sutherland occupy an area larger than the combined estuaries of the Wash, Severn, Exe, Solway and Morecambe Bay. The Bern Convention requires governments to protect endangered natural habitats and those which include species listed in its Appendices 1 and 2. Blanket peatland is an endangered natural habitat and so meets this criterion of the Convention. However, the list of important species includes rather few that are present in important numbers in Caithness and Sutherland. Black- and Red-throated Divers, Dunlins and Merlins are the four species for which the most convincing case could be made that conservation measures within the Caithness and Sutherland peatlands would be effective ways of furthering their conservation. However, for species such as Little Grebes, Peregrines, Ringed Plovers, Wood Sandpipers, Common Sandpipers, Temminck's Stints, Red-necked Phalaropes, Pied Wagtails, Grey Wagtails, Wrens, Whinchats and Stonechats the Flow Country would not be the most suitable place to achieve conservation benefits. One more species on the list, the Short-eared Owl, might actually be helped more by afforestation than by moorland conservation.

The EC Directive on the Conservation of Wild Birds identifies certain species as requiring habitat protection. Relevant species in Caithness and Sutherland are both Black-throated and Red-throated Divers, Golden Eagles, Hen Harriers, Peregrines, Merlins, Golden Plovers, Wood Sandpipers, Red-necked Phalaropes and Short-eared Owls. Again, if one was considering how the UK Government should meet its obligations under the Birds Directive, it is difficult to see that, except for Golden Plovers, Golden Eagles, Merlins and the two diver species, stopping afforestation in Caithness and Sutherland would be likely to have very great effects. And Golden Eagles have so far almost certainly been unaffected by afforestation on the blanket peatlands of Caithness and Sutherland. All in all, the arguments about international importance appear to be the result of political manoeuvring rather than of sound evaluation of the conservation value of Caithness and Sutherland. The choice by the NCC to consider the whole of Caithness and Sutherland as the area under consideration may have resulted from a sound grasp of the tactics of political influence rather than from the logic of the conservation arguments that could be based on it. This does not mean, we reiterate, that we would support further afforestation in the best parts of the Flow Country, but it does mean that we think that the conservation arguments which are based on the area's national importance are sufficiently convincing, and much more so than a weak argument based on the international importance of an enormous area most of which is unthreatened by forestry.

The initial arguments over afforestation in the Flow Country arose when it was realized by conservationists that the private forestry company Fountain Forestry had acquired large areas of land in western Caithness and eastern Sutherland; and that planting was going ahead rapidly. Approximately 32 600 ha have been either planted or received forestry grant scheme approval from the FC (Royal Society for the Protection of Birds, 1987). Fountain Forestry's holdings are mostly in the area on either side of Strath Halladale, which is in the centre of the area which the RSPB defined as the Flow Country. The public outcry over afforestation in this area derives from this fact; that it is right in the middle of the best part of the best area for moorland birds in Britain. The NCC has widened the debate considerably by choosing to consider the whole of the two districts of Caithness and Sutherland as being the area under threat; this is only true in the sense that most of the British uplands are potentially afforestable. There are two consequences of widening the issue in this way. The first is that, since a larger area is considered, the importance of the area's bird community is apparently enhanced. The figures given for the percentage of the British populations of some of the species in Table 7.2 are much higher for the larger area considered by the NCC than they would be for the smaller area which is under real threat from afforestation. For example, 6% of Britain's Golden Eagles nest in Caithness and Sutherland but there are very few pairs nesting in the Flow Country; a look at a map shows that nesting sites are rare in the flat peatlands that characterize the area and which hold many of the best birds. Thus by enlarging the area considered, and neglecting to say that only part of it is under real threat, the NCC has provided a misleading and exaggerated picture of what is happening within the whole of Caithness and Sutherland. However, widening the issue also has the effect of failing to stress the size of the effect of afforestation in the smaller area over which most of the arguments have occurred. This is particularly serious when, as we have shown, this smaller area includes the areas of the very highest conservation value. Plantations have covered a higher proportion of the core Flow Country than of Caithness and Sutherland generally because it is all flat, and with the exception of the actual dubh lochan systems, ploughable; the most recent plantings have been the ones which have been most damaging to bird populations because they have been concentrated on the very best areas for birds. The NCC's data (Stroud *et al.*, 1987; Pienkowski *et al.*, 1987) show that the most recent afforestation in the area around Strath Halladale, since 1980, has had a much more damaging effect per unit area than did the earlier afforestation, because it is mostly on landforms A and B, which are the best for

TABLE 7.2: *Percentages of British populations of some moorland birds which nest in Caithness and Sutherland (from Stroud et al., 1987).*

Red-throated Diver	14
Black-throated Diver	20
Common Scoter	39
Wigeon	20
Hen Harrier	5
Golden Eagle	6
Merlin	5
Peregrine	5
Golden Plover	18
Dunlin	39
Greenshank	66

wading birds. This reinforces the points made in the previous chapter that the positioning of afforestation is very important in determining its effects. Although forestry in Caithness and Sutherland will have greater harmful effects on moorland birds than forestry elsewhere in the country, it is still true that within this exceptional area some areas are more exceptional than others. The debate which has been centred on the activities of Fountain Forestry has been so great because conservationists feel, and our analysis has supported this feeling, that the areas planted recently have been some of the best areas of this exceptionally rich area for moorland birds.

What of the birds of the young plantations? This aspect has not been well-studied but these young forests are certainly not birdless. At this stage, when most of the plantations are very young, many open-country birds remain to breed. We have seen several pairs of Greenshanks leading their chicks through young plantations (as have Nethersole-Thompson and Nethersole-Thompson, 1979, 1986) and even found Lapwings, Redshanks and Golden Plovers nesting on the rides and at the edges of plantations. So it would be misleading to suggest that these birds simply disappear from the plantations as soon as the trees are planted. As pointed out earlier, many open-country species will persist for a few years before the trees become large enough for canopy closure to occur. These open-country species will be excluded from the growing forest as it ages but they are not all immediately displaced. Hen Harriers and Short-eared Owls hunt over these massive new plantations, and the owls certainly appear to be commoner inside the plantations than outside.

These huge new plantations may now be at their best for birds, and it will be downhill all the way from here. Many open-country species are, apparently, lost immediately the trees are planted. We do not know of any nests of Greylag Geese or Arctic Skuas within the planted areas and suspect that they are absent. Dunlins, too, seem to avoid nesting within the planted areas, although we have found nests within 20 m of ploughlines on the edge of plantations. It is curious that in *Birds, Bogs and Forestry*, the NCC provides a list of the birds which are found in plantations in Caithness and Sutherland (without explaining where these data come from) yet all the species they name are the species of the oldest forests and none of them are characteristic of the early scrub stages. Many of the rarer species to be found in the forests are missed from the list. However, even if a full list of forest species is compiled, the balance of the conservation value is much greater in the bird communities of the open moorland than of any stage of the commercial forest. The bird communities of some parts of the Caithness and Sutherland flows are by far the richest moorland bird communities in Britain, whereas the commercial forests will always lack many of the more southern woodland birds and hold few nationally rare species. However, from a purely regional point of view, bird diversity will undoubtedly be increased. This provides a good example of how little use is the concept of diversity in measuring conservation consequences. In Caithness and Sutherland no moorland species will be completely lost in the next 20 years due to afforestation, but there is a good chance that many species will increase their ranges considerably. In Caithness, for example, we would expect to see increases in the numbers of Buzzards, Tawny Owls,

Siskins, Crossbills and Redwings and probably colonization by Ospreys and Goldeneyes. There is even the hope that the northern forests could attract new breeding species to Scotland but this remains an optimistic longshot.

Could forestry have been accommodated in Caithness and Sutherland with less impact on moorland bird communities than it has actually had? The answer to this question is definitely yes. The NCC's analysis of the distribution of wading birds shows that they are extremely patchily distributed, and that the good areas for waders contain high proportions of the total numbers of birds. There is no doubt that had forestry been planned so that these wader hotspots were avoided then the same amount of forestry could easily been fitted into this area with much less damage to conservation interests. Such a situation is a pipe-dream under present conditions, since the last thing that a forestry company would do in the present climate is to ask a conservation organization for advice on where to plant. Also, it is unlikely that any conservation body would have the necessary information to provide sound advice on the conservation value of different areas for remote parts of Britain such as the Flow Country. Some would argue that because the damaging effects of afforestation can be so large, it would be reasonable to make an environmental impact assessment a condition of the granting of forestry grant scheme approval. Recent EC legislation makes this a possibility.

The Flow Country has been the scene of the most heated arguments between conservation and forestry. Although there have been exaggerated claims of international importance and some very dubious juggling with facts and figures on both sides, it is inescapable that serious harm has been done to Britain's most diverse moorland bird communities. Although conservationists can reasonably be thought to have overreacted to the effects of forestry in some other parts of the country, the Flow Country is different: there was so much more here to harm. The gains from forestry are likely to be similar to those in other areas, although perhaps it might be shorter of woodland species because it is so far north. Arguably welcome on the poorest uplands for birds, the forest assemblage can never substitute for what has gone in the Flow Country. We have argued that the problem of afforestation and birds is largely about what was there before: much of the uplands are unexceptional but the Flow Country is unique. The density and value of the moorland birds in the Flow Country proper meant that any planting would have harmed something of value. This is one place where it is possible to justify the maintenance of the existing habitat purely on bird conservation grounds regardless of how well the trees might grow. The Flow Country is entirely about whether there should be forestry, not what sort of forestry. These are quite separate issues; in fact, Fountain's planting has been carefully done, incorporating broadleaves and protecting watercourses, but this is not the issue. However carefully planted, these forests cannot substitute for the Greenshanks, Dunlins, Golden Plovers, Arctic Skuas and Scoters they have displaced.

Few organizations come out of the Flow Country with great credit. Foresters, both private and public, have to accept that their protestations of care for environmental issues ring hollow in the face of the evidence on the ground; thousands of hectares of nationally important moorland have been lost already. The NCC can take little credit from its performance in the Flows, either; the same thousands of hectares of planted land form a testament to its inability to rapidly muster a convincing argument against afforestation of the area. The RSPB emerges, we feel, with its reputation enhanced since it was largely responsible for building up a large public reaction against forestry in this area.

The arguments about birds have certainly dominated the debate over the future of the Flow Country. The NCC has had the difficult task of trying to provide recommendations for the conservation needs of other taxa, including plants and invertebrates (Lindsay *et al.*, 1988). One conclusion of that report is that the needs of groups such as freshwater invertebrates are not necessarily the same as those of birds. When, as is usually the case, the information on birds is so much more detailed, even if often still inadequate, than that for other taxa, there is a very great danger that the incompletely known needs for bird conservation may unintentionally result in harm being done to other groups. Without further work it will be difficult to know whether this has happened in the Flow Country or not.

We may now be seeing the end, or at least the beginning of the end, of the Flow Country debate. In January 1988 the Secretary of State for Scotland announced limits to the amount of land which would be afforested and also set areas which could be notified as SSSIs. While neither foresters nor conservationists will be likely to admit to being happy about the result, it is now certain that the area protected for moorland birds will be massively extended, hopefully to incorporate most of the raised boglands of the Flow Country proper.

A long-awaited report of the The Highland Regional Council working group on the Flow Country should be published during 1989. It looks like being a typical British compromise which will leave both foresters and conservationists with something to continue to moan about. Whether it will actually clarify an increasingly complex situation remains to be seen. The report divides Caithness and Sutherland into four categories of land with different expectations of forestry occurring in them. Land is classed as Unplantable, Undesirable, Possible or Preferred for forestry. The Unplantable class contains much of the higher ground to the west of the Flow Country itself which was probably never in very great danger of being planted (but which incidentally contributes much to the NCC's case that the whole area is of international importance). The Undesirable class appears to have been loosely based on the RSPB's Flow Country boundary, which must be regarded as a victory for the conservation cause; and for the birds. The Preferred land class is mostly coastal and includes areas of agricultural land, whereas the Potential class includes remaining areas of moorland outside of the Flow Country proper. It remains to be seen how closely subsequent events will follow these recommendations. If no more of the land in the Undesirable land class is planted, then conservationists can feel pleased with the way that conservation interests were in the end recognized and taken into account. However, two regrets will remain. First, it took large losses of land which is now recognized as being of high conservation quality before the implications of what was happening were realized let alone any restriction on afforestation imposed. Second, it may take similar amounts of time money and effort in other areas before conservation interests are given a place in forestry planning.

Paradoxically, because of the damage that has been done to important bird populations, conservationists come out of the Flow Country debate with more to feel happy about than do foresters. Despite the losses of birds, the Flow Country may be seen as a turning point in the arguments between conservationists and foresters. Because the harmful effects of forestry on the upland birds of Caithness and Sutherland have been so powerfully advertised by conservationists' use of the media, many people's attitudes to forestry have been shaped by this debate. There is now a much greater public interest in forestry than there was a mere five years ago, and to many people forestry is now a villain of the upland scene. The Flow Country debate has done considerable damage to forestry's image, with serious implications for future new planting. It has been largely responsible for the major changes in the taxation arrangements for forestry which were so important to the Flow Country planting. Forestry has big and obvious problems: it has been a clear loser. Whether the industry has the ability and vision to sort itself out remains to be seen.

It will be easier for the conservation organizations which have achieved much of what they set out to do to ignore the equally pressing problems the issue raises for them. Most important of all is how did they miss the key importance of the Flow Country in the first place? The standard excuse of 'lack of resources' simply will not wash: vast amounts of money have been spent on upland research and, as we have shown, the value of the Flow Country is not subtle. Any experienced upland birder should have been able to recognize its value in a matter of a few days' fieldwork. The NCC in particular has a duty to ensure that this sort of thing does not happen again. The other big problem is what to do next. The intensity of the Flow Country dispute has pushed both the NCC and RSPB towards an anti-forestry policy, not through any firm intent but simply because it is easier to deal in black and white rather than the grey areas. They have not been able to fill the vacuum left by their success, and their enhanced power in the upland debate brings with it an increasing responsibility to participate in finding solutions.

Forestry in the Flow Country has undoubtedly done significant harm to Britain's best upland bird habitats. Large areas of the best land for birds do remain untouched and are well worth saving. Although what has happened in the Flow Country may set a political pattern for the future, in factual terms this area is without doubt unique in Britain and although there are limited areas of equal importance nowhere else combines the scale and significance of the Flow Country.

CHAPTER 8

Money and power in the uplands

The interaction between forestry and birds is a complex biological issue which is more complicated politically and economically than we have so far described. Forestry concerns money, power, vast areas of land and a massive investment of taxpayers' money. In this chapter we look at the forces which have shaped both forestry and conservation in the uplands, concentrating in particular on the complex situation surrounding recent conflicts.

It is possible to portray the conflicts between forestry and conservation as a simple confrontation between the irresistible power of the forestry dragon and the plucky St George of conservation. There is a grain of truth in the idea: conservation is a latecomer to the power politics of the uplands and is still less powerful than traditional ownerships and land uses, although conservation is likely to be the eventual winner. What is quite wrong is the portrayal of a simple issue with only two players. In British politics a lot goes on beneath the surface away from the brash rhetoric of the media. Yet there are many who favour a concentration of power on environmental issues. They see the subtle changes of power between ministries and conflicting legislation as messy and would like to sweep it all away and replace it with something neat, streamlined and all powerful; perhaps a Ministry of the Countryside which takes over the decisions about forestry, conservation and agriculture.

A disadvantage with such an all-powerful body is that it is much more difficult for competing opinions to gain influence; one lobby is always likely to be pre-eminent. Conservationists might welcome such a super-Ministry if it were on their side, but it is unlikely that it would favour conservation over agriculture or forestry. In the present system, the procedural maze which is so

The bland face of upland power: since 1945 agriculture has been the most important upland industry and also the power to which forestry had to go for land.

frantically frustrating when a superb wildlife site is on the verge of destruction also allows a minority voice to be heard and new ideas to develop, albeit painfully slowly. Despite their unique central government support, since 1919 the new forests have been just as subject to the realities of our political system as has any other activity.

The Second World War swept away the world in which the FC began. The way in which land moved to forestry is of little relevance to what happened later, because in the 1920s and 1930s agriculture was weak and land ownership unprofitable. Many of the vast estates established with the industrial money of the 19th century collapsed and state forestry benefited, buying the lands of both ancient and more recently ennobled families. Other estates survived by leasing part of their land to the FC and, especially in England, large areas of FC land are still leasehold, often at 1930's rents of as little as two shillings and sixpence an acre (30p/ha). Forestry did not appear out of the blue. Its main supporters were the improving landowners of their day, many of them Scottish. This support reached the highest levels of the establishment: not only were crown forests like the New and the Dean transferred to the FC but also King George V donated bare land for planting, including the first area of Dalby Forest in the North Yorks Moors. The patronage of the people who began the drive for the new forests remains important to this day. Their successors remain powerful, influential and interested, and continue as one of the most significant forces in the development of British forestry, as any House of Lords forestry debate shows. The FC was the product of such people and it has stayed close to them, becoming one of the better-trusted parts of country-side government. This underlying support has been crucial to forestry, especially in the last few years. It represents a largely hidden power easily forgotten and hard for conservation to match.

The factor that changed the world of the countryside after 1945 was a new approach to agriculture: Britain had come close to starvation in the early years of the war and no responsible government could allow that to happen again. Agriculture was to be supported in a way not seen since the Corn Laws were repealed. Forestry played a role in this revival but it was always subsidiary to the needs of farming. The first priority for land, even overriding urban and industrial development, was to grow food. Through many changes of detail this policy survived for just about 40 years and it has dominated the way in which forestry developed during that time. It has shaped attitudes which a generation of farmers, foresters, central government administrators and even conservationists take for granted because they have never known anything else.

From the conservationists' standpoint the FC of the 1960s and 1970s looked like an impregnable, all-powerful monolith. In fact, from 1945 onwards it has always had to consult other interests before planting land. The first and most important of these were the three ministries of agriculture for England, Wales and Scotland. It was through negotiation with them that the standards for land deemed appropriate for forestry were thrashed out and there was never any real doubt that it was the agriculture ministries that were the more powerful: basically, no large areas of land deemed capable of carrying viable agriculture were to be planted. The only factor that varied through the years was the definition of 'viable', which changed as economic, political and social priorities altered. There has not been any universal standard except that agriculture has the priority: the worst land of some southern Pennine or Dartmoor hill farms looks like good in-bye land in the northwest Highlands of Scotland, so rules were made to suit local requirements.

For much of the post-1945 period these standards were decided virtually in private, because it was the FC that was buying most of the land to be planted, and the government departments could negotiate without involving outside bodies. The private sector tended to follow the lead set by the FC although most public attempts to test the limits, whether it be in obtaining permission to plant better-quality land or over environmental issues, have been through private applications to plant. In the past, successful approval for a private scheme might lead to FC acquisition of the land, but FC's delicate position as both the approver and potential planter of land meant that it rarely pushed politically sensitive cases on its own behalf.

The land that has been planted reflects this policy, as do our two post-1945 case studies, Galloway and the Flow Country. Forestry was seen as a major pillar of rural development where agriculture was so unproductive that even high levels of support could not give people an adequate standard of living; the dead sheep that littered the Galloway hills and fed the Golden Eagles and Ravens signalled a bankrupt agriculture to even the most ardent supporter of farming. It was not in any way the fault of the people: the land and the climate were against them and forestry was one way of bringing in government money to improve living conditions, which were harsh by today's standards. It is because of this that forestry still enjoys great public support at a local level in the remotest parts of the uplands, especially the Highland region of Scotland. That forestry got only the land that was left after agriculture had taken first pick is illustrated by the fact that by the early 1980s land cleared for planting in Scotland had a sheep stocking density of only one sheep to 4 ha. The planting of the Flow Country fitted into this pattern; farming had been completely abandoned over large areas and the Department of Agriculture and Fisheries for Scotland (DAFS) has as little interest in the area now as it did when it was happy to clear it for forestry in the first place.

Search the thousands of column inches written on the Flow Country, watch again the television coverage and you will be lucky to hear the agricultural factor mentioned but it was the shortage of land for forestry elsewhere as much as anything else that led to Fountain Forestry's initial purchases in the area. As environmental concern over afforestation grew, a quite separate struggle was going on behind the scenes. The area of land being planted was dropping and the main reason was lack of land. The market demand was there but the DAFS, now the agriculture department with the majority of new planting in its area, was if anything tightening its interpretation of criteria for clearing land for forestry. Prices rocketed to the point where land cleared for forestry was worth twice as much as the slightly higher quality land across the fence for which clearance was refused. The situation could not last: companies looked for land where this constraint would not exist, and Fountain found the Flow Country. Others followed a different route: the differential in land prices got so high that it almost equalled the value of grant aid, so they planted without grant, completely avoiding the voluntary consultative process which would have resulted in a refusal from the DAFS to clear the land for afforestation. Once again, the two disputes converged because the conservation lobby latched onto these cases as examples of the inability of statutory afforestation procedures to guarantee protection of the environment. To understand the full implications of this it is necessary to look in more detail at how the consultation process works.

First, however, we should bring the story of forestry and agriculture up to date. Within a year of the 1985 Flow Country explosion, in a statement by the Secretary of State for Scotland on March 24 1986 (Forestry Commission

1986b), the DAFS criteria for afforestation were relaxed and it became easier to get forestry clearance for better quality land. This took place against a background of accelerating change which has seen the first really major rethink of agricultural policy since the 1947 Agriculture Act. Conditions have changed forever; only three years on, the first farm woodland schemes are being planted on arable land. Again inconceivable five years ago, tax changes have radically altered the way the private forestry sector approaches its business. We will try to predict some of the immediate effects of these recent changes.

Both state and private forestry are supported by government funding. Almost everything in the uplands, including agriculture, tourism, industry and fisheries, is in one form or another subsidized. Long lines of communication, harsh climate and infertile soil mean that the uplands simply cannot compete with the rest of the country on an equal footing, and without support people would leave. Even with support, the populations in most upland areas continue to decline. State planting is financed by the allocation of money from central funds at a level to allow land purchase for an agreed planting programme. This investment is publicly reported in the FC's accounts. It is clear and above board and also of rather little importance to the conservation debate which has been a major issue in the recent years of declining FC planting and rising private planting.

The funding of private forestry has attracted much attention. Private forestry planting is funded by direct grant from the FC but did, until the 1988 budget, also enjoy certain tax allowances introduced as long ago as 1915. An owner could select under which schedule he wished his woods to be taxed. Under schedule B the sale of timber was tax-free. A nominal tax of one third the rental value of the land before afforestation had to be paid and establishment and maintenance costs were not eligible for tax relief. However, they were free of tax under the normal business schedule, schedule D. Both allowances made sense: forestry is a long-term business yielding a low return compared to many other enterprises, and tax allowances compensated for this.

However, whenever land changed hands a new owner could re-select the schedule under which he wished to be taxed. The realization that this switch could be used not simply as and when it occurred fortuitously but as a deliberate business tactic to minimize tax liability was the foundation of post-1945 investment forestry. Pioneered by the Economic Forestry Group, it is the foundation of all the big afforestation companies and was responsible for the majority of private upland afforestation. By switching ownership, advantage could be taken of the benefits of both schedules; D at the start of the rotation when costs are high, and B at the end when costs are low and income is hoped to be high. The initial investment was particularly attractive to individuals paying income tax at the highest rate, which has been as high as 90%. A galaxy of stars of sport and screen were attracted to forestry as an investment because of the tax advantages. The involvement of many household names has proved a tabloid headline-grabber cleverly exploited by conservation publicists. At the lower 60% rate current when the rules were changed in 1988, it had been estimated that a high-rate taxpayer could claim 47% of establishment costs against tax in addition to receiving grant aid amounting to a further 21.5% of total costs. The initial investor typically sold after about 10 years when the establishment costs had been paid and the last instalment, five years after planting, of forestry grant received. The buyer, which may be a financial institution not eligible for personal income tax relief, then elected for schedule B and paid no tax on the timber harvest. The seller paid no tax on the increased value of the trees since growing trees are exempt from capital gains tax. Thus three different sets of tax relief could be obtained from one forestry rotation (Bainbridge *et al.*, 1987).

Under this system the initial schedule D investor could expect to make a gross equivalent return of somewhere between 20% and 30% which is impressive by any standards: the precise amount depends on the eventual selling price. The schedule B investor could expect to make about 5% (net of inflation), much less impressive, but comparable to agriculture and with the similar security of land ownership and a sensible investment as a component of a wide-ranging pension fund portfolio. However, despite this incentive, private new planting still only reached 19 147 ha in the year ending March 1987, and total new planting was well below the theoretical national target of 30 000 ha at 24 489 ha (Forestry Commission 1988). On the other hand, the trend in private planting was the opposite to what might have been predicted on purely fiscal grounds: private new planting rocketed from only 6322 ha in 1978 during a period when the highest rates of taxation have fallen consistently. The cut from 90% to 60% was matched by a government call for more private forestry and a greater confidence in business and private ownership which, although failing to meet the target, resulted in rapid growth apparently against the fiscal trend.

Outside designated areas like SSSIs it remains legal in Britain to plant trees without asking anyone. What is more, tax relief was not conditional on any sort of consultation or environmental standards. This is not as exceptional as it might seem: the Treasury's job is to collect money, and other departments have responsibility for other aspects of life, whether it be landscape and wildlife or health and safety. As ever, there are checks and balances. In the case of forestry there is an unwritten understanding that the price of what is effectively self regulation is some voluntary discipline, and this is exercised through the FC's grant aid procedures. It has been quite exceptional for major planting to take place without grant aid despite the fact that far more fiscal benefit used to come from tax relief than grant aid.

Conservationists have accused the system for approval of forestry grant applications of being secretive and undemocratic, with the FC acting as judge and jury in camera. As we will see there is some justice in this charge. However, there are problems from the private foresters' point of view too, as over the years a system rather like a game of snakes and ladders has evolved. Year by year there are more additions to meet new needs and public concerns, usually to the snakes rather than the ladders. The first step is the FC assessment of the silvicultural merits of the application. This has rarely caused serious difficulties in the uplands because the FC standard has tended to be the industry standard. In recent years the FC has increasingly assessed other aspects of applications, especially landscape design, before agreeing to send a scheme out for wider consultation. The FC can turn a scheme down at this stage if it does not meet its basic silvicultural criteria.

The scheme then goes to consultation. Formal consultation gives the consultee the right to object to part or all of a scheme in which case it must, if the objection is sustained, eventually be put before the Forestry Commissioners for decision. The Forestry Commissioners invariably seek ministerial advice on cases which reach them, so sustained opposition effectively lifts consideration of a scheme from an administrative to a political level. It is not, as the mildness of the word 'consultation' might suggest, simply a case of seeking comment which can then be ignored. The right to formal consultation, or even the opportunity to be involved in consultation discussions, has played a large part in the arguments between forestry and conservation because formal consultation brings with it the power to force a political decision on disputed schemes. Who is consulted is laid down by ministerial directive and the list of consultees has expanded over the years. The most important consultees up to the 1970s were the Agriculture Departments. Most other consultations involved special cases. Water Authorities were consulted over water catchment areas, the NCC over SSSIs and the Red Deer Commission over applications within the range of Scottish Red Deer. Individual arrangements of varying degrees of compulsion were made with National Parks.

The greatest change in consultation arrangements took place in 1974 when statutory consultation was extended to include local authorities, and the Regional Advisory Committees (RACs) originally set up in 1967 were given the task of arbitrating disputed schemes. The initial role of local authorities was seen mainly as commenting on 'landscape and amenity'. The interest local authorities took varied greatly. Some did not take up the offer of consultation at all whilst others set high area thresholds so that they only saw large schemes. Interest tended to be greatest the further south and the more urban an area was, so that large forestry schemes in Scotland, which have the greatest repercussions for upland birds have been amongst the least actively discussed. Subsequently, local authorities have become more active consultees and have extended their interests to water resources, the impact of timber haulage on roads and nature conservation, but it has been a surprisingly slow development. The local authority role is especially important because in the last five years they have been encouraged to subconsult anyone they choose, an appropriate power for the democratic representatives of local people.

However, widespread consultation is not a right which authorities have used lightly. They have correctly surmised that if they subconsult and then refuse to object on the basis of advice received from a body like the RSPB they will find themselves picking up the political flak at present reserved for the administrators of the consultation system, the FC. Also, the greening of the local authorities has been as long delayed as that of bodies like the FC, MAFF or DAFS. As Brotherton (1987) points out the problem is not so much that local authorities do not have the power to push the conservation case as that they do not have the inclination. Conservation bodies have, however, had increasing success in becoming subconsultees. The RSPB has persuaded the majority of local authorities in Wales to consult it but its success has so far been limited in Scotland, which is, of course where most problems now occur.

In practice, only a tiny proportion, well under 1%, of the many thousands of schemes submitted to the FC have been formally objected to by statutory consultees since 1974. This does not mean that there are few problems: many schemes will be amended with the agreement of the owner following comments by consultees and it is the job of FC staff running the grant scheme to ensure that only schemes where fundamental disagreement exists are taken to the next step in the formal process. The inevitable delay this involves plus the increased chance of a whole scheme being rejected are valuable allies in persuading owners to compromise. If compromise is not possible, the scheme goes to the RAC. The committee is composed of distinguished people from within the relevant conservancy (FC region). Representation on the RACs has a strong bias towards members of the forestry industry, although farming and rural trade union interests are also represented. The representation of environmental interests, or rather the lack of it, has been a source of considerable discontent in the last few years. Even though individual members of committees are sometimes knowledgeable and sympathetic to the environment, they were usually selected for other reasons and the advice of environmental bodies, the professionals in the field, has not been taken. Deciding which environmental groups should have access to the RACs is clearly problematical when there are interests as diverse as insects and orienteering to consider.

The RACs' primary function is reconciliation. Their meetings are held in private and involve only owners and statutory objectors. FC staff are not involved in the discussions. An RAC does not publish its proceedings, nor any justification for its findings. If it cannot resolve the dispute it reports to the Forestry Commissioners. Again, what it reports is not published. These are sensible arrangements for an arbitration service but the position of the RACs has become increasingly uncomfortable as environmentalists in particular look for greater public participation in key cases. This resulted in proposals being put forward in 1986 (Forestry Commission, 1986b) to expand membership to allow for a greater environmental and farming representation and to further reduce the forestry flavour of the committees by appointing more obviously independent chairmen. However, proposals for opening up the way the committee's business was carried out fell far short of the demands of the environmental lobby and they failed to be ratified in parliament.

The number of cases going to RACs did not exceed 10 per annum until 1983 but have subsequently increased greatly, reflecting a hardening of opposition by some statutory consultees. The success rate for solving problems has always been modest, probably reflecting the effectiveness of FC in only allowing really intractable cases to get this far. Just over one quarter of cases are resolved by the RACs but some others are dropped by owners, so that just over half the cases considered up to 1986 reached the Forestry Commissioners. The Commissioners then consult the appropriate government minister, although it is their eventual decision that is final. Again, what goes on between the Commissioners and the minister is confidential but the belief that the Commissioners always accept the minister's advice is almost certainly correct.

Two thirds of cases that reached Commissioners in the decade to 1985 were refused a forestry grant.

The new consultation arrangements seem to have worked well in the 1970s but by 1985 were looking stretched: pressure on them was sufficient to precipitate the review of RAC procedures and the more liberal attitude to the role of local authorities. There has been an increasing clamour from a wide range of new bodies for inclusion as statutory consultees. Although the demands of voluntary bodies have been resisted, consultation with a number of extra central or local government organizations, in particular water authorities concerned about acidification, has been conceded. The new idea of notification was introduced for consultation over archaeology: this allows for the provision of advice but not statutorily binding objection. There have been territorial extensions, too, for National Scenic Areas in Scotland, the new Environmentally Sensitive Areas designation and the special extension of consultation with the NCC to the whole of Caithness and Sutherland.

The problem faced by the conservation lobby is obvious: up to 1985 nature conservation's only firm statutory locus in the process was the consultation of the NCC over proposals to carry out afforestation on SSSIs. Simply getting to see private forestry schemes other than on SSSIs depended on the willingness of the local authority to make the information it received as a statutory consultee more generally available. Until very recently this involved very few cases indeed because most effort was being directed towards protecting existing SSSIs.

Conservation was only a small part of the whole scene: to view this situation from the specialized position of this book or most of its readers would be like looking down the wrong end of a telescope. The foresters had bigger battles to fight. Compared to the problems of obtaining clearance of land from agriculture, conservation was a side-issue to be contained with as little loss to forestry as possible. The SSSI system provided the most obvious approach because site protection was the nature conservationists' chosen method of protecting wildlife interest, and SSSIs were the statutory mechanism for doing so. The NCC was as obviously smaller and weaker than the FC as the Commission itself was compared to the Agriculture Departments in the game of upland power politics. The FC's treatment of the NCC was as ruthless as its own treatment by the Agriculture Departments. In its dealings with SSSIs its duty to promote forestry dominated any other consideration, including its own weakly defined responsibilities towards the environment. It was up to the NCC, not the FC, to fight the conservation corner. The problems they faced in the uplands were severe, especially prior to the 1981 Wildlife and Countryside Act.

The SSSI system was the accepted mechanism of protection. Administered originally by the Nature Conservancy and, following its reorganisation, the NCC, the series of roughly 6000 SSSIs aims to include the best representatives of the whole suite of natural and semi-natural habitats of the British Isles. Everything of value cannot be protected, so some selection has to be made. For some habitats this is not difficult: there is so little left that all should be protected, but for most this is not the case. There is not only the problem of selecting within a habitat type but between habitats; where there is high wildlife value spread across a range of species and habitats this may cause particular problems, because such a high proportion of the land should be SSSIs; the Scottish island of Islay is a prime example, its interest ranging from Barnacle Geese on improved farmland, through rare unglaciated peat bogs to birds of prey and Choughs in the wilder hills and on the sea coasts.

It is difficult to set objective, practical cut-off points for site selection. One of the problems is that any hard and fast rules can be made to look inappropriate in particular cases and it is difficult to predict what area of land would meet the SSSI standard for any given set of rules. The NCC has rightly tended to concentrate on conserving representative areas of different vegetation types, since by conserving plant communities the animals that depend on them should also be protected. However, popular perception usually focuses on rarities, often individual bird species. This has given the NCC real difficulty because it has often made its arguments appear intellectually elitist, even to those who agree with them. For example, in the Flow Country debate the RSPB concentrated on the conflict between forestry and the needs of three breeding wader species, Greenshanks, Golden Plovers and Dunlins, whereas the NCC first stressed the national importance of "The large area and diversity of blanket bog as a physiographic/vegetation feature and the relative lack of disturbance in many places, giving the greatest extent of actively growing mire in Britain and one of the few areas of extensive natural terrestrial vegetation now remaining." (Stroud *et al.*, 1987.)

However, *A Nature Conservation Review* (Ratcliffe, 1977b), provides a superb foundation to the SSSI system. It describes Britain's main habitat types, the basis for the selection of the best sites and the sites themselves. It does not include all SSSIs but rather the national series of key sites, the best that must be conserved if we are not to lose forever whole parts of our natural and semi-natural heritage. At the time of its publication, however, it looked like becoming simply a milestone in the road to the destruction of Britain's wildlife, to be looked back at in 50 years time rather like a Victorian explorer's journal to see what had once been but had disappeared irretrievably.

Before 1981 SSSIs were far from sacrosanct. Herb rich grassland was ploughed up, heathland developed for housing and broadleaved woodlands replanted with non-native conifers or grubbed up altogether. Permission was given for afforesting parts of several SSSIs in the uplands against NCC advice (Nature Conservancy Council, 1986). So the NCC could not simply schedule a site and go away and leave it, happy in the knowledge that its future was secure. The NCC's ability to put money into SSSIs was extremely limited: apart from anything else, its national budget at the time of the 1981 Wildlife and Countryside Act was only £12 million per annum; the value of 30 000 ha (the annual national afforestation target) at the £400/ha one might then have expected to pay for good forestry land!

SSSIs extending to a few thousand hectares, or larger on land that was less valuable for other uses, might be acceptable but sites larger than that start to impose real constraints on economic development that before 1981 the NCC could do little to compensate for. The SSSI concept, which works well enough in the lowlands, is not so appropriate for the uplands, where it is harder to identify natural boundaries for sites and where some species are extremely wide-ranging. Not surprisingly the SSSI system is easier to operate to protect orchids than it is to protect eagles. As always, lack of information is a problem. To pick sites correctly, complete detailed knowledge of distribution is essential. The more you know, the more precise you can be. Where little is known, it would be safer from a scientific and conservation point of view to select larger areas. Sufficient bird survey information has rarely been available to make the quality of decision that would ensure the best possible location of SSSIs. Larger SSSIs are more vulnerable to attack because they have more impact on other land users and they are also harder to defend because detailed information for the whole site is harder to gather.

It was a no-win situation which came home to roost in the late 1970s over the Berwyn mountains in Wales. As with most such disputes, what really happened is shrouded in mystery but even the tip of the iceberg that shows above the surface is complex enough. The dispute arose not from an existing designation but from the NCC's attempts to greatly extend SSSI coverage following the results of RSPB surveys showing parts of the area to hold some of the most important moorland bird populations in Wales (Lowe *et al.*, 1986). The sites recommended by the RSPB as essential for protection were extensive, and as soon as the NCC's intention to designate large additional areas became known, opposition from local farmers, afraid of losing the option to improve agricultural land or sell to forestry, developed. This opposition included both the FC and the Welsh Office Agriculture Department (WOAD). An independent review, carried out by Reggie Lofthouse, a former chief surveyor with ADAS (Agricultural Development and Advisory Service), resulted in a land use plan which was effectively a compromise between the different interests. The report clearly identified the problem of the contestability of SSSI designation which has reappeared on a far larger scale in Caithness and Sutherland. Regardless of what the legislation says, designation of areas amounting to tens of thousands of hectares inevitably contains a strong sociopolitical aspect which is ignored in the name of scientific objectivity at the peril of the wildlife that designation aims to protect. Lofthouse also identified a manifestation of the point we have made above about relative power and standing of the organisations involved: the WOAD and the FC presented technical information which they expected to be, and was, accepted as fact. Both, however, felt free to question the NCC's factual basis for notification.

The Lofthouse report may not, however, have been quite as objective as it would have been if a proposal for afforestation had not arisen at Llanbrynmair Moors in the south of the area whilst the report was being prepared. The proposal was to afforest 1780 ha of the highest land of the Llanbrynmair estate to release capital for tenant farmers to buy the freehold of the better land from the existing owners who wished to sell. The local political pressure generated by such a proposal was obviously great and the NCC eventually dropped its objection to the proposal, almost certainly because it could not sustain it without doing something and did not have the money to compensate the farmers by buying the site. Only 40 ha were preserved for botanical interest and the area's bird value was skimmed over. What we will never know is to what extent the existing forestry proposal influenced Lofthouse's zonation of an admittedly substantial area for conservation in the Berwyns. What has become clear with hindsight is that Llanbrynmair was the best place in the Berwyns for moorland birds, holding 15% of Welsh Golden Plovers as well as Dunlins, Hen Harriers, Merlins and wintering Greenland White-fronted Geese.

The next major clash was of a very different nature. It involved a limited area of minimal socio-economic importance which was already designated as an SSSI, but in less than a year after the end of the Berwyn affair the Wildlife and Countryside Act 1981 had been passed. It at last gave real teeth and resources to the SSSI system: damage to an SSSI became an indictable offence subject to fines, and the NCC was given the power to pay compensation to an owner deprived of the financial benefit from development of the economic potential of a notified site.

Like the afforestation process, designating an SSSI is a long and complicated business. It is as unpopular with landowners as afforestation is with conservationists. Proposals to carry out forestry work, whether planting or felling, are initiated by the owner of the trees. Designation of an SSSI is initiated by a government body, the NCC, and imposes very great limitations on the freedom of the owner to use his land as he wishes. The administrative process does not help. Site selection is based on available facts and is carried out internally by the NCC: whilst advice may be accepted from other conservation bodies, no-one else has a statutory role in this phase of the process. In the case of large or especially sensitive sites the NCC may consult ministers or other government departments before formally notifying the site. The wording of the Wildlife and Countryside Act makes it clear that even the NCC's own minister cannot stop designation of an SSSI; in contrast to the FC which hovers around the margins of being a government department, the NCC is further removed from central government, with a greater degree of autonomy.

Consultation with landowners ahead of formal notification is limited because of the real risk of an informal warning leading to action to destroy the site: owners are likely to know that their land is under consideration because permission to survey it will have been sought, but their next intimation is likely to be a very formal document landing on their doormat listing a formidable range of 'potentially damaging operations' (PDO) for which prior permission must be sought from the NCC. Attempts to soften the wording and add more explanation of the implications of designation as an SSSI have been made but had to be withdrawn because they compromised the legal validity of the notification. Similarly, it seems likely to be difficult to add PDOs at a later date, which means that the list is made up by eliminating those that simply could not cause damage to the biological interest of the site; improbable but possible future events such as a craze for keeping elephants in ancient woodlands must, however, be catered for, leading to a long and forbidding (and sometimes slightly ludicrous) list. Most landowners perceive PDOs as activities which are forbidden, which is not in fact so but in shaping attitudes it is the perception which is more telling than the fact.

The next move is the owner's, to either object to the designation or seek permission to carry out one or more PDOs. Where the request is to carry on with operations such as sheep grazing or stalking deer on an upland site, permission is likely to be granted and no money will change hands. However, the proposition of an unacceptable operation such as afforestation leads to negotiation for a 'management agreement', involving payment of money by the NCC to the owner. Sometimes this is to cover the costs of the positive management of the site for conservation, but more often it is to compensate for operations which the owner can no longer carry out on his land. He may never have planned to until the SSSI was notified but this makes no difference. Negotiations for management agreements, the payment for which is based on expectation value, are usually lengthy but successful in reaching agreement. If they are not, the owner is able to carry on with his planned work; only a Nature Conservation Order made by the appropriate minister, the Secretary of State for the Environment in England and for Scotland and Wales in those countries, can reimpose the full rigours of the Wildlife and Countryside Act, so, as with forestry, the final decision remains a political one. As for forestry, there are different institutions and legislation for Northern Ireland. Whilst apparently further removed from government, the NCC must, like the FC, always be looking over its shoulder at the political acceptability of its actions. This may not be easy to gauge, because there is even less public consultation in the SSSI procedures than in those for forestry grants.

Over much of the uplands the most obvious development conflicting with a site's nature conservation value is afforestation. It was inevitable that the new powers of the 1981 Act would soon be put to the test either by intent or accident. The crunch came at Creag Meagaidh in central Inverness-shire in 1983. Fountain Forestry, who bought the site, knew it was an SSSI, and subsequent press portrayal of the dispute suggested that this was a direct challenge to the NCC and the Wildlife and Countryside Act.

Creag Meagaidh was an extraordinary battleground, rock-strewn slopes rising to solid scree. Its main conservation value lies in the montane altitude zone, well above the upper limit for tree planting (Ratcliffe, 1977). A limited area of Birch wood, much vaunted in the press, immediately struck the visitor as poor and insignificant. The area under threat was not one of the front rank SSSIs, clearly justifiable through rare and exciting species. At the same time it was hardly the finest forestry prospect in the Scottish Highlands. Creag Meagaidh was about principles and power, especially the NCC's new position in the order of things following the 1981 Act. The forestry proposal ran its full course and ended up with the Secretary of State for Scotland, who advised the Forestry Commissioners to approve a reduced afforestation scheme, which they did. There, were our legislative and political system as neat as many seem to wish it, the matter would have rested. In fact, the NCC stepped in and informed Fountain Forestry that were they to go ahead with ploughing and planting they would be prosecuted under the Wildlife and Countryside Act

1981 for damaging the SSSI. Whatever political risks the NCC took, they had the law on their side and the affair ended with the NCC buying the site for rather more than Fountain had paid for it.

Taken in isolation, the Secretary of State's decision was probably correct. The part of Creag Meagaidh in dispute was not so important as an SSSI that some forestry would have been a disaster for conservation. But it was not a case in isolation and it was vital for the future of the SSSI system that afforestation did not go ahead in this particular case, the first since the 1981 Act. The NCC held firm; their defiance of the elected rulers of Scotland was political brinkmanship of the first order, carrying considerable risk, but in this case it was vital and they won. The significance of that victory is that not a single forestry scheme has been passed on an SSSI against NCC advice anywhere in upland Britain since the conclusion of the Creag Meagaidh affair. The NCC also acted with pragmatism and good sense in the aftermath of their victory in buying the site. Many hard-line conservationists still see the paying of compensation, or in this case purchase of land prohibited from development, as morally wrong, especially when it is a company rather than individuals which benefits. However, compensation gave the NCC the financial clout to compete with the main economic land users and avoid the problems encountered in the Berwyns, where they had nothing but fresh air to offer the farmers who wished to sell Llanbrynmair.

The compensation idea certainly does have severe flaws. It is a basically negative idea, aimed at stopping people doing things. It is backed by compulsion which, if nothing else, stimulates owners into going for the maximum payable. It transfers conservation money to people who are often already rich and to landowners who have no intention of spending the money anywhere near the land for which compensation is paid. Unlike the Creag Meagaidh case, several subsequent forestry cases on SSSIs have almost certainly been brought without any intention of actually planting trees, simply to reap the immediate financial reward of compensation. An owner merely has to say that he wishes to carry out a PDO in order to get compensation for not being allowed to carry out his sometimes empty threat. However, for all its imperfections, compensation has broadly worked so far. Claims have actually involved only a tiny fraction of all SSSIs, for 1988–1989 totalling only £7.10 million paid out for management agreements. Many people feared that government would not finance the commitments entered into through the 1981 Act. The rapid increase in the NCC budget, which now stands at £40.29 million per annum (Nature Conservancy Council, 1988) is in stark contrast to general cuts in public expenditure since 1981.

At best, this is money that the NCC can put into upland communities to support conservation in exactly the same way that government funds support agriculture and forestry. The money that came with the legislation should represent a positive power, just as the giving of grants does to the other administrators of land use. No-one, however, knows where the limit for the use of SSSIs as a land use and economic measure is: between the area designated now and the whole of the uplands there must be one, and the ongoing saga of the Flow Country is really exploring this boundary. Its conclusion is really part of the future and we shall come back to it in the next chapter.

There are alternative forms of designation to the strict regime of the SSSI for upland bird protection. National Parks in England and Wales have provided a degree of protection to the open uplands and the recently introduced Environmentally Sensitive Areas (ESA) provide a more liberal way of directing and supporting traditional agriculture based around incentive payments to enter into voluntary agreements to maintain traditional farming practices. The National Parks have slowed forestry development over large parts of several key upland areas, for example in the North Yorks Moors. It is too soon to know how well the ESA designation will work; it is just one of the wide range of new measures whose success or failure is part of the future. One thing is certain; site designations do not provide a full solution to the problems of the uplands. Still only a small fraction of the afforestable uplands is subject to any designation, and opposition to restrictive designation, not just from national land-using bodies but also from local people, limits the rate at which new designated areas could be introduced even were that the desirable route to follow.

The problem of the weakness of the SSSI designation prior to the 1981 Act was not confined to the uplands or to afforestation. The main avenue for the guaranteed protection of prime sites taken by the conservation bodies was land purchase. Both the NCC and the RSPB built up substantial holdings of nature reserves, whilst county wildlife trusts purchased or leased a myriad of smaller sites. However, in contrast to the situation in the lowlands, land purchase for nature reserves has not been such a major factor in the uplands, and particularly not with respect to the afforestation debate. A number of important upland sites are owned by the NCC in particular, but many are not even technically afforestable so they play no role in protecting the uplands from afforestation. Land purchase has not been used as a way of stopping land going to forestry in the way it has been used to prevent undesirable development in the lowlands. The purchase of Creag Meagaidh was almost a postscript to that dispute rather than a key element in saving the site. We cannot think of a single case where a conservation organization has competitively outbid forestry to save a site.

The main reason is probably a combination of the sheer extent of the problem and the scattered, low-density nature of wildlife distribution in the uplands. Sites would have to be large to be significant: 1780 ha, which the NCC could not afford to buy, were planted at Llanbrynmair, and vastly greater areas in the Flow Country. We do not know what was paid for the land at Llanbrynmair, but at £400/ha purchase would have come to over £700 000, at the time more than the RSPB had paid for any reserve; the 8000 ha of Abernethy, purchased in 1987, cost the RSPB £800 000 and the public purse £1 000 000. By far and away the largest land purchase for conservation during the year, this was only one third the area of the year's afforestation at a price well below the forestry minimum of about £150/ha for which land might have been changing hands in the Flow Country. The Abernethy reserve amounts to only double the peak annual planting programme achieved in the Flow Country alone.

Few upland areas have densities of birds that justify spending huge sums on sites. Even in the core of the Flow Country, an area the size of the RSPB's Minsmere reserve (600 ha) might contain only one pair of Greenshanks; this might seem an expensive £90 000 worth to many RSPB members. This, as we have shown, is the crème de la crème. There would, of course, also be Dunlins, Golden Plovers and possibly rarities like Arctic Skuas and Common Scoters. Further south, densities and variety of species would be even lower and land prices higher. To protect just 1% of Britain's Golden Eagle population would require a land purchase of around 25 000 ha; and, of course, doing this would hardly cramp the foresters' options for land purchase at all, so that any effect on reducing the impact of afforestation on Golden Eagles would be negligible. Few upland species lend themselves very readily to protection by land purchase.

The one place where land purchase, as a spoiling tactic if nothing else, might have been expected was in the Flow Country. It has not happened because the RSPB got there too late. The large estates that made up the main Fountain Forestry purchases obviously had to be bought before afforestation could start and before the threat to birds became apparent. Since 1985 far less land has changed hands and forestry organizations have been very careful to ensure that any land they have bought could be planted before handing over the money. It is common practice for companies to act as free consultants to an owner in organizing an application for forestry grant and then buying the land once approval to plant is granted. Where there is doubt as to whether a scheme will be passed, this approach is the rule rather than the exception. Several schemes have subsequently been passed by the Secretary of State for Scotland, but despite the expected protest of principle against new planting, conservation bodies have made no attempt to buy them from under foresters' noses. This is not lack of opportunity: the dramatic fall in business caused by tax changes has left several hanging on the market. It is rather that the assessment of conservation value which went into approving some and rejecting other schemes was accurate and that the properties approved for forestry are genuinely of low conservation value. Those not passed are expected to be made

SSSIs and owners will certainly wait to see what sort of compensation deal they might get from the NCC before considering selling.

So, whilst increasing security for special sites has been achieved, conservation bodies, even the government's own NCC, are still out in the cold as far as statutory consultation over undesignated private afforestation proposals is concerned. Quite apart from the standards applied to non-SSSI applications within the grant scheme, there are also fears about the voluntary nature of the grant approval arrangements and the ability of a landowner simply to ignore them if conditions for grant become too arduous. Faced with an inaccessible system, conservation bodies have taken yet another of the diverse routes to power available within British politics: lobbying for the change of that system. The RSPB in particular amongst the conservation bodies has led a growing campaign over forestry. We will discuss the way in which this campaign has developed in more detail in the next chapter because it has the most profound implications for the future of forestry. In this chapter we will look only at the way it has affected the practical disposition of power in the uplands to date.

It is first important to realize that the major effects of conservation lobbying on the way in which upland land use decisions are made are very much late-comers to the scene: as we have shown, the vast majority of afforestation, both FC and private sector, has been approved and carried out with minimal modification resulting from environmental pressure. Conservation as a real practical force is really a feature of the 1980s. The methods and tactics employed to force change have expanded as the dispute between forestry and conservation has grown. By now it has covered a full spectrum both of topics attacked and means of putting those attacks across. Effective use of the media has characterized the campaign and has proved most influential. As time has gone on, the attack has spread from its beginnings in wildlife to focus on the whole range of forestry activity, from the fiscal approach to funding afforestation to the susceptibility of conifers to windthrow and insect attack.

By the mid-1980s this was having some practical effect on the standards applied to forestry grant applications. By 1985 the advice of Mills (1980) on hydrological management was being widely applied as a standard. Whilst the government's 1985 broadleaves policy (Forestry Commission, 1985a) had most impact in the lowlands, it also led to the requirement for some broadleaves, normally about 5%, to be included in all upland afforestation schemes. Landscape standards were also tightening, though less easy to define than the distance conifers should be kept back from streamsides. Despite their lack of direct involvement in the approval of individual afforestation schemes, the conservation bodies had provided the impetus for improving practice and more clearly defined standards within the industry. They had had a clear effect on what the forests should be like but much less success with the crucial and quite separate question of where these forests should go.

Of all the ideas for eroding forestry's power the one that proved most successful was the attack on the tax-switch system which formed the foundation of private-sector investment-forestry. It was the naming of the rich and famous that got forestry into the tabloids and colour supplements. It was tax relief that took the brunt of the Survival Anglia TV programme on the Flow Country. Publicity for the tax relief issue which reached its peak in late 1987 and early 1988 was the culmination of the increasingly strident media criticism of afforestation.

What probably surprised even its strongest proponents was that it worked. In his March 1988 budget the Chancellor of the Exchequer took forestry out of the taxation system, abolishing in the process the schedule-switching trick on which investment forestry by high earners depended. The Chancellor may well have had other good reasons for taking this action, in particular his declared wish to tidy up and simplify the taxation system. However, environmental pressure was undoubtedly an important factor and it is unlikely that the issue would have had the same priority for action without the extensive adverse publicity of the previous year. This is what the conservation lobby had demanded, but they had demanded it with the intention of stopping, or at least greatly reducing, upland afforestation. This was not the government's primary intent: a week later the new rates of forestry grant, under the new Woodland Grant Scheme (WGS) (Forestry Commission, 1988b), were announced. The key conifer grant band was more than doubled from £240/ha to £615/ha. Other grant bands were increased by the same flat rate of £375/ha. This gave the biggest proportionate increase to the area worst hit by the removal of tax relief, large-scale conifer planting. The lower rate of proportionate increase for smaller schemes and broadleaves was less important because they were never as dependent on tax relief as an incentive. There was a further important change; the Chancellor also reduced top-rate income tax again, from 60% to 40%. This would probably have reduced the incentive of the old tax-switch system considerably by itself, although it is worth remembering that private planting increased greatly after 1979 despite the reduction in the top rate of tax.

The Chancellor's action has, at a stroke, changed the face of upland forestry for ever. With changes to agricultural support and the growing voice of the environmental movement, the uplands face a period of unprecedented upheaval and the opportunity to create a better countryside for people, industry and birds. That future is the subject of our next chapter.

CHAPTER 9

The future

The Flow Country was a watershed for the uplands. The environment, with birds in the lead, has become a real force alongside agriculture and forestry. The battle left forestry critically weakened, major questions hanging over the assumptions and practices of the last 30 years.

The same trend is common to other developed countries. The state of Oregon in the USA has one of the world's biggest timber industries but its population is largely urban, earning its living from industrial pursuits. Awareness of the countryside is higher in Oregon than in Britain with over 100 000 people going deer hunting every year. Increasingly the urban majority's general interest is outweighing the more direct concern of rural people in the battle over the felling of old growth forest. This is primary native forest never felled for timber. The role of the rare Northern Spotted Owl (United States Department of Agriculture Forest Service, 1986) as wildlife totem of the issue closely parallels that of the Greenshank in the Flow Country. Larger and larger areas are now protected from felling and all forests must have detailed plans open to wide public debate.

In New Zealand lack of confidence in the forest service's environmental record has led to a split; native forests of high environmental value will be managed by a new government agency and the planted commercial forests sold on the open market. Neither solution is ideal: are the New Zealand commercial forests doomed to environmental second class citizenship? The American planning system is commendably open—but spectacularly costly.

In the British lowlands intense environmental pressure led in 1985 to the government's broadleaves policy and a previously inconceivable level of accord between foresters and environmentalists: it can be done. In the uplands a lull in hostilities seems to have developed while both sides lick the wounds of the Flow Country.

The culture shock of radical new ideas goes well beyond forestry and upland birds: a new green consciousness has developed against a background of concrete change in European agriculture which is already affecting all aspects of the countryside. Against such shifting sands predicting the future is impossible, so in this chapter we try and identify some of the developments which might help the birds of our forested and open uplands and the sorts of mechanisms needed to achieve them.

EXISTING FORESTS

The new forests cover nearly 5% of Britain. Chapter 2 outlined their birds. Some species they have made common; they provide habitat for many already numerous birds but also for some that are scarcer and of high conservation value. They have been deprecated by conservationists both for their very real failings and because any admission of their value would certainly have been used to justify the further expansion of new planting on the open uplands.

It is the identified failings of the new forests that also makes them such a good prospect for positive management; so what are the main improvements that could help birds in the future?

Diversity

Chapter 3 looked at how forestry technology ironed out the often limited variety from open-land habitats. Chapter 4 discussed current thinking on creating greater habitat diversity in the new forests. Intensive even-aged planting reduces the number of bird assemblages that can use the forest; at one extreme this may be to only those using a particular stage of pure conifer forest. Streams are shaded out and open land disappears as trees grow to close rides. There were no broadleaves on the open land before the forest was planted and none were introduced at planting. By clearing trees back from streamsides and nesting crags, making glades and planting broadleaves, there can be birds of the riparian zone, the forest edge and broadleaved woodland as well as of the productive conifer compartment.

The main objective, we suggest, should be to allow a wide variety of birds to have sustainable populations in all the new forests. This does not mean large proportions of habitats like broadleaved woodland: a 5% content of broad-leaves in an upland forest may allow birds like Pied Flycatchers and Wood Warblers to become established, a clear conservation gain, but adding a further 5% probably will not bring in any further new species. The 5% broadleaves introduced over several years to a large upland forest could extend to thousands of hectares — a far greater area than that of new broad-leaves planted annually in the whole of Scotland during the 1970s.

The diversity of the main conifer production stands is equally important: with each stage holding its own assemblage of birds, to maintain the widest range of birds it is vital not to lose the Goshawks by felling all the old stands or

Pine and larch between moorland and better farmland in the North Yorks Moors; one of the possibilities for future upland forests.

the Nightjars through all the clearfells growing into thicket stage. Greater age class variation must be a central objective for the future.

Speed of change is as important as scale in the all-important cost-effectiveness relationship. The end of the first rotation is the checkpoint: no streams should be shaded out by trees planted too close, all blocks should have some broadleaves and there should be planned open space. There may well be a case for bringing forward some work, especially to improve the youngest plantations where problems will persist longest if left.

Chapter 4 explained why a perfect age-class distribution would be prohibitively expensive in the short term. We would suggest that the managers main aim now should be to maintain the presence of birds already present and allow absent species to colonise and become well established: sheer population is less important. Although our suggestions may seem a rather mild prescription it can be applied to the whole forest and through its scale have more effect than a limited number of special showpieces.

This is the starting point: research and imagination will extend the options for diversity in the future. Non-clearfell silvicultural systems offer the possibility of productive stands with a continuous stock of old trees. Mixed conifer broadleaved woodland on better upland sites has potential as yet little explored—including the prospect of an as yet unknown bird assemblage which may be different from that found in either pure stand type.

Scarce species

Birds like Woodlarks, Nightjars, Goshawks and Ospreys are national rarities heavily dependent on forests. In every case special management, whether it be maintaining a constant supply of clearfells or putting up nesting platforms, can help maintain or expand their populations. These birds' scarcity increases their perceived conservation value and they are an enjoyable challenge, not least because they are so exciting, attractive or interesting.

Where scarce species are present, management priorities may need to be adjusted. For example, maintaining sufficient areas of old trees, including insurance for windthrow, is more important when Goshawks are present than when they are absent. Similarly, looking after Woodlarks and Nightjars may need rather more than the minimum of felling age alteration to keep enough clearfell to prevent a decrease in the bird population.

Birds of native pinewoods, including the Scottish Crossbill, Crested Tit and Capercaillie, are especially important. They also live in Scots Pine plantations, but mainly in older, open stands. The experimental work described in Chapter 4 to help Capercaillies live in some FC plantations is set against a general trend towards earlier felling and replacement by Sitka Spruce. The success of the new government native pinewoods policy, including much higher grants for planting native origin Scots Pine, will be of importance to this special group of birds.

Forestry has a big responsibility to these scarcer species; without help there is a good chance birds like Woodlarks or Capercaillies could follow the Red-backed Shrike into extinction in Britain. Foresters have the chance to play a part in expanding the ranges of exciting species now returned after past extinction from Britain like Ospreys. Foresters' ability to apply management on a large scale is proven: if these skills can be brought to bear on some of the scarcer forest birds, their futures could be rosy.

Seclusion

One of the forest's virtues for rare and vulnerable birds is its scale and seclusion, far better than moorland because potential human predators cannot see so far and trees shield birds from unintentional disturbance. As recreational use increases, it will become increasingly important to zone some 'quiet' areas in the forest. Also, some of our bigger birds could spread back into more populous parts of the country if they can find safe places to nest. Is there any good ecological reason why Ospreys could not nest near the coast in Dumfrieshire or even, for that matter, in Devon? The forests need to have quiet spots ready for such a spectacular reinvasion.

Unplanting

Deforesting parts of the new forests is a touchy subject with foresters for obvious reasons. Some areas have been felled and left unplanted to improve the shape of ugly forest blocks. In exactly the same way that forests planted without planning for landscape may not be very beautiful, planting without conservation planning has inevitably covered areas which in hindsight might have been better left. However much some might wish for it, removal of large areas of forest does not make a lot of sense—once ploughed and planted most sites are unlikely to recover to their former state, and if there are still untouched areas of the same habitat, money and effort are better spent looking after them. Similarly, trying to make a surrogate moor out of the forest by creating larger glades is also likely to be rather unprofitable for birds.

Against the background of a growing forest estate there may be some cases where the return of limited areas to open ground is justified by exceptional conservation value. Small features of high value like bogs are one example, habitats that have become nationally scarce another. However, consider first whether cyclical felling could achieve the same effect for birds of open ground. Forest clearance must be a last resort and not a substitute for proper management of existing semi-natural habitat; we would strongly oppose any suggestion of clearing forest in Breckland, for example, whilst there are large areas of degraded semi-natural heathland, much of it in the same national ownership as the forest.

But money remains an underlying problem for this sort of adjustment, not so much the cost of tree removal or restoration of the cleared ground as the revenue foregone from not growing a tree crop. The FC cannot make its 3% from not growing trees and every bit of land which is not productive pressurizes the rest.

WHAT PRACTICAL MEASURES ARE NEEDED TO ENSURE THE FUTURE FOR BIRDS IN THE NEW FORESTS?

Economics

There is a growing tension within both public and private forestry between the need to achieve economic targets and pressure to incorporate better environmental practices into forests. The future of birds in the new forests is tied closely to the targets set for forestry because the improvements that could help them in the future do cost money, if not as straight cash then as future revenue foregone by altering felling ages, leaving land unplanted, etc.

Foresters have been blamed for, and have vigorously defended, a situation for which we are all in part responsible. The FC has always been under pressure to produce more for less money. There is little difference between the forests of the 1950s and 1960s and the appalling tower blocks that 30 years later are being demolished. Both were symptomatic of their time and it is fair to say that as a society we got pretty much what we paid for.

Are we making the same mistake again? Since 1974 the FC has had a target rate of return of 3% (Forestry Commission, 1974a). It has struggled to reach it (National Audit Office, 1986) despite an allowance of 2% because of the supposed environmental benefits of forestry (primarily recreational) against the normal Treasury test discount rate of 5%. Ironically, as we discussed in Chapter 3, most of FC's worst offences against the environment have been in the name of achieving the 3%.

Recently conservationists joined criticism of forestry's failure to meet its economic targets (e.g. Bowers, 1984; Stewart, 1985, 1987). For the poorest land the case is strong: despite higher land prices the economic return from land that can grow GYC 16 trees is far higher than the GYC 12 on cheap land of the Flow Country (National Audit Office, 1986). Vast areas of Lodgepole Pine will not grow at a yield class of even 8 which means that at a discount rate of 3% it is likely to make a negative return over the rotation. The greatest risk to moorland birds is also on the poorest land. But these arguments are aimed at the whole forest industry in Britain with worrying implications: the conservation lobby seems inadvertently to have been drawn into advocating the sort of short-term stock market economics which many feel threatens the future of the planet.

A forest industry pressurized to produce an unattainable rate of return cannot also produce costly environmental goods. If the forests got into their present state with the tacit approval of society as a whole, then the present public concern can lead to the correction of some of their problems. British forests are often compared unfavourably with the continent, where longer rotations, natural regeneration and irregular silviculture are commoner. All are possible because there is no debt for the original establishment of the forest; it was there already and has never been destroyed, so regeneration is paid for out of part of the profit of the felling of the previous rotation and the forest is expected to show a year on year profit. Rate of return on capital is very low but the forest is productive, attractive, employs people and does not cost anyone anything. What it comes down to is that the choice is ours: what do we want from our forests? Are they simply pulp factories, managed until the environmental pips squeak for a still low financial return, or do we want more, including more birds?

We suggest that consideration needs to be given to changing economic targets and the overall objectives of the new forests, not so foresters can drive ahead, planting ever less rewarding sites, but so that positive environmental changes can go ahead quickly and widely. For the first time the value of 'non market' benefits is likely to play a part in assessing the value of forests to society (e.g. Pearce *et al.*, 1989) — to some extent the conservation value of birds, but largely the value of the forests for recreation. Every year millions of people visit these forests and simply working out what they spend getting there adds up to a sum of money which tranforms the economic perception of the new forests. They are effectively 'voting with their feet' — deciding to spend their disposable income on a forest visit rather than, say, a trip to the seaside or a stately home. The very large numbers of visitors to the Thetford bird trail (Forestry Commission, 1987) also indicates the value birders put on forest birds — and justify spending what, in the case of the FC at least, is their money on further helping birds. This sort of demand justifies changes in approach towards the sort of forests visitors want to see: if thousands visit every year to see Goshawks the forest manager should not fell all their potential nesting sites.

For shyer birds it is fortunate that economic methods can go further than the physical visit to a site. Thousands wrote to their MPs or joined the RSPB over the Flow Country but few ever visited it. They were prepared to take action to ensure the continued survival of something they had never seen in person. In many cases they also proved that they were prepared to pay something towards its survival through a subscription to a conservation body.

The removal of tax relief from private forestry also looks rather like a withdrawal of subscription by that body to which we all contribute, the government. The tax changes look like achieving conservation's main aim: stopping planting of the remote uplands, mediocre tree-growing land but exceptional for birds, but there are undesirable side effects: in future, forest operations will have to cost as close to the planting grant as possible. There is not the flexibility to do extra work, including conservation, that there was when the cost was discounted by 60%. Direct grants for specific conservation work over and above the minimum required under the woodland grant scheme might have to be considered for private forestry if the private forests are to see the same improvements for birds in the future as FC land; this should perhaps include allowances for keeping extra land unplanted, as the high cost of land means that the profitability of a whole scheme declines rapidly if too much is left unplanted and ineligible for grant.

Practicalities

As pundits, consultants and instant experts vie for the most outrageous headline it is easy to forget that real life includes a forester and half a dozen workers with planting spades or chainsaws outside a hut in the forest at dawn. It is how well that forester understands what he is doing and how well he instructs his workers that determine what happens in the forest. Conference resolutions are no good to a Goshawk chick if the chainsawyer does not spot the nest till after the tree is on the ground. Bringing long-term plans, not just for timber felling or birds but also for landscape, recreation, water quality and many other considerations, together and getting them to work on the ground is a formidable task. The chain of management stretches from the scientist developing new techniques through national strategic planning and right down a lengthy line of communications to the man on the ground.

The FC plays the central role in this process. As well as managing its own land it carries out research, advises and, through the grant arrangements, to a large extent controls the private sector. One of the most frequent of the instant solutions for solving all forestry's problems is to abolish it (the previous one, removing tax relief having failed to completely stop new planting). The flaw in the argument is that it is not the way the FC has worked but the objectives it has pursued that are the problem: however you look at it, the creation of the new forests is a spectacular practical achievement. The same determined practicality is gradually being turned to bird conservation in the new forests. Its effects have been positive, in some cases quite dramatically so. The sort of respect growing up between people working on the ground is evinced by the top RSPB/Esso conservation award for 1989 going to an FC forester. Where the FC and conservationists are heading in the same direction, the Commission is a welcomed and invaluable ally. No other institution has experience of landscape-scale design nor the confidence to go out and solve practical problems on the scale the new forests present.

Many other factors than birds come into arguments about who should be doing what and how in the countryside: looking solely at birds, it is beyond our scope to suggest how they might best be organized. However, what we would emphasize is that, change or no change, it is vital to the future of birds in the new forests that we do not forget the perspective of the forester on the ground. That is where the real action is; paper programmes are worth nothing and we must above all else preserve the skills and motivation needed to make good things really happen for birds in the forest.

Research

Nowhere do the above arguments apply more than to the research and information-gathering crucial to further progress in improving forests for birds. Our ignorance of what happens to birds in the new forests is profound. Perhaps the most glaring gaps in our knowledge are caused by the shortage of simple survey information. Whereas O'Connor and Shrubb (1986) could call on a very large number of amateur studies of the birds of farmland as well as the information from the organized work of amateurs participating in the Common Birds Census, there was hardly any similar information published on the birds of conifer plantations available to us. We still do not know the simplest and most basic things, like how long Golden Plovers will continue nesting after their territories have been afforested, what actually happens to the birds as trees grow to overshadow small watercourses, or how long it is until young broadleaved trees planted in a conifer forest will affect the birds that are found there. One problem is that most people live out of the uplands where the new forests are, but that is not the whole answer; it has not stopped effective national surveys of inaccessible birds like Peregrines.

With some notable exceptions, for example David Lack in the 1930s and more recently Ian Newton's Institute of Terrestrial Ecology team, academic researchers have ignored the new forests. Considering the scale of what is effectively one huge experiment this is a great pity. The FC itself has been very slow to appreciate the wildlife value of its own forests; research on wildlife that did not damage trees was extremely limited until the 1980s, and only in the last few years has it been involved in work on birds of prey, Woodlarks, Nightjars and Black Grouse. Even now the FC is rather poor at publicizing the conservation value of the birds contained within its own forest estate. An FC-commissioned national survey of Redwings, Long-eared Owls, Goshawks or Crossbills would almost certainly help raise interest in forest birds. Our experience is that once birdwatchers actually get into these plantations they wonder why they haven't spent more time there before.

The best organization to carry out this sort of work is probably the British Trust for Ornithology, through its large amateur membership. Responsible for most of the data in the O'Connor and Shrubb study mentioned above, the BTO has been notably absent from the upland forests. Although its Common Birds Census has many woodland plots, there are hardly any in conifers. Remoteness is a problem but Ken Williamson, who was interested in the birds of the new forests and carried out some useful work under the auspices of the BTO, used to organize survey expeditions to Scottish woods. Bibby's point count studies in Wales have shown just how much can be achieved quickly by good amateur birders linked to a sophisticated but easy to use field and analysis package.

The need for this sort of survey information is particularly important because of the opportunities for improving upland forests: new management techniques are being applied on a wide scale with little information on their effects. It is fair to assume that they will be better than what went before but without more monitoring fine tuning will not be possible and avoidable mistakes are sure to happen. What are the effects of clearfell size on different species and how long do open land birds stay after replanting? Are we now planting far enough back from watercourses for Dippers, Grey Wagtails and Common Sandpipers to flourish in the midst of large conifer plantations? What are the best broadleaved species to plant and are our ideas on scattering small groups throughout the forest right? How can we best provide a continuous cone supply for Crossbills and Red Squirrels? These are just a few of the questions needing answers, and each one will differ regionally, between sites and between tree species. One of the beauties of bird research in forests is that big sample sizes and complicated approaches are not always needed: frequently anything, however small, is better than nothing on a topic not covered before anywhere by anyone.

Information is at the core of good environmental management and foresters will badly need the help of organizations like the ITE, the BTO and the RSPB if they are to optimize their management work for forest birds.

NEW PLANTING

New planting will remain the major issue as far as forestry and birds are concerned. Whether or not the existing forests were planted in the right place, any harm has already been done. New planting still gives scope for major bird conservation disasters. Conflict and confrontation between forestry developers and conservationists has become habit. For the birds it is vital that this does not continue indefinitely and in this section we look at some of the major issues that will affect the future of new planting and upland birds.

As we discussed in Chapter 8, the factors which shaped afforestation have until recently had little to do with environmental values. It would be wrong to suppose that there has been no planning; the overlapping influences of agricultural policy, land quality and forestry subsidies have provided remote but quite forceful guidance to planting. They did not intentionally concentrate planting in any particular region or habitat type but in areas like Galloway or the Flow Country had much the same effect as if they had. Since 1981 the Site of Special Scientific Interest (SSSI) system administered by the NCC has become powerful and effective. It is unlikely that further planting will take place on SSSIs without NCC consent. SSSIs protect the best effectively, but in the uplands it is hard to define the boundaries of sites or to cater within them for very wide-ranging birds like Golden Eagles. Outside these sites existing consultation arrangements between forestry and other interests tend to lead to an unconnected case by case approach to planting, causing skirmish after skirmish and massive wastes of time and money on both sides.

The land planted in the Flow Country was unprotected by designation or even the knowledge of high conservation value at the time it was bought for forestry. Further new forests must take more account of bird values if they are not to cause unacceptable damage in the future. To achieve this there must be knowledge of where priorities for conservation lie and a greater environmental content to the mechanisms which direct land use.

Is further new planting likely to be ecologically justified?

Whether there should be more trees and what effect new planting has had and might have in the future have never really been discussed. The forestry case has usually been economically based and its occasional sallies on wildlife have tended to be so superficial as to lose rather than gain credibility (e.g. Forestry Commission, 1977a). Foresters have had surprising success in concentrating argument around the bottomless pit of forestry economics. Even in the Flow Country, as the dispute developed it was easy to forget that the fact that an area could be proved to return 3% from a Sitka Spruce crop was not in itself a justification to plant trees.

We discussed how the conservation value of moorland birds might be assessed in Chapter 5 but obviously the relative value of moorland to forest birds depends to some extent on personal perception. We have tried, particularly through the case studies, to develop some perspective on this. It leads to one clear conclusion: that the relative value of forestry for birds depends on what moorland birds are there before planting. Forestry undoubtedly eliminates many species of open land birds. It equally certainly provides habitat for a different assemblage of considerable attractiveness and interest in its own right. Not all uplands are equal: ecological degradation by burning and overgrazing has wrecked large areas of upland for birds and the premise that all semi-natural vegetation is more important than any artificial habitat is hard to argue, for birds at least.

On the other hand, the best of the afforestable uplands are very special. Artificial they may be, but they have been here for a very long time, hold birds that are indisputably attractive and exciting and are a significant proportion of Europe's upland heath bird habitat. We believe that maintaining established populations of upland birds has higher conservation value than increasing incomers: thus, for species of equal interest and rarity, for the sake of argument Greenshanks and Goshawks, the Greenshanks must rate a higher conservation value. In practice this means that the best areas for upland birds, the best example of which is the Flow Country, should not be afforested at all.

These best areas are limited: although there are birds everywhere in the uplands, we suggest that the loss of some of the poorest grass moorland assemblages of Skylarks, Meadow Pipits and a few Kestrels or Buzzards has been more than compensated by the introduction to many regions of new breeding species like Crossbills or Siskins. However, if some areas should have no expansion of forestry at all it does not follow that others should be entirely covered by trees. We discuss in the next section the very considerable areas potentially available for forestry. There is no justification for total loss of any habitat type, however poor for birds, on a regional scale and the regional extinction of any moorland bird species can and should be avoided. Our case studies for Galloway and the North Yorks Moors suggest that it may be the last fairly small proportion of planting in a region that can do most harm to wide-ranging moorland birds.

Most regions of Britain now have some new forests and breeding birds like Siskins and Crossbills have spread into them. There is also now some forest almost everywhere for new colonists. This means that further new forest may allow populations to increase, in some cases to become properly established, but is unlikely to bring in many new species. Without planning that takes account of the variable value of the open uplands for birds some forest will inevitably be planted on the best land. Serious loss to vulnerable species will occur and regional extinction is a possibility. National extinction is less likely now that much of the Flow Country is to remain as open land. Located on the land of lowest conservation value, however, further new planting could have a very limited effect on open-land birds and hardly touch scarcer species at all. Further new planting would have a neutral or slightly beneficial overall effect on bird conservation value.

Birds are only a small part of the future of afforestation. A particularly important factor is the effect of different human priorities in different parts of Britain. In Chapter 5 we demonstrated the very strong latitudinal trend in the diversity of moorland birds: the further north you go, the richer the moorland bird assemblage you are likely to find. In 1988 the Secretary of State for the Environment announced that planting of conifers in the English uplands would no longer be allowed. The main reason for this was conservation pressure, as much from landscape interests as wildlife conservationists. In contrast, during 1989 the Highland Regional Council called for a doubling of afforestation in the far north of Scotland. People in the north have less economic prosperity, people in the south less wilderness. The geography of human aspirations is the direct opposite to that of moorland bird diversity.

Conservation values, also, do not always overlap. Even in the Flow Country peatland and bird conservation priorities are not identical. We know most about birds and vegetation in the uplands, less about other groups.

It is not appropriate here, in a book about birds, to get too deeply involved in the tortuous arguments over the economic justification for forestry. However, the broader ecological importance of planting trees in Britain is relevant because it affects birds elsewhere. The plight of the tropical rainforest is well known as is the effect of world trade on deforestation. Britain led the way, many centuries ago, in destroying its forests. We now depend on other people for our wood and have undoubtedly contributed to deforestation. Much of our softwood comes from primary forest, the regeneration of which can be far from certain. Even though we cannot ever produce all our timber needs, surely we should take advantage of our ideal tree-growing climate to produce rather more? Is the only example we can set other countries one of the sort of dependency that follows forest destruction? More trees would also provide some cushion, if not a defence, to future timber shortages.

Is this another spurious justification for upland planting? Perhaps, but maybe we are moving back closer to the real reason for the start of the new forests in 1919. The First World War was the spur but the feeling remains that the real reason for the foundation of the FC was a deep-seated feeling that we need more trees, the same feeling that many people now have following the 1987 hurricane in southern England and from watching television of the tropical rainforest burning. In the intervening years it has been unfashionable to express that sort of emotion, just as it has been unfashionable to suggest that serious economic development should be halted for a few birds.

If new planting in the future is ecological rather than economic, it is all the more important that we do not destroy our own environment to save someone else's. Criticism of new planting has focused entirely on what has happened, in particular damaging locations, whether from a landscape or conservation viewpoint, and intensive economic use of the land. Future forests must be located to do the least environmental harm and be generous in their design. Like our existing forests we probably need to spend much more on future new forests; they may never make 3% but they might make a tiny contribution to the future of the planet.

How much more new planting should there be and where is it best located for bird conservation?

Location and area of new forest are inextricably linked; as we have shown, location is the key to the impact further afforestation has on moorland birds. The big problem is information: to plan a strategy an overview is essential. For this it is assumed that massive survey is essential and NCC research has covered only a tiny fraction of the uplands. But it is wrong that only on the ground can detailed survey give us the answers we need. In fact, it is often difficult to draw comparisons between very detailed bird surveys because of differences in method, differences between years and incompleteness of coverage. There are other approaches which give better prospects of success.

The ITE Merlewood land classification project (Bunce *et al.*, 1983; Bunce, 1984) gives a valuable and dispassionate analysis of the nature of the land most likely to be planted in future. It classifies a representative selection of areas across the whole of Britain into different land types using a wide variety of environmental variables. The importance of the classification is that it gives a broad view allowing predictions of the effect of different land use changes across the whole of Britain to be made. A simple economic model is used to look at the effect of different levels of forestry planting on vegetation types.

Taking all the upland land that under present conditions could physically be afforested, a massive 2 718 000 ha (more than double the present area of upland forest), Mitchell *et al.* (1983) found that the loss of existing moorland vegetation types was likely to be surprisingly even. Between 40% and 60% would be lost for the major vegetation types; only the very wettest bogs would be less affected. Of the largest categories *Calluna vulgaris* (58% predicted loss, 583 000 ha) and *Vaccinium myrtillus* (54%, 470 000 ha) would be affected more than grass moorland, with *Agrostis tenuis/Festuca ovina* (42%, 325 000 ha) and *Nardus stricta* (45%, 328 000 ha) and *Molinia caerulea* (57%, 267 000 ha).

Although large areas of all vegetation types would be left after such an unimaginable spread of forestry they would form small isolated remnants compared with their present distribution. In some areas moorland of any type would disappear completely. The effects on birds would obviously be immense.

However, the ITE model does not advocate this level of planting. The real situation is less dramatic. Present planting levels would see a further 6% of moorland vegetation converted to trees by the end of the century (Bunce 1988). Between 1978 and 1984 heather moorland was the most afforested, at 30 200 ha and 27% of the area planted. *Mollinia*-dominated moorland was next most affected at 22 400 ha and 20% of the total. As Bunce points out, how one sees these figures depends very much on one's outlook. On the one hand 6% is not a very large proportion of a land type but on the other the area of land concerned is enormous.

The information this study gives us is of considerable importance: the location of new forests is not just about geography. The finding that the vegetation type most at risk from forestry is *Calluna* moorland is worrying. The RSPB has carried out bird surveys over very large areas of the Pennines and south Scotland using the ITE land classification as a sampling framework. The preliminary findings confirm that many species of high conservation importance are linked to heather moorland and that the birds of grass moorland are far less interesting and numerous. This type of widespread survey, carried out within a known sampling framework, is another step in providing the information we need to see where forestry will do least harm. It will enable researchers to say that, compared with *Nardus* moorland say, heather moorland is X times as important for Curlews, Y times as important for Golden plover and Z times as important for Snipe. What this approach does not tell you is where all the different bits of heather and *Nardus* moorland are (although it can tell you how much of them there is).

This introduces the role of a major advance in identifying which moorlands are good for birds; satellite imagery. Both the NCC and RSPB are actively engaged in using satellite imagery to map moorland vegetation and to predict the numbers of moorland birds on unsurveyed land. These new high-technology approaches can produce maps of vegetation types for very large areas of land. Combined with a knowledge of the bird communities of each vegetation type from the linked ground survey projects (ground truthing) mentioned above, the satellites can produce maps of bird numbers (Avery and Haines-Young, 1990). Progress in this field has been rapid over the past two years but the main advances still lie ahead.

The third line of information which will become available at about the same time as the other two are producing clear and useable results is the new BTO breeding bird atlas. Potentially this exciting new project could provide much of the information on which to base a compromise between forestry and conservation interests. Extensive bird surveys of the types used to identify different land classes or to link in with satellite information have, by necessity, to cover large areas fairly quickly. They thus tend to provide good information on the numbers of moderately common species such as waders, certainly in enough detail to say which are good wader areas and which are not, but they miss the rarer species which are often of equal importance in assessing the overall conservation value of the area. It is this gap that the Atlas will go a long way to filling.

Thus, there is far more useful information available for making decisions than is generally acknowledged. Laborious and expensive ground survey is not the necessity sometimes claimed. Within a few years there is a good prospect of a fairly comprehensive overview of upland birds emerging and we are even now in a position to make some predictions about the sort of land on which forestry is likely to do least harm to open-land birds.

The two major trends are geographical and vegetational. The further north one goes, the more likely moorland is to hold a diverse and valuable bird fauna. On the vegetation side there are two trends—first, wetter ground is likely to be more valuable, and second, heather is likely to be more valuable than grass. These rules are, inevitably, not universal—badly degraded grass moor with high sheep mortality may be important for carrion-eating birds. Also, grass moors are not by definition badly degraded and it would be wrong to write off a vegetation type as a whole because of what is true for much of it. In practice these conclusions point to central and southern Scotland and northern England as being preferable for forestry to northern Scotland. Wales still has extensive areas of grass moor but also has the most heavily afforested uplands of the three countries.

How are these criteria likely to fit with forestry and other land use interests? A trend away from the wettest ground for planting is already established in the aftermath of the Flow Country. Planting grass rather than heather moor is likely to be better for growing trees. There is a direct conflict between the priorities for birds and people geographically, with the greatest support for further planting in the northern part of the country. Remember, however, that we are talking about trends and not absolutes: it is a question of altering the balance between heather and grass, not saying that if another heather moor is planted it will be a major disaster for birds. Similarly, there is much land in northern Scotland that is not of high value for birds and could be planted. What could be disastrous would be a higher and higher proportion of planting being pushed north by conservation pressures in the south.

This is a simplistic approach but it demonstrates that broad trends can be identified as a basis for strategic planning. Obviously there are a range of other questions that planners have to face. To what extent should we protect sites because they are typical as opposed to exceptional? Should large areas of forest or moor be favoured over the same total area split into several pieces? Should particular species be given special consideration, and if so which should these be? Is the best way to protect upland birds to concentrate on protecting particular vegetation types, perhaps without even using the bird information at all? The questions multiply but the answers are rarely clear. Different people and different organizations will see things differently.

Underlying many of forestry's present problems is the uncertainty of how much land will eventually be planted. All there has ever been is an annual target, at present 33 000 ha per annum. Only the vaguest thought has been given to the eventual size of the national forest estate. While the forests were still a tiny proportion of the uplands this did not really matter but it does now: within the present generation we could see more than half the afforestable uplands planted. Whether a set area target would be sensible, let alone practical, is open to debate, but a clearer indication of what we are aiming for in planting new forests could take much of the heat out of the present conflict between forestry and conservation. If it were accepted that forestry should occupy a certain percentage of the available upland land then there would be much greater opportunity for conservationists to make sensible, practical proposals as to where new afforestation should go and even perhaps accept some plantations in areas where they would prefer them not to be in the certain knowledge that this was not the thin end of a massive wedge.

The present uncertainty benefits no-one. Foresters are apt to see their industry under threat and regard each tree planted as a battle won, whereas conservationists see themselves besieged by the massed ranks of conifers which must be fought on all fronts if the uplands are not to be doomed. As more and more afforestable land is planted it will become less worthwhile for foresters to fight an increasingly effective and vigorous battle for the land which is left. This has already happened in England and to a large extent in Wales, where environmental constraints, National Parks and a simple shortage of freehold land for sale mean that there is little prospect for substantial new afforestation schemes. The Secretary of State for the Environment's 1988 announcement that planting of conifers in the English uplands would no longer be allowed rules out further expansion in England. It was a victory for conservation of considerable political importance but its practical impact was limited: only 536 ha of conifer new planting took place in England in the year ending March 1987 (Forestry Commission 1988a).

Being specific about the area which might be planted is impossible: a national target would almost certainly have to be built up from a combination of national overview (like the ITE study) and regional planning. However, it is possible to make some predictions. A substantial area of new planting will take place whatever happens. Approvals for grant aid on as much as 100 000 ha of land as yet unplanted may be held by private landowners at any one time. Evidence that afforestation was likely to be curtailed could well make them more valuable and more likely to be planted. However, new planting is unlikely to be stopped in its tracks and the least we would see being planted given a further rapid decline in popular support for forestry would be 250 000 ha. At the top end it is unlikely that more than 1 000 000 further hectares would be planted, and that only against a background of a radical change of attitude towards forestry. It is something we believe forestry could achieve but only by joining forces with rather than fighting conservation.

A larger area of forest can only be planted without harm to birds if it is carefully planned, but based on what is now known we believe the impact of quite large areas of carefully located forest could be suprisingly low. It looks as if it could be theoretically possible to afforest a further 500 000 ha without the loss of any Greenshanks or Dunlins and probably no Merlins. Fewer Golden Plovers should be affected than have been lost to only 30 000 ha of planting in the Flow Country. They would also maintain their geographical range because key heather in Wales and the Pennines would not be planted. Very rare species like the Chough would be given more general protection to enhance that already provided by the SSSI system. The main losses would be to numerous and widespread birds like Meadow Pipits and Skylarks. Moorland fringe birds like Curlews and Whinchats are also likely to be more affected. It is more difficult to predict the impact of this level of planting on the carrion-eaters: a concentration in south central Scotland could well bring Ravens to the brink of regional extinction but the other key species, Golden Eagles, Buzzards and Red Kites are absent or scarce from these areas. Even discounting any value attributable to forest birds, it seems that it may be possible to plant a lot more trees with remarkably little damage, certainly far less in total than has been suffered in the Flow Country.

How can newly planted forests be better for birds?

The good design of new planting cannot compensate for the forest being in the wrong place: much of the design of the Flow Country forests is generous in the extreme, with very wide riparian zones, wide deer passes and space left round lochs. However, once the decision to plant has been taken, it is important that the new forest starts with good design.

Many of the design criteria are common to existing forests, as described in Chapter 4; open space, keeping the crop conifers back from watercourses, leaving bogs and crags unplanted and planting a proportion of broadleaves. Obviously, age–class diversity is not a major factor at the new planting stage. It may be possible to make special allowances for scarce or interesting species already living in the area: as we have discussed it is possible to design the forest to retain edge birds like Black Grouse but it is more difficult to help true moorland birds like Golden Plovers.

To be ideal for birds, some aspects of design, in particular the inclusion of more broadleaves and open land, need to go a little further than the minimum. Cost is again the main barrier and it is far more obvious at this initial investment stage: there is no timber income to cushion environmental work and the investment in land is very recent. Two factors apart from the removal of tax relief may make this even more difficult: first, forestry may be moving off the worst and very cheap land. Better land is more expensive making it costly to give up land for conservation, and additionally the difference in return between broadleaves or open land and conifers is even greater. Second, if more owners are to plant their own land rather than selling it on to forestry companies, the gap between grant income and cost becomes more critical because they are less likely to have up-front investment money than the clients of the companies.

The land use problem also still hangs over forestry: forestry and agriculture still tend to be mutually exclusive large-scale land uses in the uplands. Although in the heart of the uplands it may be best for birds to have separated and unbroken areas of moorland and forest, around the edges a more diverse mixed landscape would probably be better. It would certainly look better and being able to operate two land uses side by side would make it easier to, for example, keep straths open for waders. However, forestry and farming do tend to remain very much separated and sale of part of a holding for forestry usually weakens rather than strengthens the farming on what is left.

How can future new planting best be located in practice for bird conservation?

Knowing where new forestry should best go for birds is only half the battle: there is then the question of getting it there. We discussed the background to the location of forestry in Chapter 8. Up until recently designations have been the main control available to conservation. In England and Wales this includes National Parks, covering quite a high proportion of the uplands, but in Scotland there are only SSSIs. The other option open to conservation, especially the voluntary sector, is purchase, but this has been suprisingly little used over forestry issues: in general, areas involved are too large and the bird interest too diffuse, but in the Flow Country where purchase might have worked most of the land deals had already taken place by the time the RSPB became heavily involved.

SSSIs have worked reasonably in recent years even in Scotland in protecting the most special sites. During the Flow Country dispute they were seen as the only effective conservation defence against forestry, and massive extension was advocated. The Secretary of State for Scotland decided in 1987 that the SSSI area in Sutherland and Caithness should be greatly expanded to meet the special value of the Flow Country. However, in the wider uplands SSSIs are not designed as a broad land use tool. They are seen as being a single-use designation and unduly restrictive, both points justified in practice. The NCC has a very high degree of control over SSSIs and tend to require management as for a nature reserve, which is reasonable if the area designated is one of the real conservation jewels of the nation. Owners have limited rights of appeal once a site is chosen for designation and the cost of compensation paid by government to owners can be high.

A number of alternatives have been suggested in the last few years. Most have been designed not to direct but to stop forestry, in some cases because of a basic dislike of what forestry was doing to the uplands, and in others because no avenue for compromise was available. Crippling forestry financially was an obvious target and was the subject of the tax relief campaign. Increasing control over forestry by one means or another has also had a lot of support. Most often suggested was formal local authority planning control over new planting to replace the present consultation system discussed in Chapter 8. The idea poses problems for birds: local government decisions are coloured by local attitudes, and, as we have discussed, a national overview is essential for proper strategic planning. Sympathy for conservation tends to be less the further north and more rural the region, exactly where moorland bird interest is highest. The operation of planning control depends largely on the attitude of the local authority, which as demonstrated for National Parks by Brotherton (1987) doesn't always match the opinion of conservationists as closely as the advocates of statutory planning control fondly imagine. There is also the problem of maintaining sufficient expertise at a local level to evaluate often complex conservation cases and it is very much up to each authority who they consult outside designated areas: there would be no guaranteed role for the NCC or RSPB.

An alternative is greater central government control, usually through some sort of mega-ministry for the countryside. This is an idea that never seems to develop much beyond the abolition of the FC, which seems to be the main target. But does central government really need more control to exert its influence? Because infertility and remoteness make the uplands so uncompetitive economically, almost every activity is funded from central government to some extent. It pays forestry and agricultural grants, it pays the NCC's wages and the SSSI compensation bills and it grant aids the National Parks. Through bodies like the Highlands and Islands Development Board (HIDB) it supports industrial and tourism development. Upland funding is not just with UK money: agricultural money is underpinned by the European Economic Community's (EC) Common Agricultural Policy (CAP) and roads and other development projects are supported by the EC's regional development funds. In the fragile economy of the uplands it does not take much more than a tweak on the reins to correct the balance of power or economics.

It is against this background that a ray of hope in the form of an indicative forestry strategies system has emerged. In conception it looks as if it could meet the needs of upland birds in identifying on a broad scale where planting would do least harm. The approach now planned for the whole of Scotland started with the regional council in Strathclyde and has been produced as a map showing four broad categories of land. First is land unsuitable for forestry, derived from the new Macaulay Land Use Research Institute (McLURI) land capability for forestry classification (Bibby *et al.*, 1988). This land is effectively out of contention. Potentially plantable land is then divided into areas where there is no obvious objection to planting, land where there is some constraint and finally what are effectively 'no go' areas where there are serious constraints to planting. The strategy has no statutory compulsion: simply the clear signal that there will be fierce opposition to planting in areas considered most sensitive. The scale of the map makes it difficult to pick out precise boundaries: it is not meant to replace SSSIs in protecting discrete special sites.

From a bird point of view there must be a little concern that strategies are generated at a regional rather than national level. However, the NCC is one of the participants in the project and can provide the national conservation overview. The Flow Country experience suggests that a nationally imposed system might have been unacceptable regionally. Strategies have got off to a surprisingly good start: the level of co-operation between agencies is unexpected and creditable after the acrimony of the Flow Country. It is almost certain that serious disagreements will arise as strategies are developed for further regions but we hope this will not wreck them: this initiative gives hope for the future planning of new planting almost inconceivable only a year ago.

The alteration of strategies as further conservation information becomes available could be contentious; conservationists must accept that this should be a two-way process, not simply a question of forgetting the constrained areas and putting all their effort into proving higher value on the land identified as most suitable for planting. In this organisations like ITE and McLURI could be vital: as pure research bodies they do not have the problem of the dual role as both advocates and researchers of, for example, the NCC.

A factor which a forestry strategy needs to take into account is its effect on the owners of land which it may rule out for planting. Sale for afforestation has provided an economic longstop in the uplands in the past. Forestry is only one element in the upland economy. Agriculture is most important and it too is changing rapidly at the moment. On the positive side the sheep enterprise has held up well within the CAP so far, being one of the few commodities not in surplus. The new designation of environmentally sensitive areas (ESA) is directly supporting traditional agriculture in selected farming areas of particularly high landscape or wildlife value. A British-initiated scheme, it broke new ground by making payments for maintaining a way of life and farming without requiring the agricultural 'improvements' that can do so much environmental damage.

However, SSSI or ESA payments cover only a small part of the uplands. Agriculture is economically weak at present and despite the DAFS insistence that grazing should not be permitted on set-aside land, more and more lowland farmers are keeping sheep. The risk of a serious collapse in upland farming is real and would have dire consequences for birds.

Sport, shooting Red Grouse and stalking Red Deer, is also vital to upland birds. Heather management for grouse provides ideal habitat for a range of important birds including Merlins and Golden Plovers. Red Deer, at present at record population levels, are one of the most important factors in shaping upland vegetation. Large areas of grouse moors and deer forests have been sold for afforestation. The encouragment of current management on the former and better management of the latter will be important for upland birds in the future. It may even be necessary to support these activities with public money: but if bird conservation is a worthwhile objective we must be ready to support the means to achieve our ends, however unconventional by today's standards. The message of ESAs is that it is now acceptable to pay for non-market goods like landscape or wildlife rather than having to dress up in a disguise of production of conventional agricultural goods.

NEW HORIZONS

Agricultural change in Europe is already affecting the British forestry scene. Large areas of arable land need to be taken out of production to reduce surpluses. In the uplands some of the better marginal land long denied to forestry is becoming available. Planting on the more productive lowlands is seen by many conservationists as an alternative to upland afforestation. In fact, the two are not mutually exclusive but there is a link, in particular for Britain's timber production potential. It is possible that our perspective on forestry will be very different in 10 year's time.

The growth potential of tree species is limited: a tree like Oak, for example, will not reach the GYC 12 average of upland Sitka Spruce whatever soil it is planted on. Forgetting social issues and subsidies, the marginal land where the poorest improved land and the moor meet is probably the best place to grow trees in Britain as it has little importance for the nation's food supply but is good enough for conifers to grow close to their biological potential. Marginal agricultural land contains a large proportion of Britain's Lapwings, Curlews and Snipe. Fertile valley slopes, too steep for cultivation, are the productive hunting areas for species such as Red Kites and Golden Eagles. We could even face a second battle of the Flow Country if forestry came off the moors of Caithness onto the marginal land around Halkirk and Westerdale, which from personal observation probably holds some of the densest populations in Britain of species such as Snipe, Curlews, Lapwings, Redshanks and Oyster-catchers.

We actually know very little about the value to birds of these marginal lands; there seems no reason why exactly the same circumstances as in the uplands should not arise. It is likely that planting the part of this land poorest for birds would do little harm and produce some excellent forest, but where is that? We now have some idea of the relative value of the open moorland but nothing with which to make comparisons with this habitat. Both the RSPB and NCC are doing some work on the birds of marginal land; maybe for once we can tackle a problem before it becomes a crisis.

The Farm Woodland Scheme

Fertile soils give scope for more variety but creating an interesting wood from scratch is a far greater design problem than manipulating something already there, added to which many farm woodland planters will have had little experience of woodland management. Initially fairly conventional woods are likely to be the norm, meaning that regular plantations of commercial species will dominate. However, early farm woodland schemes do look promising. The way the scheme works helps. First, it is concentrated on land that has been cultivated in the last 10 years, which means that rough corners or unploughable steep banks which often hold most of the remaining conservation value in farmland are not at risk. Second, standard rate FC grant is paid for planting broadleaves but conifers only get the much lower pre-1988 budget rate of grant. Third, annual payments designed to substitute for lost agricultural income is at a fixed rate but goes on for 20 years for conifers, 30 for most broadleaves and 40 for Oak and Beech.

Broadleaves have dominated early applications in England, with a predominance of native species and quite a good variety of minor species like Gean supporting the main crops of Oak, Ash, and Beech. Shrubs like Hazel are popular for providing game cover; field sports are an important motivation for planting and also generate valuable extra income for the farm. Rather more conifers in the mix would actually add variety and probably some extra species of birds to these plantings.

Funnily enough, there are fewer rare birds likely to move into these farm woodlands than live in upland conifer forests. The most obvious beneficiaries, thrushes, tits, Sparrowhawks and woodpeckers are all common. Rarer farm woodland birds might include Nightingales, Lesser-spotted Woodpeckers and

Hawfinches. In the long term, a much larger area of lowland woodland and improved attitudes to raptors (the former seems more likely than the latter at present) could see the spread of birds like Buzzards, Goshawks and even Red Kites throughout the lowlands.

Obtaining the maximum benefit from lowland woods will be far more difficult than most conservationists realize; there is an enormous gap between planting trees, whatever their species, in neat lines and the creation of really diverse woodland. Species mix is important; we would like to see a proportion of conifers in most plantings, a predominance of native broadleaves and a strategic scattering of minor species. Silvicultural method and structure are less obvious but equally vital: middle-aged plantations of any species lack structural diversity. A shrub layer will add great value for birds. A mix of high forest and other systems like coppice with standards will again increase habitat diversity and provide for birds like Nightingales. In the absence of coppicing a proportion of clear felling should take place, providing habitat for Nightjars, Whitethroats and Tree Pipits and an element of conifer will spread time of felling so the whole wood does not reach its economic rotation age at the same time.

It will also be necessary to develop the non-woodland aspects of new lowland plantings; rides and glades and their scrub edges are vital to lepidoptera and Sylvia warblers. Initially, most will be too narrow and too straight, developing into dank, whistling wind tunnels. Glades also make it possible to shoot deer which will be just as big a threat to trees, especially coppice regrowth, in the lowlands as in the uplands. Existing habitat features like ponds can be developed. Hedgerow trees should be incorporated into new woods rather than felled; it will be a long time before the young planting produces natural dead wood habitats of its own.

Making the most for birds of the opportunities of new planting on better land is going to take far more effort than is realised. Big new centrally funded research programmes need to be directed as much towards wildlife and other social and environmental considerations like landscape, recreation and sport as the silviculture on which the first projects concentrate. The farm woodland scheme has avoided the potentially disastrous error of allowing planting on the remaining lowland rough ground, in fact to the extent that for ecological safety and maximising savings in agricultural overproduction it excludes the vast area of improved permanent pasture that is of low conservation value. The clear message that the remaining fragments of open semi-natural habitat in the lowlands are not the place to plant trees has been heeded; it should not be forgotten.

New semi-natural forests

What about restoring extensive areas of semi-natural-type forest in the wilder parts of the Scottish Highlands? It is an idea that has been put forward by several conservationists. Obviously, it has no economic basis and the only people likely to be able to try it at present is a voluntary conservation body like the RSPB. On the surface the product of a vivid imagination, the means may not be far off in the shape of the new Native Pinewoods grant scheme which will pay broadleaved rates for planting or regenerating native original Scots Pine forest. Already the RSPB has plans for greatly extending the Caledonian Pine forest at Abernethy. The biggest problem for both them and anyone else with similar aspirations is Red Deer; is there a possibility that Scotland could one day have a few natural altitudinal tree lines?

CONCLUSIONS

It would be hard to overestimate the importance of birds in the differences between forestry and conservation. They are the sentinels of land use, the standard bearers of our feeling for the life with which we share our country and planet. We have painted a very different picture of the new forests, the open uplands and their birds from the popular mythology that has grown up around them in the heat of the last few years. Although it is a view coloured by our own experiences of both forest and moor, we have tried to analyse the rather scanty facts available as carefully, but as provocatively, as possible. That forests do hold many birds, some of them interesting and exciting, should not be the surprise it is: regardless of the fact that most of the trees are non-native, forests do provide a three-dimensional habitat closer at least in structure to our lost native climax vegetation than moorland. They provide shelter for nesting and from the bleak climate of the uplands and a far greater volume of vegetative material than the moor. It is true that often they have had value despite foresters' concentrated efforts to create uninteresting tree farms, but the changes that have now started to make themselves felt can reverse much of the harm and deserve support.

However, conservation is not only about volume of wildlife but also diversity, rarity, naturalness and beauty. Densities of moorland birds may be low, but what a superb and uplifting group of species they are, from the plaintive piping of the elegantly-attired summer Golden Plover to the soaring grandeur of the Golden Eagle or the dash of the Merlin. Whether we have forest or moorland is not an either/or; we can and must have both. Fortunately, the forests do seem to have done far less damage to moorland bird populations than might have been expected, but the close shave with our finest moorland habitat in the Flow Country must be the incentive to look for a better way of doing things in future.

It is sobering to look back at Frank Fraser-Darling's West Highland Survey (1955), completed 40 years ago. We have not done well in the interim. A strong supporter of forestry as a valuable component of the upland economy, he was critical of the single-minded excesses of commercial forestry which both the FC and the private forestry companies honed to such a fine art in the following 30 years. Nor would he have been pleased by the narrowness of the campaign launched over the last four years from the NCC's headquarters in Inverness which bears his name. 40 years on the land remains devastated, the Red Deer numbers climb as the refugee vegetation of the moors gets shorter and sparser and we still squabble over the ashes.

But there is hope. We have the environmental movement and agricultural surpluses to thank for a rapidly developing shift in attitudes. There is a real chance that we may again be able to see the land as something other than a production unit to be abused at the dictat of nett discounted revenue or gross margin. We are starting to accept that the products of land other than food or timber, including birds, have a value that is worth taking into account in allocating resources. What we have to try and achieve now is a greater unity of objective: we know we can grow trees or conserve birds but can we do both at the same time? Now we must grasp Fraser-Darling's 40-year-old vision of human and wildlife ecology as a single force in our uplands. But the importance is so much greater now: after all, if a rich country like Britain cannot tackle the test set by forestry and birds, where does it leave third-world countries struggling for survival?

References

Aldhous, J.R. (1972). *Nursery Practice*. Forestry Commission Bulletin 43. London: HMSO.

Aldhous, J.R. and Low, A.J. (1974). *The Potential of Western Hemlock, Western Red Cedar, Grand Fir and Noble Fir*. Forestry Commission Bulletin 49. London: HMSO.

Anderson, M.A. (1982). *The Effects of Tree Species on Vegetation and Nutrient Supply in Lowland Britain*. Research Information Note 75/82/SSS. Edinburgh: Forestry Commission.

Anderson, M.A. (1986). Conserving soil fertility: applications of some recent findings, in Davies, R.J. (ed.) *Forestry's Social and Environmental Benefits and Responsibilities*, pp. 65–78. Edinburgh: Institute of Chartered Foresters.

Anderson, M.A. and Carter, C.I. (1988). Shaping ride sides to benefit wild plants and butterflies. In Jardine, D.C. (ed.) *Wildlife Management in Forests*. Proceedings of a Discussion Meeting. Edinburgh: Institute of Chartered Foresters.

Anderson, M.L. (1950). *The Selection of Tree Species*. Edinburgh: Oliver and Boyd.

Andrews, J. (1979). Who can impress the forests? *Birds* **Winter**: 20–23.

Atterson, J. and Davies, E.J.M. (1967). Fertilisers—their use and method of application in British forestry. *Scottish Forestry* **21**: 222–228.

Avery, M.I. (1989). Effects of upland afforestation on some birds of the adjacent moorlands. *Journal of Applied Ecology* **26**: 957–966.

Avery, M.I., Winder, F.W.R. and Egan, V. (1989). Predation on artificial nests adjacent to forestry plantations in northern Scotland. *Oikos* **55**: 321–323.

Avery, M.I. and Haines-Young, R. (in press). Population estimates of wading birds derived from satellite imagery. *Nature*

Bainbridge, I.P., Minns, D.W., Housden, S.D. and Lance, A.N. (1987). *Forestry in the Flows of Caithness and Sutherland*. Sandy: Royal Society for the Protection of Birds.

Balda, R.P. (1975). *Vegetation Structure and Breeding Bird Diversity*. United States Department of Agriculture Forestry Service General Technical Report.

Balfour, J. and Steele, R.C. (1980). Nature conservation in upland forestry—objectives and strategy. In *Forestry and Farming in Upland Britain*. Forestry Commission Occasional Paper No. 6. Edinburgh: Forestry Commission.

BANC (1987). *Forests for Britain—The BANC Report*. Sussex: Packard Publishing.

Barnes, R.F.W. (1987). Long-term declines of Red Grouse in Scotland. *Journal of Applied Ecology* **24**: 735–741.

Batten, L.A. (1971). Firecrests breeding in Buckinghamshire. *British Birds* **64**: 473–475.

Batten, L.A. (1973). The colonisation of England by the Firecrest. *British Birds* **66**: 159–166.

Baxter, E.V. and Rintoul, L.J. (1953). *The Birds of Scotland: their History, Distribution and Migration*. Edinburgh: Oliver and Boyd.

Beedy, E.C. (1981). Bird communities and forest structure in the Sierra Nevada of California. *Condor* **83**: 97–106.

Berry, J. and Johnston, J.L. (1980). *The Natural History of Shetland.* London: Collins.

Bibby, C.J. (1973). The Red-backed Shrike: a vanishing British species. *Bird Study* **20**: 103–110.

Bibby, C.J. (1986). Merlins in Wales; site occupancy and breeding in relation to vegetation. *Journal of Applied Ecology* **23**: 1–12.

Bibby, C.J. (1988). Management in commercial forests for birds. In Jardine, D.C. (ed.), *Wildlife Management in Forests*, pp. 60–65. Edinburgh: Institute of Chartered Foresters.

Bibby, C.J., Aston, N. and Bellamy, P.E. (1989a). Effects of broadleaved trees on birds in upland conifer plantations in North Wales. *Biological Conservation* **49**: 17–29.

Bibby, C.J., Bain, C.G. and Burges, D.J. (1989b). Bird communities of highland birchwoods. *Bird Study* **36**: 123–133.

Bibby, C.J. and Nattrass, M. (1986). Breeding status of the merlin in Britain. *British Birds* **79**: 170–185.

Bibby, C.J., Phillips, B.N. and Seddon, A.J.E. (1985). Birds of restocked conifer plantations in Wales. *Journal of Applied Ecology* **22**: 619–633.

Bibby, J.S., Heslop, R.E.F. and Hartnup, R. (1988). *Land capability classification for forestry in Britain.* Aberdeen, The Macaulay Land Use Research Institute.

Blatchford, O.N. (1978) *Forestry Practice.* Forestry Commission Bulletin 14, 9th edition. London: HMSO.

Booth, T.C. (1977). *Windthrow Hazard Classification.* Research information note 22. Farnham: FC.

Booth, C.J., Cuthbert, M. and Reynolds, P. (1984). *The Birds of Orkney.* Orkney: The Orkney Press.

Bowers, J.K. (1984). *Do We Need More Forests?* Discusssion Paper Series, No. 137. Leeds: Leeds University School of Economics.

Brazier, J.D. and Hands, R.G. (1986). Structural sawnwood yields of Sitka Spruce from adjacent compartments at Fernworthy, Dartmoor, established at different planting distances. In *Oceanic Forestry.* pp. 32–36. Tring: Royal Forestry Society of England, Wales and Northern Ireland.

Brotherton, I. (1987). The case for consultation. *Ecos* **8(4)**: 18–23.

Brown, E.R. (ed.). (1985). *Management of Wildlife and Fish Habitats in Forests of Western Oregon and Washington.* Portland: US Forest Service.

Buckham, A., Cramb, T., Hudson, M.J., Leslie, R. and Scruton, R. (1982). *Ecological aspects of even-aged plantations: conservation aspects.* Dumfries, Institute of Chartered Foresters, Borders Branch.

Bunce, R.G.H. (1984). The use of simple data in the production of strategic sampling systems. In Brandt, J. and Aggar, P. (eds). pp. 45–56. *Methodology in Landscape and Ecological Research and Planning.* Vol. 4: Methodology of Evaluation/synthesis of Data and Landscape Ecology. International Association of Landscape ecology.

Bunce, R.G.H. (1988). The impact of afforestation on semi-natural vegetation in Britain. In Jardine, D.C. (ed.). pp. 54–59. *Wildlife Management in Forests.* Edinburgh: Institute of Chartered Foresters.

Bunce, R.G.H., Barr, C.J. and Whittaker, H.A. (1983). A stratification system for ecological sampling. In Fuller, R.M. (ed.). pp. 39–46. *Ecological mapping from ground, air and space.* Cambridge: Institute of Terrestrial Ecology.

Cadbury, J.C. (1987). Moorland birds: Britain's international responsibility. *RSPB Conservation Review* **1**: 59–64.

Callander, R.F. (1986). The history of native woodlands in the Scottish highlands. In Jenkins, D. (ed.). pp 40–45. *Trees and Wildlife in the Scottish Uplands.* Huntingdon: ITE.

Campbell, B. (1967). The Dean nestbox study, 1942–64. *Wildlife in the Forest,* supplement to Forestry: 13–14.

Campbell, B. (1973). *Crossbills.* London: HMSO.

Campbell, B. (1974). *Birds and woodlands.* London: HMSO.

Campbell, J.W. (1957). The rarer birds of prey: their present status in the British Isles. *British Birds* **50**.

Campbell, L.H. and Talbot, T. (1987). Breeding status of Black-throated Divers in Scotland. *British Birds* **80**: 1–8.

Campbell, L.H. (1983). *Moorland Breeding Birds in the North Yorks Moors National Park.* RSPB Internal Report. Sandy: RSPB.

Carlisle, A. (1977). The impact of man on the native pinewoods of Scotland. In Bunce, R.G.H. and Jeffers, J.N.R. (eds). pp. 70–77. *Native Pinewoods of Scotland.* Merlewood: ITE.

Centre for Agricultural Strategy. (1980). *Strategy for the UK Forest Industry.* CAS Report No. 6. Reading: Centre for Agricultural Study.

Clarke, W.G. (1925). *In Breckland Wilds.* London: Robert Scott.

Condry, W.M. (1960). *Nature in Wales.* **1**: 25–27.

Cook, M.J.H. (1982). Breeding status of the Crested Tit. *Scottish Birds* **12**: 97–106.

Cramp, S. and Simmons, K.L. (1977). *The Birds of the Western Palaearctic* Vol. 1. Oxford: Oxford University Press.

Cramp, S. and Simmons, K.L. (1980). *The Birds of the Western Palaearctic* Vol. 2. Oxford: Oxford University Press.

Cramp, S. and Simmons, K.L. (1983). *The Birds of the Western Palaearctic* Vol. 3. Oxford: Oxford University Press.

Cramp, S. (1984). *The Birds of the Western Palaearctic* Vol. 4. Oxford: Oxford University Press.

Crick, H.Q.P. and Spray, C.J. (1987). The impact of aerial applications of fenitrothion on forest bird populations, In Leather, S.R., Stoakley, J.T. and Evans, H.F. (eds) *Population Biology and Control of the Pine Beauty Moth.* Forestry Commission Bulletin 67. London: HMSO.

Currie, F.A. and Bamford, R. (1981). Bird populations of sample pre-thicket forest plantations. *Quarterly Journal of Forestry* **75**: 75–82.

Currie, F.A. and Bamford, R. (1982a). The value to wildlife of retaining small conifer stands beyond normal felling age within forests. *Quarterly Journal of Forestry* **76**: 153–160.

Currie, F.A. and Bamford, R. (1982b). Songbird nestbox studies in forests in North Wales. *Quarterly Journal of Forestry* **76**: 250–255.

Dare, P.J. (1958). *Devon Birds* **11**: 22–32.

Davies, M. (1988). *The importance of Britain's Twites.* RSPB Conservation Review 2: 91–94.

Davies, R.J. (ed.) (1986). *Forestry's Social and Environmental Benefits and Responsibilities.* Edinburgh: Institute of Chartered Foresters.

Davies, R.J. (1987). *Trees and Weeds.* Forestry Commission Handbook No. 2. London: HMSO.

Deane, C.D. (1954). *Bulletin of the Belfast Museum and Art Gallery* **1**: 120–193.

Denne, T., Bown, M.J.D. and Abel, J.A. (1986). *Forestry: Britain's Growing Resource.* London: UK Centre for Economic and Environmental Development.

Dennis, R.H. (1987a). Boxes for Goldeneyes: a success story. *RSPB Conservation Review* **1**: 85–87.

Dennis, R.H. (1987b). Osprey recolonisation. *RSPB Conservation Review* **1**: 88–90.

Dennis, R.H., and Dow, H. (1984). The establishment of a population of Goldeneyes *Bucephala clangula* breeding in Scotland. *Bird Study* **31**: 217–222.

Drakeford, T. (1979). *Survey of the Afforested Spawning Grounds of the Fleet Catchment.* Dumfries: FC unpublished report.

Drakeford, T. (1982). *Management of upland streams (an experimental fisheries management projection the afforested headwaters of the River Fleet, Kircudbrightshire).* pp. 86–92. Institute of Fisheries Management. 12th Annual Study Course, Durham.

Dutt, W.A. (1906). *Wild life in East Anglia.* London: Methuen.

Engstrom, R.T., Crawford, R.L. and Baker, W.W. (1984). Breeding bird populations in relation to changing forest structure following fire exclusion: a 15-year study. *Wilson Bulletin* **96**: 437–450.

Evans, J. (1984). *The Silviculture of Broadleaved Woodland.* Forestry Commission Bulletin No. 62, London: HMSO.

Ferguson-Lees. I.J. (1971). Studies of less familiar birds. 164. Wood Sandpiper. *British Birds* **64**: 114–117.

Ferry, C. and Frochot, B. (1980). Bird communities and the dissemination of forest plants. In Oelke, H. (ed.). pp. 136–140. *Bird Census Work and Nature Conservation.*

Flegg, J.J.M. and Bennett, T.J. (1974). The birds of oak woodland. In Morgan, M.S. and Perring, F.H. (eds). pp. 324–340. *The British Oak.* Faringdon: Botanical Society of the British Isles.

Ford, E.D., Malcolm, D.C. and Atterson J. (eds), (1979). *The Ecology of Even-aged Forest Plantations.* Cambridge: Institute of Terrestrial Ecology.

Forestry and British Timber (1984). UK softwood output set to double. *Forestry and British Timber* December 1983/January 1984: 10–13.

Forestry Commission (1943). *Post-War Forest Policy.* London: HMSO.

Forestry Commission (1965) *Forty-fifth Annual Report of the Forestry Commissioners for the Year Ended 30th September 1964.* London: HMSO.

Forestry Commission (1952). *Census of Woodlands 1947–1949.* London: HMSO.

Forestry Commission (1974a). *Fifty-Third Annual Report and Accounts of the Forestry Commission for the Year Ended 31st March 1973.* London: HMSO

Forestry Commission (1974b). *British Forestry.* London: HMSO.

Forestry Commission (1977a). *The Wood Production Outlook in Britain.* London: HMSO.

Forestry Commission (1977b). *Report on Forest Research, 1977.* London: HMSO.

Forestry Commission (1978a). *Fifty-seventh Annual Report and Accounts of the Forestry Commission for the Year Ended 31st March 1977.* London: HMSO.

Forestry Commission (1978b). *Report on Forest Research 1978.* London: HMSO.

Forestry Commission (1982). *Improving Upland Forests for Birds.* Edinburgh: Forestry Commission.

Forestry Commission (1985). *The Policy for Broadleaved Woodlands.* Edinburgh: Forestry Commission.

Forestry Commission (1986a). *The Forestry Commission and Conservation.* Policy and Procedure Paper No. 4. Edinburgh: Forestry Commission.

Forestry Commission (1986b). *Sixty-sixth Annual Report and Accounts of the Forestry Commission for the Year Ended 31 March 1986.* London: HMSO.

Forestry Commission (1986c). *British Forestry.* Edinburgh: Forestry Commission.

Forestry Commission (1987). *Thetford Forest Birds.* Santon Downham, Suffolk, Forestry Commission.

Forestry Commission (1988a). *67th Annual Report and Accounts for the Year Ended 31 March 1987.* London: HMSO.

Forestry Commission (1988b). *Woodland Grant Scheme.* Edinburgh: Forestry Commission.

Franklin, J.F., Cromack, K., Denison, W., McKee, A., Maser, C., Sedell, J., Swanson,

F., and Juday, G., (1981). *Ecological characteristics of old-growth Douglas-fir forests*, USDA Forest Service, Pacific Northwest Forest and Range Experimental station, General technical report PNW–118.

Fraser-Darling, F. (1955). *West Highland Survey*. Oxford OUP.

French, D.D., Jenkins, D. and Conroy, J.W.H. (1986). Guidelines for managing woods in Aberdeenshire for song birds. In Jenkins, D. (ed.). pp. 129–143. Huntingdon: ITE.

Fuller, R.J. (1980). A method for assessing the ornithological interest of sites for conservation. *Biological Conservation* **17**: 235–244.

Fuller, R.J. (1982). *Bird Habitats in Britain*. Calton: T. and A.D. Poyser.

Fuller, R.J. and Langslow, D. (1986). Ornithological conservation evaluation. In Usher, M.B. (ed.). *Wildlife conservation evaluation*. London: Chapman and Hall.

Fuller, R.J. and Moreton, B.D. (1989). Breeding bird populations of Kentish Sweet Chestnut (*Castanea sativa*) coppice in relation to the age and structure of the coppice. *Journal of Applied Ecology* **24**: 13–27.

Garfitt, J.E. (1983). Afforestation and upland birds. *Quarterly Journal of Forestry* **77**: 253–255.

Gibbs, J.N. and Greig, B.J.W. (1970). *Fomes annosus*. Forestry Commission leaflet 5. London: HMSO.

Gladstone, Hugh S. (1910). *The birds of Dumfriesshire*. London: Witherby.

Glowaci'nski, Z. (1972). Secondary succession of birds in an oak-hornbeam forest. *Bull. Acad. Pol. Sci. Serie Sci. Biol. CL II* **20**: 705–710.

Glowaci'ski, Z. (1975a). Succession of bird communities in the Niepolomice Forest (Southern Poland). *Ekol. Pol.* **23**: 231–263.

Glowaci'nski, Z. (1975b). Birds of the Niepolomice Forest (a faunistic-ecological study). *Acta Zool. Crac.* **20**: 1–88.

Glowaci'nski, Z. (1979). Some ecological parameters of avian communities in the successional stages of a cultivated pine forest. *Bull. Acad. Pol. Sci. Serie Sci. Biol. CL II*, **27**: 169–177.

Glowaci'nski, Z. (1980). Energetics of bird fauna in consecutive stages of semi-natural pine forest. *Ekol. Pol.* **28**: 71–94.

Glowaci'nski, Z. (1981b). Stability in bird communities during the secondary succession of a forest ecosystem. *Ekol. Pol.* **29**: 73–95.

Glowaci'nski, Z. and Järvinen, O. (1975). Rate of secondary succession in forest bird communities. *Ornis Scandinavica* **6**: 33–40.

Glowaci'nski, Z. and Weiner, J. (1975). A bird community of a mature deciduous forest: its organization, standing crop and energy balance (IBP 'Ispina project'). Bull. Acad. Pol. Sci.., *Serie Sci. Biol. CL II* **23**: 691–697.

Glowaci'nski, Z. and Weiner, J. (1977). Energetics of bird communities in successional series of a deciduous forest. *Pol. Ecol. Stud.* **3**: 147–175.

Glowaci'nski, Z. and Weiner, J. (1980). Energetics of bird fauna in consecutive stages of a semi-natural pine forest. *Ekol. Pol.* **28**: 71–94.

Gomersall, C.H., Morton, J.S. and Wynde, R.M. (1984). Status of breeding Red-throated Divers in Shetland, 1983. *Bird Study* **31**: 223–229.

Good, J.E.G. (ed.) (1986). *Environmental Aspects of Plantation Forestry in Wales*. Proc. ITE Symposium, Plas-tan-y-Bwlch. Bangor: ITE.

Goodier, R, and Bunce, R.G.H. (1977). The native pinewoods of Scotland: the current state of the resource. In Bunce, R.G.H. and Jeffers, J.N.R. (eds). pp. 78–87. *Native Pinewoods of Scotland*. Cambridge: ITE.

Granfield, E.F. (1971). *Design, Construction and Maintenance of Earth Dams and Excavated Ponds*. Forestry Commission Forest Record 75. London: HMSO.

Gribble, F.C. (1983). Nightjars in Britain and Ireland in 1981. *Bird Study* **30**: 165–176.

Gromadzki, M. (1970). Breeding bird communities of mid-field afforested areas. *Ekol. Pol.* **18**: 307–350.

Grove, R. (1983). *The Future for Forestry.* British Association of Nature Conservationists.

Haapanen, A. (1965a). Bird fauna of the Finnish forests in relation to forest succession. I. *Annales Zoologica Fennici* **2**: 153–196.

Haapanen, A. (1965b). Bird fauna of the Finnish forests in relation to forest succesion. II. *Annales Zoologica Fennici* **3**: 176–200.

Haila, Y., Järvinen, O. and Väisänen, R.A. (1980). Habitat distribution and species associations of land bird populations on the Åland Islands, SW Finland. *Annales Zoologica Fennica* **17**: 87–106.

Hamilton, G.A., Hunter, K. and Ruthven, A.D. (1981). Inhibition of brain acetylcholinesterase activity in songbirds exposed to fenitrothion during aerial spraying of forests. *Bulletin of Environmental Contamination and Toxicology* **27**: 856–863.

Hamilton, G.J. (ed.) (1974). *Aspects of Thinning.* Forestry Commission Bulletin No. 55. London: HMSO.

Hamilton, G.J. and Christie, J.M. (1971). *Forest Management Tables (metric).* Forestry Commission Booklet No. 34. London: HMSO.

Hamilton, G.J. and Christie, J.M. (1974). *Influence of Spacing on Crop Characteristics and Yield.* Forestry Commission Bulletin No. 52. London: HMSO.

Hamilton, R.B. and Noble, R.E. (1975). *Plant Succession and Interactions with Fauna.* United States Department of Agriculture Forestry Service General Technical Report W01:96115.

Hansson, L. (1983). Bird numbers across edges between mature conifer forest and clearcuts in Central Sweden. *Ornis Scandinavica* **14**: 97–103.

Harriman, R. (1978). Nutrient leaching from fertilised forest watersheds in Scotland. *Journal of Applied Ecology* **15**: 933–942.

Harriman, R. and Morrison, B.R.S. (1982). Ecology of streams draining forested and non-forested catchments in an area of central Scotland subject to acid precipitation. *Hydrobiologia* **88**: 251–263.

Heinselman, M.L. (1973). Fire in the virgin forests of the Boundary Waters Canoe area, Minnesota. *Quartenary Research* **3**: 329–382.

Helle, P. (1984). Effects of habitat area on breeding bird communities in Northeastern Finland. *Annales Zoologica Fennici* **21**: 421–425.

Helle, P. (1985a). Habitat selection of breeding birds in relation to forest succession in Northeastern Finland. *Ornis Fennica* **62**: 113–123.

Helle, P. (1985b). Effects of forest regeneration on the structure of bird communities in northern Finland. *Holarctic Ecology* **8**: 120–132.

Henderson, D.M. and Faulkner, R. (eds). (1987). Sitka Spruce. *Proceedings of the Royal Society of Edinburgh:* **93B**(v): 1–234.

Hibberd, B.G. (1985). Restructuring of plantations in Kielder forest district. *Forestry* **58**: 119–129.

Hibberd, B.G. (1986). *Forestry Practice.* Forestry Commission Bulletin No. 14, 10th edition. London: HMSO.

Hill, M.O. (1979). The development of a flora in even-aged plantations, in Ford, E.D., Malcolm, D.C. and Atterson, J. (eds). *The Ecology of Even-Aged Forest Plantations.* Cambridge: Institute of Terrestrial Ecology.

Hino, T. (1985). Relationships between bird community and habitat structure in shelterbelts of Hokkaido, Japan. *Oecologia (Berlin)* **65**: 442–448.

H.M. Treasury (1972). *Forestry in Great Britain; an Inter-departmental Cost Benefit Study.* London: HMSO.

Holtam, B.W., Chapman, E.S.B., Ross, R.B. and Harker, M.G. (1967). *Forest Management, and the Harvesting and Marketing of Wood in Sweden*. Forestry Commission bulletin no 41. London: HMSO.

Holtam, B.W. (ed.) (1971). *Windblow in Scottish Forests in January 1968*. Forestry Commission Bulletin No. 45. London: HMSO.

Hope Jones. P. (1972). Succession in breeding bird populations of sample Welsh oakwoods. *British Birds* **65**: 291–299.

Hope Jones, P. and Colling, A.W. (1984). Breeding and protection of Montagu's Harriers in Anglesey, 1955–64, *British Birds* **77**: 41–46.

Hope Jones, P. (1987a). *A History of Black Grouse in Wales*. RSPB Report to the Forestry Commission. Sandy.

Hope Jones, P. (1987b). *Black Grouse Habitat and Food in North Wales, 1986*. RSPB report to the Forestry Commission. Sandy.

Hope Jones, P. (1987c). *Black Grouse-specific Management Suggestions: General Principles, and Particular Recommendations for Five Sites in North Wales*. RSPB report to the Forestry commission, Sandy.

Hornbuckle, J. and Herringshaw, D. (1985). *Birds of the Sheffield Area*. Sheffield: Sheffield Bird Study Group.

Hornung, M. and Newson, M.D. (1986). Upland afforestation influences on stream hydrology and chemistry. *Soil Use Management* **2**: 61–65.

Hudson, P.J. (1986). *Red Grouse: the Biology and Management of a Wild Gamebird*. Fordingbridge: Game Conservancy Trust.

Husqvarna. undated. *Organised Felling*. 11 page booklet—Husqvarna Forestry Technique series. Sweden Husqvarna.

Innes, R.A. and Seal, D.T. (1971). Native Scottish pinewoods. In *Wildlife Conservation in Forest Management* ed Jones, E.W and Edwardson, T.E. Supplement to Forestry: 66–73.

Irvine, J. (1977). Breeding birds in New Forest broadleaved woodland. *Bird Study* **24**: 105–111.

Jalkanen, P. (1960). Pälkäneen linnustota. (Zusammenfassung: Uber die Vogelfauna von Pälkäne.). *Pälkäneseuran julk.* **2**: 1–122.

Jardine, D.C. (ed.) (1988). *Wildlife Management in Forests*. Proceedings of a Discussion Meeting. Edinburgh: Institute of Chartered Foresters.

Jenkins, D. (ed.) (1985). *Trees and Wildlife in the Scottish Uplands*. Proc. ITE symposium, Banchory. Banchory: ITE.

Jessop, R.M. (1982). The impact of afforestation on the avifauna of a Scottish moor. *Arboricultural Journal* **6**: 107–119.

Johnston, D.R., Grayson, A.J., and Bradley, R.T. (1967). Forest Planning. London: Faber and Faber.

Johnston, D.W. and Odum, E.P. (1956). Breeding bird populations in relation to plant succession on the Piedmont of Georgia. *Ecology* **37**: 50–62.

Jones, A.M. (1982). Capercaillie in Scotland—towards a conservation strategy. In Lovel, T.W.I. (ed.). *Proceedings 2nd International Symposium on Grouse*. Paris: International Council for Game and Wildlife Conservation.

Jones, A.M. (1984). Nesting habitat of Capercaillie in Scottish plantations. In Hudson, P.J. and Lovel, T.W.I. (eds). pp. 301–316. *Proceedings 3rd International Grouse Symposium*. Paris: International Council for Game and Wildlife Conservation.

Jones, A.T., and Smith, R.O. (1980). *Harvesting Windthrown Trees*. Forestry Commission Leaflet No. 75. London: HMSO.

Knystautas, A. (1987). *The Natural History of the USSR*. London: Century Hutchinson.

Lack, D. (1933). Habitat selection in birds, with special reference to the effects of afforestation on the Breckland avifauna. *Journal of Animal Ecology* **2**: 239–262.

Lack, D. (1939). Further changes in the Breckland avifauna caused by afforestation. *Journal of Animal Ecology* **8**: 277–285

Lack, D. and Lack, E. (1951). Further changes in bird-life caused by afforestation. *Journal of Animal Ecology* **20**: 173–179.

Langslow, D.R. (1983). The impacts of afforestation on breeding shorebirds and raptors in the uplands of Britain. In Evans, P.R., Hafner, H. and Hermite, P.L. pp. 17–25. *Shorebirds and Large Waterbirds Conservation.* Brussels: Commission of European Communities.

Lay, D.W. (1938). How valuable are woodland clearings to birdlife. *Wilson Bulletin* **45**: 254–256.

Lean, G. and Rosie, G. (1988). Forests of Money. *The Observer Magazine*, 14 February 1988: 34–41.

Leather, S.R. and Barbour, D.A. (1987). Associations between soil type, lodgepole pine (*Pinus contorta*) provenance, and the abundance of the pine beauty moth, *Panolis flammea. Journal of Applied Ecology* **24**: 945–952.

Leather, S.R., Stoakley, J.T., Evans, H.F. (eds). (1987). *Population Biology and Control of the Pine Beauty Moth.* Forestry Commisssion Bulletin No. 67. London: HMSO.

Leslie, R. (1981). Birds of the northeast England forests. *Quarterly Journal of Forestry* **75**: 153–158.

Leslie, R. (1985). The population and distribution of Nightjars (*Caprimulgus europaeus*) on the North Yorks Moors. *Naturalist* **110**: 23–28.

Leslie, R. (1986). *Forestry and Conservation—The Baseline* In Davies, R.J. (eds) *Forestry's Social and Environmental Benefits and Responsibilities.* Edinburgh: The Institute of Chartered Foresters.

Likens, G.E., Bormann, F.H., Johnson, N.M., Fisher, D.W. and Pierce, R.S. (1970). Effects of forest cutting and herbicide treatment on nutrient budgets in the Hubbard Brook watershed ecosystem. *Ecological Monographs* **40**: 23–47.

Lindsay, R.A., Charman, D.J. Everingham, F., O'Reilly, R.M., Palmer, M.A., Rowell, T.A. and Stroud, D.A. (1988). *The Flow Country: the peatlands of Caithness and Sutherland.* Peterborough: NCC.

Lines, R. (1976). The development of forestry in Scotland in relation to the use of Pinus contorta. In *Pinus contorta Provenance Studies.* Forestry Commission Research and Development Paper **114**: 2–5. Edinburgh: Forestry Commission.

Lines, R. (1987). *Choice of Seed Origins for the Main Forest Species in Britain.* Forestry Commission Bulletin No. 66. London: HMSO.

Locke, G.M.L. (1970). *Census of Woodlands 1965—67.* London HMSO.

Locke, G.M.L. (1987). *Census of Woodlands and Trees 1979–82.* London: HMSO.

Lockie, J. (1955). The breeding habits of Short-eared Owls after a vole plague. *Bird Study* **2**: 53–69.

Lovenbury, G.A., Waterhouse, M. and Yalden, D.W. (1978). The status of Black Grouse in the Peak District. *Naturalist* **103**: 3–14.

Low, A.J. (ed.) (1985). *Guide to Upland Restocking Practice.* Forestry Commission leaflet no 84. London: HMSO.

Low, A.J. (1986). *Use of Broadleaved Species in Upland Forests.* Forestry Commission leaflet 88. London: HMSO.

Lowe, P., Cox, G., MacEwan, M., O'Riordan, T. and Winter, M. (1986) *Countryside Conflicts,* Aldershot: Gower/Maunice Temple.

MacArthur, R.H. and MacArthur, J.W. (1961). On bird species diversity. *Ecology* **42**: 594–598.

MacArthur, R.H., MacArthur, J.W. and Preer, J. (1962). On bird species diversity: II. Prediction of bird census from habitat measurements. *American Naturalist* **96**: 167–174.

MacArthur, R.H., Recher, H.F. and Cody, M. (1966). On the relation between habitat selection and species diversity. *American Naturalist* **100**: 319–332.

MacArthur, R.H. and Wilson, E.O. (1967). *The Theory of Island Biogeography.* Princeton: Princeton University Press.

McIntosh, R. (1984). *Fertiliser Experiments in Established Conifer Stands.* Forestry Commission Forest Record 127. London: HMSO.

McIntosh, R. (1988). Integrating Management Objectives in an Upland Forest. In Jardine, D.C. (ed.) pp 1–5. *Wildlife Management in Forests.* Proceedings of a Discussion Meeting. Edinburgh: Institute of Chartered Foresters.

Mackenzie, J.M., Thomson, J.H. and Wallis, K.E. (1976). *Control of Heather by 2,4-D.* London: HMSO.

Malcolm, D.C. and Cuttle, S.P. (1983). The application of fertilisers to drained peat. 1. Nutrient losses in drainage. *Forestry* **56**: 155–174.

Mannan, R.W., Meslow, E.C. and Wight, H.M. (1980). Use of snags by birds in Douglas fir forests, Western Oregon. *Journal of Wildlife Management* **44**: 787–797.

Marquiss, M. and Newton, I. (1982). The Goshawk in Britain. *British Birds* **75**: 243–260.

Marquiss, M., Newton, I. and Ratcliffe, D.A. (1978). The decline of the Raven (Corvus corax) in relation to afforestation in southern Scotland and northern England. *Journal of Applied Ecology* **15**: 129–144.

Marquiss, M., Ratcliffe, D.A., and Roxburgh, R. (1985). The numbers, breeding success and diet of golden eagles in southern Scotland in relation to changes in land use. *Biological Conservation* **34**: 121–141.

Massey, M.E. (1974). The effect of woodland structure on breeding bird communities in sample woods in south central Wales. *Nature in Wales* **14**: 95–105.

Mather, J.R. (1986). *The Birds of Yorkshire.* Kent: Croom Helm.

Mearns, R. (1983). The status of the Raven in southern Scotland and Northumbria. *Scottish Birds* **12**: 211–218.

Meek, E.J. and Little, B. (1977). The spread of the Goosander in Britain and Ireland. *British Birds* **70**: 229–237.

Meek, E.J. The breeding ecology and decline of the Merlin Falco columbarius in Orkney. *Bird Study* **35**: 209–218.

Miller, K.F. (1985). *Windthrow Hazard Classification.* Forestry Commission leaflet 85. London: HMSO.

Mills, D.H. (1980). *The Management of Forest Streams.* London: HMSO.

Mitchell, C.P., Brandon, O.H., Bunce, R.G.H., Barr, C.J., Tranter, R.B., Downing, P., Pearce, M.L. and Whittaker, H.A. (1983). Land availability for production of wood energy in Britain. In Strub, A., Chartier, P. and Schleser, G. (eds). pp 159–163. *Energy from Biomass* (2nd EC Conference, Berlin, 1982). London: Applied Science.

Mitlin, D.C. (1987). *Price-size Curves for Conifers.* Forestry Commission Bulletin No. 68. London: HMSO.

Moeller, G.H. and Seal, D.T. (eds). (1984). *Technology Transfer in Forestry.* Forestry Commission Bulletin No. 61. London: HMSO.

Moore, N.W. (1987). *The Bird of Time.* Cambridge: CUP.

Moore, N.W. and Hooper, M.D. (1975). On the number of bird species in British woods. *Biological Conservation* **8**: 239–250.

Morrison, B.R.S. and Wells, D.E. (1981). The fate of Fenitrothion in a stream environment and its effect on the fauna following aerial spraying of a Scottish forest. *Science of the Total Environment* **19**: 223–252.

Morrison, M.L., and Meslow, E.C. (1984). Effects of the herbicide Glyphosate on bird community structure, Western Oregon. *Forest Science* **30**: 95–106.

Moss, D. (1978a). Diversity of woodland song-bird populations. *Journal of Animal Ecology* **47**: 521–527.

Moss, D. (1978b). Song-bird populations in forestry plantations. *Quarterly Journal of Forestry* **72**: 4–14.

Moss, D. (1979). Even-aged plantations as a habitat for birds. In Ford, E.D., Malcolm, D.C. and Atterson, J. (eds). pp. 413–427. *The Ecology of Even-aged Forest Plantations.* Cambridge: ITE.

Moss, D., Taylor, P.N. and Easterbee, N. (1979a). The effects on song-bird populations of upland afforestation with spruce. *Forestry* **52**: 129–150.

Moss, R., Weir, D., and Jones, A. (1979). Capercaillie management in Scotland. In Lovell, T.W.I. (ed.). pp. 140–155. *Woodland Grouse 1978.* Bures, Suffolk: World Pheasant Association.

Mowle, A. (1986). *Nature Conservation in Rural Development.* Focus on Nature Conservation No. 18. Peterborough: Nature Conservancy Council.

National Audit Office (1986). *Review of Forestry Commission Objectives and Achievements.* London: HMSO.

Natural Resources (Technical) Committee. (1957). *Forestry, agriculture and marginal land.* London: HMSO.

Nature Conservancy Council. (1986). *Nature conservation and Afforestation in Britain.* Peterborough: NCC.

Nelson, T.H. (1907). *The Birds of Yorkshire.* 2 vols. London: A. Brown and Sons.

Nethersole-Thompson, D. (1951). *The Greenshank.* London: Collins.

Nethersole-Thompson, D. (1966). *The Snow Bunting.* Edinburgh: Oliver and Boyd.

Nethersole-Thompson, D. (1971). *Highland Birds.* Inverness: Highlands and Islands Development Board.

Nethersole-Thompson, D. (1973). *The Dotterel.* London: Collins.

Nethersole-Thompson, D. (1975). *Pine Crossbills.* Calton: T. and A.D. Poyser.

Nethersole-Thompson, D. and Nethersole-Thompson, M. (1979). *Greenshanks.* Calton: T. and A.D. Poyser.

Nethersole-Thompson, D. and Nethersole-Thompson, M. (1986). *Waders: Their Breeding, Haunts and Watchers.* Calton: T. and A.D. Poyser.

Neustein, S.A. (1976a). A history of plough development in British forestry — II Historical review of ploughing on wet soils. *Scottish Forestry* **30**: 89–111.

Neustein, S.A. (1976b). A history of plough development in British forestry — III Historical review of ploughing drier soils. *Scottish Forestry* **30**: 253–274.

Neustein, S.A. and Seal, D.T. (1984). Technology transfer: lessons of the British experience. In Moeller, G.H. and Seal, D.T. (eds). (1984). *Technology Transfer in Forestry.* Forestry Commission Bulletin No. 61. London: HMSO.

Newson, M.D. (1985). Forestry and water in the uplands of Britian: the background of hydrological research and options for harmonious land-use. *Quarterly Journal of Forestry* **79**: 113–120.

Newton, I. (1972). *Finches.* London: Collins.

Newton, I. (1983). Birds and forestry. In Harris, E.H.M. (ed.). pp. 21–36. *Centenary Conference on Forestry and Conservation.* Tring: Royal Forestry Society of England, Wales and Northern Ireland.

Newton, I. (1984). Raptors in Britain — A review of the last 150 years. *BTO News* **131**: 6–7.

Newton, I. (1986). Principles underlying bird numbers in Scottish woodlands. In Jenkins, D. (ed.). pp. 121–128. *Trees and Wildlife in the Scottish Uplands.* Huntingdon: ITE.

Newton, I. and Moss, D. (1981). Factors affecting the breeding of Sparrowhawks and the occurrence of their song-bird prey in woodlands. In Last, F.T. and Gardiner, A.F. (eds). pp. 125–131. *Forest and Woodland Ecology.* Cambridge: ITE.

Newton, I., Meek, E. and Little, B. (1978). Breeding ecology of the Merlin in Northumberland. *British Birds* **71**: 376–398.

Newton, I., Davies, P.E and Moss, D. (1981). Distribution and breeding of red kite in relation to land use in Wales. *Journal of Applied Ecology* **18**: 173–186.

Newton, I., Meek, E. and Little, B. (1986a). Further observations on the population and breeding ecology of the Merlin in Northumbria. *British Birds* **79**: 155–170.

Newton, I., Wyllie, I. and Mearns, R. (1986b). Spacing of sparrowhawks in relation to food supply. *Journal of Animal Ecology* **55**: 361–370.

Newton, I., Davis, P.E. and Davies, J.E. (1982). Ravens and Buzzards in relation to sheep farms and forestry in Wales. *Journal of Applied Ecology* **19**: 681–706.

Newton, I., Robson, J.E. and Yalden, D.W. (1981). Decline of the Merlin in the Peak District. *Bird Study* **28**: 225–234.

Niles, J. (1971). Forestry and conservation. 2. Strictly for the birds. *Journal of the Devon Trust for Nature Conservation* **3**: 80–85.

North Yorks Moors National Park Committee. (1977). *North Yorks Moors National Park Plan.* Helmsley: North Yorks Moors National Park.

Norris, C.A. (1960). The breeding distribution of thirty bird species in 1952. *Bird Study* **7**: 129–184.

O'Connor, R.J, and Shrubb, M. (1986). *Farming and Birds.* Cambridge: CUP.

Odum, E.P. (1950). Bird populations of the Highlands (North Carolina) Plateau in relation to p-lant succession and avian invasion. *Ecology* **31**: 587–605.

Orford, N. (1973). Breeding distribution of the twite in central Britain. *Bird Study* **20**: 51–62, 121–126.

Ormerod, S.J. (1985). The diet of breeding dippers *Cinclus cinclus* and their nestlings in the catchment of the River Wye, mid-Wales: a preliminary study by faecal analysis. *Ibis* **127**: 316–331.

Ormerod, S.J. and Edwards, R.W. (1985). Stream acidity in some areas of Wales in relation to trends in afforestation and the usage of agricultural limestone. *Journal of Environmental Management* **20**: 189–197.

Ormerod, S.J., Boilstone, M.A. and Tyler, S.J. (1985a). Factors influencing the abundance of breeding dippers, *Cinclus cinclus* in the catchment of the River Wye, mid-Wales. *Ibis* **127**: 332–340.

Ormerod, S.J., Tyler, S.J, and Lewis, J.M.S. (1985b). Is the breeding of Dippers influenced by stream acidity? *Bird Study* **32**: 32–39.

Ormerod, S.J., Allinson, N., Hudson, D. and Tyler, S.J. (1986a). The distribution of breeding dippers (*Cinclus cinclus* Aves) in relation to stream acidity in upland Wales. *Freshwater Biology* **16**: 501–507.

Ormerod, S.J., Mawle, G.W., and Edwards, R.W. (1986b). The influence of forests on aquatic fauna. In Good, (ed.). pp . *Environmental Aspects of Plantation Forestry in Wales.* Bangor: ITE.

Ormerod, S.J., Weatherley, N.S., and Boole, P. (1987). Short-term experimental acidification of a Welsh stream: comparing the biological effects of hydrogen ions and aluminium. *Freshwater Biology.*

Parslow, J.L.F. (1973). *Breeding Birds of Britain and Ireland.* Berkhamstead: Poyser.

Parsons, A.J. (1976). Birds of a Somerset wood. *Bird Study* **23**: 285–293.

Pawsey, R.G. and Gladman, R.J. (1965). *Decay in Standing Conifers Developing from Extraction Damage.* Forestry Commission Forest Record 54. London: HMSO.

Pearce, D.W., Markandya A. and Barbier, E. (1989). Sustainable development. *Report to the UK Department of the Environment.*

Pearson, D.L. (1975). The relation of foliage complexity to ecological diversity of three Amazonian bird communities. *Condor* **77**: 453–466.

Peet, R.K. (1974). The measurement of species diversity. *Annual Review of Ecology and Systematics* **5**: 285–307.

Peterken, G.F. (1981). *Woodland Conservation and Management.* London: Chapman and Hall.

Peterken, G.F. (1986). Commercial forests and woods — the nature conservation opportunities. In Davies, R.J. (ed.). *Forestry's Social and Environmental Benefits and Responsibilities*: 53 — 64. Edinburgh: Institute of Chartered Foresters.

Peterken, G.F. (1987). Natural features in the management of upland conifer forests. In Henderson, D.M. and Faulkner, R. (eds). Sitka Spruce. *Proceedings of the Royal Society of Edinburgh:* **93B**(v): 223–234.

Petty, S.J. (1979). *Breeding Biology of the Sparrowhawk in Kielder Forest, 1975–1978.* Tyneside Bird Club Occasional Report no 2: 1–18.

Petty, S.J. (1983). A study of Tawny Owls *Strix aluco* in an upland spruce forest. *Ibis* **125**: 592.

Petty, S.J. (1987a). The design and use of a nestbox for tawny owls *Strix aluco* in upland forests. *Quarterly Journal of Forestry* **81**: 103–109.

Petty, S.J. (1987b). Breeding of tawny owl *Strix aluco* in relation to their food supply in an upland forest in D.J. Hill (ed.) *Reproduction and Management in Birds of Prey.* University of Bristol: 167–179.

Petty, S.J. (1988). The management of raptors in upland forests. In Jardine, D.C. (ed.) pp. 7–24. Wildlife Management in Forests. *Proceedings of a Discussion Meeting.* Edinburgh: Institute of Chartered Foresters.

Petty, S.J. and Anderson, D. (1986). Breeding by hen harriers *Circus cyaneus* on restocked sites in upland forests. *Bird Study* **32**: 194–195.

Petty, S.J., (1989a), Goshawks: their status, requirements and management. *Forestry Commission bulletin 81.* London: HMSO.

Petty, S.J., (1989b). Productivity of Tawny Owls (*Strix aluco*) in relation to the structure of a spruce forest in Britain. *Ann. Zool. Fennici* **26**: 227–233.

Petty, S.J. and Avery, M.I. (1990). *Forestry and Birds.* Forestry Commission.

Phillips, J.S. (1973). Stonechats in young forestry plantations. *Bird Study* **20**: 82–84.

Picozzi, N. and Hewson, R. (1970). Kestrels, Short-eared Owls and Field Voles in Eskdalemuir in 1970. *Scottish Birds* **6**: 185–190.

PIEDA (1986). *Forestry in Great Britain — an Economic Assesment for the National Audit Office.* London: National Audit Office.

Pielou, E.C. (1966). Species diversity and pattern diversity in the study of ecological succession. *Journal of Theoretical Biology* **10**: 370–383.

Pienkowski, M.W., Shroud, D.A. and Reed, T.M. (1987). Problems in maintaining breeding habitat, with particular reference to peatland waders. *Wader Study Group Bulletin 49*: Suppl. IWRB, Special publication 7: 95–101.

Piersma, T. (1986). Breeding waders in Europe: a review of population size estimates and a bibliography of information sources. *Wader Study Group Bulletin 48 Supplement.* Wader Study Group.

Pleasance, B. (1982). Nightjar enquiry 1981. The Breckland. Unpublished.

Poxton, I.R. (1986). Breeding Ring Ouzels in the Pentland Hills. *Scottish Birds* **14**: 44–48.

Poxton, I.R. (1987). Breeding status of the Ring Ouzel in Southeast Scotland 1985–86. *Scottish Birds* **14**: 205–208.

Pyatt, D.G. and Low, A.J. (1986). *Forest Drainage.* Forestry Commission Research Information Note 103–86-SILN. Roslin Forestry Commission.

Pyatt, D.G., Anderson, A.R., Stanndard, J.P. and White, I.M.S. (1985). A drainage experiment on a peaty gley soil at Kershope Forest, Cumbria. *Soil Use and Management* **1**: 89–94.

Rackham, O. (1976). *Trees and Woods in the British Landscape*. London: Dent.

Rackham, O. (1980). *Ancient Woodland*. London: Arnold.

Rackham, O. (1986). *The History of the Countryside*. London: Dent.

Rands, M.R.W., Hudson, P.J. and Sotherton, N.W. (1988). Gamebirds, ecology, conservation and agriculture. In Hudson, P.J. and Rands, M.R.W. (eds). pp. 1–17. *Ecology and Management of Gamebirds*. Oxford: Blackwell Scientific Publications.

Rankin, G.D. and Taylor, I.R. no date. *Changes Within Afforested but Unplanted Ground: Birds*. NCC Contract number HF3/03/175.

Ratcliffe, D.A. (1976). Observations on the breeding of the Golden Plover in Great Britain. *Bird Study* **23**: 63–116.

Ratcliffe, D.A. (1977a). Uplands and birds—an outline. *Bird Study* **24**: 140–158.

Ratcliffe, D.A. (ed.) (1977b). *A Nature Conservation Review*. Cambridge: CUP.

Ratcliffe, D.A. (1980). *The Peregrine Falcon*. Calton: T. and A.D. Poyser.

Ratcliffe, D.A. (1984a). The Peregrine breeding population of the United Kingdom in 1981. *Bird Study* **31**: 1–18

Ratcliffe, D.A. and Thompson, D.B.A. (1988). The British uplands: their ecological character and international significance. In Usher, M.B. and Thompson, D.B.T. (eds). In *Ecological Change in the Uplands*. Oxford: Blackwell Scientific Publications.

Ratcliffe, P.R. (1984b). Population dynamics of red deer (Cervus elephas) in Scottish commercial forests. *Proceedings of the Royal Society of Edinburgh* **82B**: 291–302.

Ratcliffe, P.R. (1985a). Population density and reproduction of red deer in Scottish commercial forests. *Acta Zooligca Fennica* **172**: 191–193.

Ratcliffe, P.R. (1985b). *Glades for Deer Control in Upland Forests*. Forestry Commisssion Leaflet no 86. London: HMSO.

Ratcliffe, P.R. (1987a). *The Management of Red Deer in Upland Forests*. Forestry Commission Bulletin No. 71. London: HMSO.

Ratcliffe, P.R. (1987b). The distribution and current status of Sika deer, *Cervus nippon*, in Great Britain. *Mammal Review* **17**: 39–58.

Ratcliffe, P.R. (1988). The management of Red Deer populations resident in upland forests. In Jardine, D.C. (ed.) pp 44–50. *Wildlife Management in Forests. Proceedings of a Discussion Meeting*. Edinburgh: Institute of Chartered Foresters.

Ratcliffe, P.R. and Rowe, J.J. (1985). A biological basis for managing red and roe deer in British commercial forests. *Transactions International Union Game Biology* **17**: 917–925.

Ratcliffe, P.R., Hall, J. and Allen, J. (1986). Computer predictions of sequential growth changes in commercial forests as an aid to wildlife management, with reference to red deer. *Scottish Forestry* **40**: 79–83.

Reed, T.M., Langslow, D.R. and Symonds, F.L. (1983). Breeding waders of the Caithness flows. *Bird Study* **12**: 180–186.

Richards, W.N. (1984). Problems of water management and water quality arising from forestry activities. In Harding, D.L. and Fawell, J.F. (eds) *Woodlands, Weather and Water* pp. 67–85. London: Institute of Biology.

Robbins, C.S. (1979). *Effect of Forest Fragmentation on Bird Populations*. United States Department of Agriculture Forestry Service General Technical Report.

Roberts, M.E. and James, D.B. (1972). Some effects of forest cover on nutrient cycling and river temperature. In Taylor, J.A. (ed.) *Research Papers in Forest Meteorology* pp. 100–108. Aberystwyth: Cambrian News.

Robertson, A.W.P. (1954). *Bird Pageant*. London: Batchworth Press.

Robinson, M. and Blyth, K. (1982). The effect of forestry drainage operations on upland sediment yields: a case study. *Earth Surface Processes and Landforms* **7**: 85–90.

Robinson, M. (1980). *The effect of pre-afforestation drainage on the streamflow and water quality of a small upland catchment*. Wallingford: Institute of Hydrology.

Rogers, E.V. (1975). *Ultra Low Volume Herbicide Spraying*. Forestry Commission Leaflet No. 62. London: HMSO.

Roth, R.R. (1976). Spatial heterogeneity and bird species diversity. *Ecology* **57**: 773–782.

Røv, N. (1975). Breeding bird community structure and species diversity along an ecological gradient in deciduous forest in western Norway. *Ornis Scandinavica* **6**: 1–14.

Royal Society for the Protection of Birds (1985). *Forestry in the Flow Country — the Threat to Birds*. Sandy: RSPB.

Ryle, G.B. (1966). *The Impact of the Twentieth Century Forest upon Nature Conservation*. Handbook of the Society for the Promotion of Nature Reserves.

Ryle, G.B. (1969). *Forest Service: The First 45 Years of the Forestry Commission*. Newton Abbot: David and Charles.

Sale, J.S.P., Tabbush, P.M., and Lane, P.B., (1986). *The Use of Herbicides in the Forest — 1986*. Forestry Commission Booklet No. 51. Edinburgh: Forestry Commission.

Sammalisko, L. (1957). The effect of woodland-open peatland edge on some peatland birds in South Finland. *Ornis Fennica* **34**: 81–89.

Scott, T.M. and King, C.J. (1974). *The Large Pine Weevil and Black Pine Beetles*. Forestry Commission Leaflet No. 58. London: HMSO.

Seago, M.J. (ed.) (1971). *Norfolk Bird Report — 1970*. Norwich: Norfolk and Norwich Naturalists Trust.

Shannon, C.E. and Weaver, W. (1949). *The Mathematical Theory of Communication*. Urbana III: University of Illinois Press.

Sharrock, J.T.R. (1972). Habitat of Redwings in Scotland. *Scottish Birds* **7**: 208–209.

Sharrock, J.T.R. (1976). *The Atlas of Breeding Birds in Britain and Ireland*. Calton: T. and A.D. Poyser.

Shawyer, C.R. (1987). *The Barn Owl in the British Isles*. London: The Hawk Trust.

Sills, N. (1988). Transformation at Titchwell: a wetland reserve management case history. *RSPB Conservation Review* **2**: 64–68..

Simms, E. (1971). *Woodland Birds*. London: Collins.

Simpson, L.M. and Henderson-Howat, D.B. (1985). *Thetford Forest Management Plan: a Conservation Review*. Forestry Commission Forest Record No. 130. London: HMSO.

Sitters, H.P. (1986). Woodlarks in Britain, 1968–83. *British Birds* **79(3)**: 105–116.

Sitters, H.P. (1988). *Breeding Birds of Devon*. Plymouth: Devon Bird Watching and Preservation Society.

Smith, B.D. (1980). The effects of afforestation on the trout of a small stream in southern Scotland. *Fish Management* **11**: 39–57.

Smith, J.J. (1955). Rabbit clearance in Kings Forest, 1947 to 1951. *Journal of the Forestry Commission*. **23**: 70—71

Smith, K.W. (1983). The status and distribution of waders breeding on wet lowland grassland in England and Wales. *Bird Study* **30**: 177–192.

Smith, K.W. (1987). The ecology of the Great spotted Woodpecker. *RSPB Conservation Review* **1**: 74–77.

Springthorpe, G.D. and Myhill, N.G. (1985). *Forestry Commission Wildlife Rangers' Handbook*. Chester: Forestry Commission.

Stafford, J. (1962). Nightjar Enquiry, 1957—58. *Bird Study 9*. **2**: 104–115.

Staines, B.W. and Welch, D. (1984). Habitat selection and impact of Red (*Cervus elaphus* L.) and Roe (*Capreolus capreolus* L.) deer in a Sitka spruce plantation. *Proceedings of the Royal Society of Edinburgh* **82B**: 303–319.

Staines, B.W., Petty, S.J. and Ratcliffe, P.R. (1987). Sitka Spruce forests as a habitat for birds and mammals, in Henderson, D.M. and Faulkner, R. (eds) Sitka Spruce. *Proceedings of the Royal Society of Edinburgh* **93B**(v): 1–2340.

Steele, R.C. (1972). Wildlife conservation in woodlands. London: HMSO.

Stewart, P.J. (1985). British forestry policy: time for a change? *Land Use Policy* **2**: 16–29.

Stewart, P.J. (1987). *Growing Against the Grain: United Kingdom Forestry Policy*. A report commissioned by the Council for the Protection of Rural England, London.

Stirling Maxwell, Sir J. (1929). *Loch Ossian plantations 1929*. 2 vols, privately published.

Stoner, J.H., Gee, A.S and Wade, K.R. (1984). The effects of acidification on the ecology of streams in the upper Tywi catchment in west Wales. *Environmental Pollution (Series A)* **35**: 125–157.

Stoner, J.H. and Gee, A.S. (1985). Effects of forestry on water quality and fish in Welsh rivers and lakes. *Institution of Water Engineers Journal* **39**: 27–45.

Stoner, J.H. (co-ordinator) (1987). *Llyn Brianne Acid Waters Project—First Technical Summary Report*. Welsh Water Authority.

Stowe, T.J. (1987). Management of sessile oakwoods for pied flycatchers. *RSPB Conservation Review* **1**: 78–83.

Stroud, D.A, Reed, T.M., Pienkowski, M.W., Lindsay, R.A. (1987). *Birds, Bogs and Forestry*. Peterborough: Nature Conservancy Council.

Stroud, D.A. and Reed, T.M. (1986). The effects of plantation proximity on moorland breeding waders. *Wader Study Group Bulletin* **46**: 25–28.

Sykes, J.M., Lowe, V.P.W. and Briggs, D.R. (1985). *Changes in Plants and Animals in the First 10 Years after Upland Afforestation*, pp. 16–21. NERC/ITE Annual Report for 1984.

Tabbush, P.M. (1988) Silvicultural principles for upland restocking. *Forestry Commission bulletin 76*. London HMSO.

Taylor, I.R., Dowell, A. Irving, T., Langford, I.K. and Shaw, G. (1988). The distribution and abundance of the Barn Owl (*Tyto alba*) in south-west Scotland. *Scottish Birds* **15**: 40–43.

Thom, V.S. (1986). *The Birds of Scotland*. Calton: T. and A.D. Poyser.

Tompkins, S.C. (1986). *The Theft of the Hills: Afforestation in Scotland*. London: Ramblers Association.

Tubbs, C.R. (1974). *The Buzzard*. London: David and Charles.

Turcek, F.J. (1957). The bird succession in the conifer plantations on mat-grass land in Slovakia (SSR). *Ibis* **99**: 587–593.

Turner, D.J. (1977). *The Safety of the Herbicides 2,4-D and 2,4,5,-T*. Forestry Commission Bulletin No. 57. London: HMSO.

Tyler, S.J. (1987). River birds and acid water. *RSPB Conservation Review* **1**: 68–70.

Ulfstrand, S. (1975). Bird flocks in relation to vegetation diversification in a South Swedish coniferous plantation during winter. *Oikos* **26**: 65–73.

Ulmanen, I. and Valste, J. (1965). Sodankylän Koitelaiskairan linnustosta 1962–64. *Lintumies* **1**: 51–56.

United States Department of Agriculture Forest Service (1986). *Spotted Owl Guidelines*. Portland, Oregon: USFS Pacific Northwest Region.

Village, A. (1981). The diet and breeding of long-eared owls in relation to vole numbers. *Bird Study* **28**: 213–224.

Village, A. (1982). The home range and density of Kestrels in relation to vole abundance. *Journal of Animal Ecology* **51**: 413–428.

Village, A. (1986). Breeding performance of kestrels at Eskdalemuir, south Scotland. *Journal of Zoology (London)* **208**: 367–378.

Watson, A. (1969). Preliminary counts of birds in central highland pine woods. *Bird Study* **16**: 158–163.

Watson, A. and Rae, R. (1987). Dotterel numbers, habitat and breeding success in Scotland. *Scottish Birds* **14**: 191–198.

Watson, D. (1977). *The Hen Harrier*. Berkhamstead: Poyser.

Watson, J. (1979). Food of merlins in young conifer forests. *Bird Study* **26**: 253–258.

Watson, J., Langslow, D.R. and Rae, S.R. (1986). *The Impact of Land-use Changes on Golden Eagles in the Scottish Highlands*. Peterborough: Nature Conservancy Council.

Webb, N. (1986). *Heathlands*. London: Collins.

Williamson, K. (1972). The conservation of bird life in the new coniferous forests. *Forestry* **45**: 87–100.

Williamson, K. (1973). Habitat of Redwings in Wester Ross. *Scottish Birds* **7**: 268–269.

Williamson, K. (1975). Bird colonisation of new plantations on the moorland of Rhum, Inner Hebrides. *Quarterly Journal of Forestry* **69**: 157–168.

Williamson, K. (1976). Bird-life in the wood of Cree, Galloway. *Quarterly Journal of Forestry* **70**: 206–215.

Willson, M.F. (1974). Avian community organization and habitat structure. *Ecology* **55**: 1017–1029.

Wood, R.F. (1974). *Fifty Years of Forest Research*. Forestry Commission Bulletin No. 50. London: HMSO.

Woolhouse, M.E.J. (1983). The theory and practice of the species–area effect, applied to the breeding birds of British woods. *Biological Conservation* **27**: 315–332.

Yalden, D.W. (1979). *Decline of the Black Grouse in Derbyshire — critical comments*. Derbyshire Bird Report 1978: 7–9.

Yapp, W.B. (1953). The birds of Welsh high level oakwoods. *Nature in Wales* **1**: 161–166.

Yapp, W.B. (1955). The succession of birds in developing *Quercetum petraeae*. *North Western Naturalist* **26**: 58–67.

Yapp, W.B. (1962). *Birds and Woods*. Oxford: Oxford University Press.

Yapp, W.B. (1974). Birds of the northwest Highland birchwoods. *Scottish Birds* **8**: 16–31.

Young, C., Herger, B., Marx, G. and Scabrook, K. (1985). *The Forests of British Columbia*. North Vancouver: Whitecap Books.

Zehetmayr, J.W.L. (1954). *Experiments on Tree Planting on Peat*. Forestry Commission Bulletin No. 22. London: HMSO.

Zehetmayr, J.W.L. (1960). *Afforestation of Upland Heaths*. Forestry Commission Bulletin No. 32. London: HMSO.

Appendix

*Scientific names of birds, plants and fungi, mammals and invertebrates
mentioned in the text*

BIRDS

Avocet *Recurvirostra avocetta*
Blackbird *Turdus merula*
Blackcap *Sylvia atricapilla*
Blackcock *Lyrurus tetrix*
Bluethroat *Luscinia svecica*
Brambling *Fringilla montifringilla*
Bullfinch *Pyrrhula pyrrhula*
Bunting, Cirl *Emberiza cirlus*
    Snow *Plectrophenax nivalis*
Bustard, Great *Otis tarda*
Buzzard *Buteo buteo*
Capercaillie *Tetrao urogallus*
Chaffinch *Fringilla coelebs*
Chiffchaff *Phylloscopus colybita*
Chough *Pyrrhocorax pyrrhocorax*
Cowbird, Brown-headed *Molothrus ater*
Crossbill, Common *Loxia curvirostra*
    Parrot *Loxia pytyopsittacus*
    Scottish *Loxia scotica*
    Two-barred *Loxia leucoptera*
Crow, Carrion or Hooded *Corvus corone*
Cuckoo *Cuculus canorus*
Curlew *Numenius arquata*
    Stone *Burhinus oedicnemus*
Dipper *Cinclus cinclus*
Diver, Black-throated *Gavia arctica*
    Red-throated *Gavia stellata*
Dotterel *Eudromias morinellus*
Dove, Stock *Columba oenas*
    Turtle *Streptopelia turtur*
Dunlin *Calidris alpina*
Dunnock *Prunella modularis*
Eagle, Golden *Aquila chrysaetos*
    White-tailed *Haliaeetus albicilla*
Eider *Somateria mollissima*
Fieldfare *Turdus pilaris*
Firecrest *Regulus atricapillus*
Flycatcher, Pied *Ficedula hypoleuca*
    Spotted *Muscicapa striata*
Goldcrest *Regulus regulus*
Goldeneye *Bucephala clangula*
Goshawk *Accipiter gentilis*
Grebe, Little *Tachybaptus ruficollis*
    Slavonian *Podiceps nigricollis*
Greenfinch *Carduelis chloris*
Greenshank *Tringa nebularia*

Goosander *Mergus merganser*
Goose, Barnacle *Branta leucopsis*
    White-Fronted *Anser albifrons*
    Greylag *Anser anser*
Grosbeak, Pine *Pinicola enucleator*
Grouse, Black *Lyrurus tetrix*
    Red *Lagopus lagopus*
    Willow *Lagopus lagopus scoticus*
Gull, Black-headed *Larus ridibundus*
    Common *Larus canus*
Harrier, Hen *Circus cyaneus*
    Montagu's *Circus pygargus*
Hawfinch *Coccothraustes coccothraustes*
Heron, Grey *Ardea cinerea*
Jay, Siberian *Perisoreus infaustus*
Kestrel *Falco tinnunculus*
Kingfisher *Alcedo atthis*
Kite, Red *Milvus milvus*
Lapwing *Vanellus vanellus*
Mallard *Anas platyrhynchos*
Merganser *Mergus serrator*
Merlin *Falco columbarius*
Moorhen *Gallinulla chloropus*
Nightingale *Luscinia megarhyncos*
Nightjar *Caprimulgus europaeus*
Nutcracker *Nucifraga caryocatactes*
Nuthatch *Sitta europea*
Oriole, Golden *Oriolus oriolus*
Osprey *Pandion haliaetus*
Ousel, Ring *Turdus torquatus*
Owl, Barn *Tyto alba*
    Eagle, *Bubo bubo*
    Long-eared *Asio otus*
    Northern Spotted *Strix occidentalis caurina*
    Pygmy *Glaucidium passerinum*
    Short-eared *Asio flammeus*
    Tawny *Strix aluco*
    Tengmalm's *Aegolinus funereus*
    Ural *Strix uralensis*
Oystercatcher *Himantopus ostralegus*
Partridge, Common *Perdix perdix*
    Red-legged *Alectoris rufa*
Peregrine *Falco peregrinus*
Phalarope, Red-necked *Phalaropus lobatus*
Pheasant, Common *Phasianus colchicus*
    Golden *Chrysolophus pictus*
Pigeon, Wood *Columba palumbus*

Pintail *Anas acuta*
Pipit, Meadow *Anthus pratensis*
    Tree *Anthus trivialis*
Plover, Golden *Pluvialis apricaria*
    Ringed *Charadrius hiaticula*
Ptarmigan *Lagopus mutus*
Raven *Corvus corax*
Redpoll *Carduelis flammea*
Redshank *Tringa totanus*
Redstart *Phoenicurus phoenicurus*
Redwing *Turdus iliacus*
Robin *Erithacus rubecula*
Sanderling *Crocethia alba*
Sandpiper, Common *Tringa hypoleucos*
    Pectoral *Calidris melanotos*
    Wood *Tringa glareola*
Scaup *Aythya marila*
Scoter, Common *Melanitta nigra*
Shrike, Red-backed *Lanius colluris*
Siskin *Carduelis spinus*
Skua, Arctic *Stercorarius parasiticus*
    Great *Stercorarius skua*
Skylark *Alauda arvensis*
Snipe *Gallinago gallinago*
Sparrowhawk *Accipiter nisus*
Starling *Sturnus vulgaris*
Stint, Temminck's *Calidris temmincki*
Stonechat *Saxicola torquata*
Swallow *Hirundo rustica*
Swan, Whooper *Cygnus cygnus*
Teal *Anas crecca*
Tern, Roseate *Sterna dougallii*
Thrush, Mistle *Turdus viscivorous*
    Song *Turdus philomelos*
Tit, Bearded *Panurus biarmicus*
    Blue *Parus caeruleus*
    Coal *Parus ater*
    Crested *Parus cristatus*
    Great *Parus major*
    Marsh *Parus palustris*
    Willow *Parus montanus*
Twite *Carduelis flavirostris*
Wagtail, Grey *Motacilla cinerea*
    Pied *Motacilla alba*
Warbler, Dartford *Sylvia undata*
    Garden *Sylvia borin*
    Grasshopper *Locustella naevia*
    Greenish *Phylloscopus trochiloides*
    Willow *Phylloscopus trochilus*
    Wood *Phylloscopus sibilatrix*
Waxwing *Bombycilla garrulus*
Wheatear *Oenanthe oenanthe*
Whinchat *Saxicola rubecula*
Whitethroat *Sylvia communis*
Wigeon *Anas penelope*
Woodcock *Scolopax rusticola*

Woodlark *Lullula arborea*
Woodpecker, Black *Dryocopus martius*
    Great-spotted *Dendrocopus major*
    Green *Picus viridis*
    Lesser-spotted *Dendrocopus minor*
    Three-toed *Picoides tridactylus*
Wren *Troglodytes troglodytes*
Wryneck *Jynx torquilla*
Yellowhammer *Emberiza citrinella*

PLANTS AND FUNGI

Alder *Alnus glutinosa*
Ash *Fraxinus excelsior*
Aspen *Populus tremula*
Bark canker fungus *Lachnellula willkommii*
Beech *Fagus sylvatica*
Birch *Betula pendula*
Bog Cotton *Eriophorum vaginatum*
Bracken *Pteridium aquilinum*
Bramble *Rubus fruticosus*
Cherry, Wild *Prunus avium*
Chestnut, Sweet *Castanea sativa*
Fir, Douglas *Pseudotsuga menziesii*
    Grand *Abies grandis*
    Noble *Abies procera*
    Silver *Abies alba*
Gean *Prunus avium*
Grass, Purple moor- *Molinia caerulea*
Hazel *Corylus avellana*
Heather *Calluna vulgaris*
Hemlock, Western *Tsuga heterophylla*
Holly *Ilex aquafolium*
Juniper *Juniperus communis*
Larch, European *Larix decidua*
    Hybrid *Larix eurolepsis*
    Japanese *Larix kaempferi*
Oak, Common *Quercus petraea*
    Red *Quercus robur*
Pine, Corsican *Pinus nigra*
    Lodgepole *Pinus contorta*
    Scots *Pinus silvestris*
Resin top fungus *Peredermium pini*
Root rot fungus *Heterobasidium annosus*
Rowan *Sorbus aucuparia*
Sedge, Sand-dune *Carex arenaria*
Snowberry *Symphoricarpos rivularis*
Spruce, Norway *Picea abies*
    Sitka *Picea sitchensis*
Sycamore *Acer pseudoplatanus*
Willow *Salix* sp.
Willowherb *Epilobium angustifolium*

MAMMALS

Deer, Fallow *Dama dama*
    Red *Cervus elaphus*
    Roe *Capreolus capreolus*
    Sika *Cervus nippon*
Fox *Vulpes vulpes*
Horse *Equus* sp.
Otter *Lutra lutra*
Pine marten *Martes martes*
Rabbit *Orycotolagus cuniculus*
Squirrel, Grey *Sciurus carolinensis*
    Red *Sciurus vulgaris*
Stoat *Mustela ermina*
Vole, Water *Arvicola amphibius*

INVERTEBRATES

Aphids *Adelges nordmannianae*
    Adelges laricis
Beetle, Bark *Dendroctonus micans*
    Black Pine *Hylaster* sp.
    Pine Shoot *Tomicus piniperda*
Moth, Pine Beauty *Panolis flammea*
    Pine Looper *Bupalus piniaris*
Sawfly, Pine *Neodiprion sertifer*
Weevil, Large Pine *Hylobius abiertis*

Index